LINEAR PROGRAMMING

For Undergraduate Three Year Degree General Course Students

LINEAR PROGRAMMING

P M Karak

Selection Grade Lecturer

Department of Mathematics, Sree Chaitanya College

Habra, West Benagal

New Central Book Agency (P) Ltd

LONDON

HYDERABAD ERNAKULAM BHUBANESWAR

NEW DELHI KOLKATA PUNE GUWAHATI

NCBA

REGD OFFICE

8/1 Chintamoni Das Lane, Kolkata 700 009, India

email: ncbapvtltd@ncbapvtltd.com

OVERSEAS

NCBA (UK) Ltd, 149 Park Avenue North, Northampton, NN3 2HY, UK

email: ncbauk@yahoo.com

EXPORT

NCBA Exports Pvt. Ltd, 212 Shahpur Jat, New Delhi 110 049

email: ncbaexports@ncbapvtltd.com

BRANCHES

2/27 Ansari Road, 1st Floor
Daryaganj, **New Delhi** 110 002
email: ncbadel@ncbapvtltd.com

Shop Nos. 3 & 4, Vinayak Towers
681/B Budhwar Peth, Appa Balwant Chowk
Pune 411 002
email: ncbapune@ncbapvtltd.com

Shop No. 15, Bharati Towers, Ground Floor
Forest Park
Bhubaneswar 751 001
email: ncbabub@ncbapvtltd.com

House No. 3-1-315, 1st Floor
Nimboliadda, Kachiguda, **Hyderabad** 500 027
email: ncbahydb@ncbapvtltd.com

GSS Shopping Complex, 1st Floor
Opposite College Post Office, Convent Road
Ernakulam 682 035
0484-4021054 / 53

Radhamayee Complex [Second Floor]
Opp. Axis Bank, Panbazar, Jaswanta Road
Guwahati 781 001
email: ncbaguwahati@ncbapvtltd.com

LINEAR PROGRAMMING [General] ● P M Karak

First Published: 1979

Thoroughly Revised, Enlarged and Updated Central Edition: 1999

Reprinted: 2001, 2002, 2008, 2011, 2013, 2015, 2019

PUBLISHER AND TYPESETTER

New Central Book Agency (P) Ltd
8/1 Chintamoni Das Lane, Kolkata 700 009

PRINTER

New Central Book Agency (P) Ltd
Web-Offset Division, Dhulagarh, Sankrail, Howrah

PROJECT SUPERVISOR

Abhijit Sen

PROJECT TEAM

Abhijit Sen and Subir Pal

ISBN: 978 81 7381 367 2

Price ₹ 385.00

PREFACE TO THE CENTRAL EDITION

This enlarged revised central edition of the book is meant for the students of three-year degree pass course of all Indian Universities particularly for the Universities of West Bengal. The chapters of the book have been rearranged throughly to make the development of the subject more consistent and to solve the LPP by simplex method the computations have been made in such a way that it will definitely save both time and labour and it is hoped that it will be easily acceptable for them.

As an author of the textbook of this subject and from the teaching experience for more than two decades in the subject, it has been noticed that there is some confusion and ambiguity regarding the development of the subject, particularly in the chapter *Simplex Method or Simplex Algorithm.* To remove such ambiguity and confusion, I have tried to explain the subject matter, explicitly, perfectly and scientifically. Whether I have succeeded or not, is a matter to be judged by the readers.

Actually, it is too difficult for the students reading in Degree Pass Course to understand the theoretical development of the chapter. That is why initially I have discussed the working principle clearly step by step and solved the problems accordingly, avoiding the general difficult discussions. The detailed discussions have been made in the *Appendix.*

At the end, I convey my sincere thanks to Mr Amitabha Sen and the employees of NCBA (P) Ltd for their whole hearted cooperation in the publication of the book in such a nice presentable manner.

Any suggestions from any corners for further improvement of the book will be highly appreciated.

Samvabana
AH 59, Salt Lake City
Calcutta 700091

P M Karak

PREFACE TO THE FIRST EDITION

Minimum Cost and Maximum Profit is the fundamental theme in the realm of Applied Economics, Business and Commerce. In the past, business communities tried to minimize the cost of production by using the trial and error method. In the early part of the sixth decade George Dantzig invented a scientific device based on Mathematics for minimizing the cost of production. This device is called *Programming*. In it, *Linear Programming* has an immense important.

Now the business world is coming to realize the efficacy of the application of the device in different fields of Applied Economics. As a natural consequence, its theoretical developments and practical applications have grown at a rapid rate in the last 25 years.

Keeping in mind the needs of the day, Calcutta University has rightly introduced the subject in the Pass and Honours syllabus in the newly introduced curriculum of Mathematics in the Degree Course. Burdwan University has also introduced the subject in Mathematics Honours Course. The present volume is written according to the Pass Course syllabus of Calcutta University and the Honours Course syllabus of Burdwan University. Each chapter included in the syllabus has been discussed properly and in detail. Again, as the book is written mainly for the Pass Course students, some complex theoretical discussions have been excluded. In the *Appendix*, proofs of some theorems including replacement theorem regarding vectors have been given in detail. Interested students may go through the proofs of these theorems. An alternative method of computing $z_j - c_j$ is also given in the appendix.

I am indebted to the authors and publishers of those books whose help I have taken at the time of the preparation of the manuscripts. I appeal to my colleagues to help me in making the book a better one by constructive criticism of the same. Any detection of printing mistakes, inaccuracies or factual mistakes will be gracefully acknowledged.

July 1979
Śree Chaitanya College P M Karak

Contents

Chapter 1

Mathematical Background

1.1 Introduction

The method of solving a *Linear Programming Problem* (Simply L.P.P.) mainly lies in finding out a solution set of m simultaneous linear equations with $n(n > m)$ unknowns $x_j (j = 1, 2, \cdots, n)$

$$\left.\begin{array}{l} a_{11}x_1 + a_{12}x_2 + \cdots + a_{1n}x_n = b_1 \\ a_{21}x_1 + a_{22}x_2 + \cdots + a_{2n}x_n = b_2 \\ \cdots\cdots\cdots\cdots\cdots\cdots\cdots\cdots\cdots\cdots \\ \cdots\cdots\cdots\cdots\cdots\cdots\cdots\cdots\cdots\cdots \\ a_{m1}x_1 + a_{m2}x_2 + \cdots + a_{mn}x_n = b_m \end{array}\right\} \quad (1.1.1)$$

subject to the restrictions $x_j \geq 0$, $j = 1, 2, \cdots, n$, which makes a *linear function* $z = c_1x_1 + c_2x_2 + \cdots + c_nx_n$ of all the variables (unknowns), either a maximum or minimum. The function z is known as the *objective function.* The quantities a_{ij}, $[i = 1, \cdots, m;\ j = 1, 2, \cdots, n]$; b_i's $[i = 1, 2, \cdots, m]$ and c_j's $[j = 1, 2, \cdots, n]$ are known constants. In order to solve the problem some fundamental knowledges about matrix and vector algebra are essential.

1.2 Matrices

Matrix: m, n being positive integers, $m \times n$ numbers a_{ij} $(i = 1, 2, \cdots, m;$ $j = 1, 2, \cdots, n)$ arranged in a rectangular block or array with m rows and n columns as given below

$$\begin{bmatrix} a_{11} & a_{12} & \cdots & \cdots & a_{1n} \\ a_{21} & a_{22} & \cdots & \cdots & a_{2n} \\ \cdots & \cdots & \cdots & \cdots & \cdots \\ a_{m1} & a_{m2} & \cdots & \cdots & a_{mn} \end{bmatrix}$$

is called a *matrix of order* $m \times n$ or m by n matrix. It is generally denoted by A or $[a_{ij}]_{m\times n}$. All a_{ij}'s are the elements of the matrix. A matrix is said to be real if all the elements a_{ij} are real.

Square matrix: When $m = n$, i.e., the number of rows and columns are equal, the matrix is known as a *square matrix* of order n and when $m \neq n$, the matrix is known as a *rectangular matrix.*

Diagonal matrix: In a square matrix, if all off-diagonal elements are zero, i.e., if $a_{ij} = 0$ for $i \neq j$, the matrix is known as *diagonal matrix.*

Identity matrix: In a diagonal matrix, if all non-zero elements be unity, the matrix is known as *Identity matrix* or *unit matrix.* It is usually denoted by I_n or simply by I.

Null matrix: If all elements of a matrix be zero, the matrix is known as *null matrix.*

Transpose of a matrix or **transposed matrix:** A matrix, formed by interchaning the elements of the rows and columns of a matrix is known as *transposed matrix* of the original one. The transpose of a $m \times n$ matrix A is denoted by A' or A^t and A^t will be a $n \times m$ matrix. The transpose of a square matrix is also a square matrix of the same order. The transpose of the transposed matrix is the original matrix, i.e., $(A')' = A$ etc.

Row matrix: Matrix having a single row is known as a *row matrix.* $1 \times n$ matrix is a row matrix.

Column matrix: A matrix having a single column is known as a *column matrix.* $m \times 1$ matrix is a column matrix.

From the definition, it is clear that the transpose of a row matrix is a column matrix and vice versa.

Equality of two matrices: Two matrices $A = [a_{ij}]$, $B = [b_{ij}]$ of same order $m \times n$ are said to be equal if $a_{ij} = b_{ij}$ for all i and j.

Addition and subtraction of matrices: The sum of two matrices $A = [a_{ij}]$, $B = [b_{ij}]$ of same order $m \times n$ is also a matrix $C = [c_{ij}]$ of the same order, where $c_{ij} = a_{ij} + b_{ij}$ for all i and j etc.

From the above definition, it is evident that

$$\begin{aligned} A + B &= B + A \quad \text{(Commutative law)} \\ (A + B) + C &= A + (B + C) \quad \text{(Associative law)} \\ (A + B)' &= A' + B'. \end{aligned}$$

The difference of two matrices $A = [a_{ij}]$, $B = [b_{ij}]$ of same order $m \times n$, is also a matrix $C = [c_{ij}]$ of the same order, where $c_{ij} = a_{ij} - b_{ij}$.

Multiplication of matrix with scalar quantity: If $A = [a_{ij}]$ be a $m \times n$ matrix and k be any scalar quantity then $kA = Ak = [ka_{ij}]$ is also a matrix of the same order.

From this we can say $k(A + B) = kA + kB$ and so on.

Multiplication of matrices: If $A = [a_{ij}]_{m \times n}$ and $B = [b_{jk}]_{n \times p}$ be two matrices of order $m \times n$ and $n \times p$ respectively, the product of the two matrices A and B which is denoted as AB is a matrix C of order $m \times p$ such that $C = [c_{ik}]_{m \times p}$, where

$$c_{ik} = \sum_{j=1}^{n} a_{ij} b_{jk}.$$

For matrix product to be feasible, the number of columns of the first matrix must be equal to the number of rows of the second matrix. From the definition, it is evident that in general BA does not exist though AB exists, and even if both AB and BA exist, they are not equal in general. For two square matrices A and B, of same order AB and BA both exist but $AB \neq BA$ in general.

Properties of matrix multiplication: If A, B and C are all matrices then

$$\begin{aligned} (A+B)C &= AC+BC \\ C(A+B) &= CA+CB \\ (AB)C &= A(BC) \\ a(AB) &= A(aB) = (aA)B, \quad \text{where } a \text{ is a scalar.} \end{aligned}$$

$(AB)' = B'A'$ and $AI = IA = A$.

It is assumed in all the cases that the matrix product is possible.

$AB = 0$, does not necessarily follow that either A or B be zero or both of them be zero.

With the knowledge of the matrix product, the set of simultaneous linear equations given in (1.1.1) can be written with the help of matrix notation as given below in a more compact form

$$\mathbf{Ax} = \mathbf{b}, \tag{1.2.1}$$

where A is a matrix of order $m \times n$ given by

$$A = \begin{bmatrix} a_{11} & a_{12} & \cdots & a_{1n} \\ a_{21} & a_{22} & \cdots & a_{2n} \\ \dots & \dots & \dots & \dots \\ a_{m1} & a_{m2} & \cdots & a_{mn} \end{bmatrix} \tag{1.2.2}$$

$\mathbf{x}$ is column matrix of order $n \times 1$, having elements $x_1, x_2, \cdots, x_n$. $\mathbf{b}$ is column matrix of order $m \times 1$ having elements $b_1, b_2, \cdots, b_m$. A is known as the *co-efficient matrix.*

Sub-matrix: Matrix formed by omitting some rows and columns of a matrix, is known as a *sub-matrix* of the original matrix.

Determinant of a matrix: The determinant formed with the elements of a square matrix is known as *determinant of the matrix.*

If A be a square matrix of order n, $A = [a_{ij}]_{n\times n}$, then the determinant of the matrix which is denoted by $|A|$ or det A, is given by

$$|A| = \begin{bmatrix} a_{11} & a_{12} & \cdots & a_{1n} \\ a_{21} & a_{22} & \cdots & a_{2n} \\ \dots & \dots & \dots & \dots \\ a_{n1} & a_{n2} & \cdots & a_{nn} \end{bmatrix}. \tag{1.2.3}$$

Note: $|A|$ is here different from the modulus of a quantity A, where the same notation has been used.

The method of simplification of a determinant and the properties of a determinant are known to the students of the degree course. We shall not here discuss all the properties. We discuss only about the **Minor** and the **Cofactor** of an element of a determinant.

Minor: The minor of an element a_{ij} of a determinant $|A|$ is a determinant formed by omitting the ith row and the jth column of the determinant $|A|$. It is usually denoted by M_{ij}.

Cofactor: The cofactor of an element a_{ij} of a determinant $|A|$ is denoted by C_{ij} and given by $C_{ij} = (-1)^{i+j} M_{ij}$.

Singular and non-singular matrix: If the vlaue of the determinant of a square matrix be a non-zero quantity, then the matrix is said to be a *a non-singular matrix* and if the vlaue of the determinant of the matrix be zero then the matrix is said to be a *a singular matrix*. We know that det A = det A^t.

Thus if a square matrix be non-singular, then its transpose is also non-singular.

Rank of a matrix: The rank of a $m \times n$ matrix A is will be r, if all square sub-matrices of order $(r+1)$ are singular and at least one square sub-matrix of order r is non-singular. From the definition, it is clear that $r_{\max} \leq \min(m, n)$. The rank of a matrix A is usually denoted by $r(A)$.

(i) $A = \begin{bmatrix} 2 & -3 \\ 3 & 4 \end{bmatrix}$, $|A| = \begin{vmatrix} 2 & -3 \\ 3 & 4 \end{vmatrix} = 2 \times 4 - (-3) \times 3 = 17 \neq 0.$

Thus $r(A) = 2$.

(ii) $A = \begin{bmatrix} 5 & 10 \\ 3 & 6 \end{bmatrix}$, $|A| = \begin{vmatrix} 5 & 10 \\ 3 & 6 \end{vmatrix} = 5 \times 6 - 10 \times 3 = 0.$

Thus $r(A) \neq 2$.

Now one square sub-matrix of order 1 is $B_1 = [5]$ and $|B_1| = 5 \neq 0$.

Thus $r(A) = 1$.

(iii) $A = \begin{bmatrix} 2 & 3 & -1 \\ 1 & 2 & 0 \end{bmatrix}$, one square sub-matrix of order 2 is $B_1 = \begin{bmatrix} 2 & 3 \\ 1 & 2 \end{bmatrix}$,

$|B_1| = \begin{vmatrix} 2 & 3 \\ 1 & 2 \end{vmatrix} = 2 \times 2 - 3 \times 1 = 1 \neq 0.$

Thus $r(A) = 2$.

(iv) $A = \begin{bmatrix} 2 & 3 & 5 \\ 4 & 6 & 10 \end{bmatrix}$, one square sub-matrix of order 2 is $B_1 = \begin{bmatrix} 2 & 3 \\ 4 & 6 \end{bmatrix}$,

$|B_1| = 0$, similarly $|B_2| = \begin{vmatrix} 2 & 5 \\ 4 & 10 \end{vmatrix} = 0$, $|B_3| = \begin{vmatrix} 3 & 5 \\ 6 & 10 \end{vmatrix} = 0.$

Thus $r(A) \neq 2$.

Now one square sub-matrix of order 1, $B_4 = [2]$ and $|B_4| = 2 \neq 0$.

Thus $r(A) = 1$.

(v) $A = \begin{bmatrix} 1 & 3 & 4 \\ 2 & -2 & 2 \\ 3 & 1 & 1 \end{bmatrix}$,

$$|A| = \begin{vmatrix} 1 & 3 & 4 \\ 2 & -2 & 2 \\ 3 & 1 & 1 \end{vmatrix} = 1 \begin{vmatrix} -2 & 2 \\ 1 & 1 \end{vmatrix} - 3 \begin{vmatrix} 2 & 2 \\ 3 & 1 \end{vmatrix} + 4 \begin{vmatrix} 2 & -2 \\ 3 & 1 \end{vmatrix} = 40 \neq 0.$$

Thus $r(A) = 3$.

(vi) $A = \begin{bmatrix} 4 & 3 & 2 \\ 2 & 1 & 4 \\ 2 & 3 & -8 \end{bmatrix}$,

$$|A| = \begin{vmatrix} 4 & 3 & 2 \\ 2 & 1 & 4 \\ 2 & 3 & -8 \end{vmatrix} = 4\begin{vmatrix} 1 & 4 \\ 3 & -8 \end{vmatrix} - 3\begin{vmatrix} 2 & 4 \\ 2 & -8 \end{vmatrix} + 2\begin{vmatrix} 2 & 1 \\ 2 & 3 \end{vmatrix} = 0.$$

Thus $r(A) \neq 3$.

Now one square sub-matrix of order 2 is $B_1 = \begin{bmatrix} 4 & 3 \\ 2 & 1 \end{bmatrix}$, $|B_1| = -2 \neq 0$.

Thus $r(A) = 2$.

(vii) $A = \begin{bmatrix} 1 & 2 & 3 \\ 3 & 6 & 9 \\ 2 & 4 & 6 \end{bmatrix}$, $|A| = 0$.

Thus $r(A) \neq 3$.

Now one square sub-matrix of order 2 is $B_1 = \begin{bmatrix} 1 & 2 \\ 3 & 6 \end{bmatrix}$, $|B_1| = 0$.

Another square sub-matrix of order 2 is $B_2 = \begin{bmatrix} 2 & 3 \\ 4 & 6 \end{bmatrix}$, $|B_2| = 0$.

Similarly, it can be shown that all square sub-matrices of order 2 are singular. Thus $r(A) \neq 2$. Last of all, one square sub-matrix of order 1 is [1] which is non-singular. Thus $r(A) = 1$.

Note: I_n **has rank** n **and a null matrix has rank 0.**

Adjoint of a matrix: The adjoint of a square matrix A is a matrix of the same order, which is the transpose of the matrix B, where the elements of B are the cofactors of the corresponding elements of the determinant of the matrix A. It is denoted by Adj (A).

If $A = [a_{ij}]_{n\times n}$ then

$$\text{Adj}\,(A) = \begin{bmatrix} c_{11} & c_{21} & \cdots & c_{n1} \\ c_{12} & c_{22} & \cdots & c_{n2} \\ \cdots & \cdots & \cdots & \cdots \\ c_{1n} & c_{2n} & \cdots & c_{nn} \end{bmatrix}, \tag{1.2.4}$$

where c_{ij} etc. are the cofactors of the corresponding elements a_{ij} of $|A|$ given in (1.2.3).

Inverse of a matrix: If there exist two square matrices A and B of same order such that $AB = BA = I$ (identity matrix of same order) then B is called as the *inverse matrix* of A and vice versa. The inverse matrix of A which is denoted by A^{-1} is given by

$$A^{-1} = \frac{1}{|A|} \cdot \text{Adj}\,(A). \tag{1.2.5}$$

From the definition, it is clear that $AA^{-1} = A^{-1}A = I$. It is also clear that inverse exists only in the non-singular matrices. The inverse matrix of an identity matrix is the same matrix.

Determination of the inverse of a 2 × 2 non-singular matrix:

$$A = \begin{bmatrix} a_{11} & a_{12} \\ a_{21} & a_{22} \end{bmatrix}; \quad |A| = a_{11}a_{22} - a_{12}a_{21} \neq 0.$$

Here the cofactors of $a_{11} = c_{11} = a_{22}$, $c_{12} = -a_{21}$, $c_{21} = -a_{12}$ and $c_{22} = a_{11}$; it follows that

$$\text{Adj}\,(A) = \begin{bmatrix} c_{11} & c_{21} \\ c_{12} & c_{22} \end{bmatrix} = \begin{bmatrix} a_{22} & -a_{12} \\ -a_{21} & a_{11} \end{bmatrix}$$

and thus

$$A^{-1} = \frac{1}{|A|}\,\text{Adj}\,(A) = \frac{1}{|A|}\begin{bmatrix} a_{22} & -a_{12} \\ -a_{21} & a_{11} \end{bmatrix}. \tag{1.2.6}$$

Note: Readers should remember the formula for A^{-1}.

▶ **Example 1.2.1** *Find the inverse of the matrix* $A = \begin{bmatrix} 2 & -1 \\ 4 & 3 \end{bmatrix}$.

Solution: $\det A = |A| = 2 \times 3 - (-1) \times 4 = 10 \neq 0$. Hence the matrix is non-singular and therefore the inverse exists. Thus using (1.2.6) we have,

$$A^{-1} = \frac{1}{10}\begin{bmatrix} 3 & 1 \\ -4 & 2 \end{bmatrix} = \begin{bmatrix} \frac{3}{10} & \frac{1}{10} \\ -\frac{2}{5} & \frac{1}{5} \end{bmatrix}$$

and finally we can verify that

$$\begin{aligned} AA^{-1} &= \begin{bmatrix} 2 & -1 \\ 4 & 3 \end{bmatrix}\begin{bmatrix} \frac{3}{10} & \frac{1}{10} \\ -\frac{2}{5} & \frac{1}{5} \end{bmatrix} \\ &= \begin{bmatrix} 2 \times \frac{3}{10} + (-1)\left(-\frac{2}{5}\right) & 2 \times \frac{1}{10} + (-1) \times \frac{1}{5} \\ 4 \times \frac{3}{10} + 3 \times \left(-\frac{2}{5}\right) & 4 \times \frac{1}{10} + 3 \times \frac{1}{5} \end{bmatrix} = \begin{bmatrix} 1 & 0 \\ 0 & 1 \end{bmatrix} = I_2. \end{aligned}$$

Determination of the solution set of m linear equations with m variables by matrix method: Let the set of equations be

$$A\mathbf{x} = \mathbf{b},$$

where A is a square matrix of order m. We make an assumption that A is non-singular, i.e., $|A| \neq 0$. The solution $\hat{\mathbf{x}}$ is given by $\hat{\mathbf{x}} = A^{-1}\mathbf{b}$ and the solution set is unique.

▶ **Example 1.2.2** *Solve the linear equations by matrix method.*

$$\begin{aligned} x_1 + 2x_2 + 3x_3 &= 6 \\ 3x_1 + 2x_2 + 9x_3 &= 14 \\ 2x_1 + 4x_2 + x_3 &= 7. \end{aligned}$$

Solution: $A = \begin{bmatrix} 1 & 2 & 3 \\ 3 & 2 & 9 \\ 2 & 4 & 1 \end{bmatrix}$, $|A| = 20 \neq 0$.

$$A^{-1} = \frac{1}{|A|}\begin{bmatrix} -34 & 10 & 12 \\ 15 & -5 & 0 \\ 8 & 0 & -4 \end{bmatrix} \text{ by (1.2.5)}$$

$$\hat{\mathbf{x}} = A^{-1}\begin{bmatrix} 6 \\ 14 \\ 7 \end{bmatrix} = \frac{1}{20}\begin{bmatrix} -34 & 10 & 12 \\ 15 & -5 & 0 \\ 8 & 0 & -4 \end{bmatrix}\begin{bmatrix} 6 \\ 14 \\ 7 \end{bmatrix}, \text{ or, } \hat{\mathbf{x}} = \begin{bmatrix} 1 \\ 1 \\ 1 \end{bmatrix}$$

and the solution set is unique.

1.3 Vectors

All row and column matrices are known as *row* and *column* vectors respectively. The vectors are usually denoted by lower case bold type as $\mathbf{a}, \mathbf{b}, \mathbf{x}$ etc.

Row vector: A row matrix with n elements $a_1, a_2, \cdots, a_n$ is known as *n-component row vector.* It is usually denoted by $\mathbf{a}$ and $\mathbf{a} = (a_1, a_2, \cdots, a_n)$. To represent row vector first bracket "( )" has been used. The elements a_1, a_2 etc. are known as the *components of the vector.*

Column vector: A column matrix with n elements $b_1, b_2, \cdots, b_n$ is known as *n-component column vector* $\mathbf{b}$ and

$$\mathbf{b} = \begin{bmatrix} b_1 \\ b_2 \\ \cdots \\ b_n \end{bmatrix}.$$

In matrix notation column matrix is written in a vertical column. But a system is used to write the column vector in a horizontal row. The vector $\mathbf{b}$ is written as $\mathbf{b} = [b_1, b_2, \cdots, b_n]$ and third bracket "[ ]" is used to denote it as a column vector.

Note: Both row and column vectors are written in a horizontal row. And for row vector the first bracket "( )" is used and for column vector, the third bracket "[ ]" is used.

Analytically, both the row and column vectors may be considered as points in n-dimensional space which is known as *vector space* or *n-dimensinoal Eucledean space* which is usually denoted by $V_n(F)$ or E^n or E_n etc. Geometricaly, all vectors may be considered as the position vector of points with respect to the origin. In a vector space all vectors must be either row vectors or column vectors.

Unit vector: A vector is said to be a unit vector, if all its components are zero *except one with unit value.* $\mathbf{e}_i$ is a unit vector in n-dimensional vector space whose i-th component is 1. $\mathbf{e}_1 = (1, 0, \cdots, 0)$, $\mathbf{e}_2 = (0, 1, 0, \cdots, 0)$, $\mathbf{e}_3 = (0, 0, 1, \cdots, 0)$, $\mathbf{e}_n = (0, 0, \cdots, 1)$ are all unit vectors in n-dimensional space.

Null vector: *A* vector is said to be a null vector, if all the components of the vector be equal to zero. It is usually denoted by $\mathbf{0}$. An n-component null vector is written as $\mathbf{0} = (0, 0, \cdots, 0)$ with n zeros.

Equality of two vectors: Two n-component row vectors are said to be equal, if the corresponding components of both the vectors are equal. Similar is the case for column vectors. Two n-component row vectors $\mathbf{a}$ and $\mathbf{b}$ are equal only when $a_i = b_i (i = 1, 2, \cdots, n)$, where $\mathbf{a} = (a_1, a_2, \cdots, a_n)$ and $\mathbf{b} = (b_1, b_2, \cdots, b_n)$.

Sum of two vectors: The sum of two n-component row vectors is also a row vectors, whose i-th component is the sum of the ith components of the constituent vectors. If $\mathbf{a} = (a_1, a_2, \cdots, a_n)$ and $\mathbf{b} = (b_1, b_2, \cdots, b_n)$ be two n-component row vectors then

$$\mathbf{a} + \mathbf{b} = \mathbf{c},$$

where $\mathbf{c} = (a_1 + b_1, a_2 + b_2, \cdots, a_i + b_i, \cdots, a_n + b_n)$.

Similar is the case for two n-component column vectors.

Difference of two vectors: The difference of two n-component row vectors is also a row vector, whose ith component is the difference of the ith components of the constituent vectors. If $\mathbf{a} = (a_1, a_2, \cdots, a_n)$, $\mathbf{b} = (b_1, b_2, \cdots, b_n)$ be two n-component row vectors then $\mathbf{a} - \mathbf{b} = \mathbf{c}$, where $\mathbf{c} = (a_1 - b_1, a_2 - b_2, \cdots, a_n - b_n)$. Similar is the case for column vectors. The difference of two equal vectors is a null vector.

If $\mathbf{a} = (a_1, a_2, \cdots, a_n)$ than $\lambda\mathbf{a} = (\lambda a_1, \lambda a_2, \cdots, \lambda a_n)$, where λ is a scalar.

Inner product of two vectors: If $\mathbf{a}$ and $\mathbf{b}$ are two n-component row and column vectors respectively, then the product of the vectors $\mathbf{a}$ and $\mathbf{b}$ is known as the *inner product* (actually matrix product) of the vectors denoted by $\mathbf{ab}$ and given by $\mathbf{ab} = a_1b_1 + a_2b_2 + \cdots + a_nb_n$, where $\mathbf{a} = (a_1, a_2, \cdots, a_n)$ and $\mathbf{b} = [b_1, b_2, \cdots, b_n]$ and $\mathbf{ab} \neq \mathbf{ba}$.

Inner product is also termed as scalar product. Inner product is a scalar quantity. Thus

$$(\mathbf{ab})' = \mathbf{b}'\mathbf{a}' = \mathbf{ab} \text{ etc.}$$

Note: In the case of inner product, the first vector must be row vector and the second vector must be column vector.

Length of a vector: If $\mathbf{a} = (a_1, a_2, \cdots, a_n)$ be an n-component row vector then the length of the vector $\mathbf{a}$ which is usually denoted by $|\mathbf{a}|$ is the square root of the inner product $\mathbf{aa}'$ or $|\mathbf{a}| = \sqrt{\mathbf{aa}'} = \sqrt{a_1^2 + a_2^2 + \cdots + a_n^2}$.

If the vector $\mathbf{b} = [b_1, b_2, \cdots, b_n]$ be an n-component column vector then

$$|\mathbf{b}| = \sqrt{\mathbf{b}' \cdot \mathbf{b}} = \sqrt{b_1^2 + b_2^2 + \cdots + b_n^2}.$$

Distance between two points: If $\mathbf{a}$ and $\mathbf{b}$ are vectors (which are considered as points) of the same vector space then the distance between the points is denoted by $|\mathbf{a} - \mathbf{b}|$ and given by

$$|\mathbf{a} - \mathbf{b}| = \sqrt{(a_1 - b_1)^2 + (a_2 - b_2)^2 + \cdots + (a_n - b_n)^2},$$

where $\mathbf{a} = (a_1, a_2, \cdots, a_n)$ and $\mathbf{b} = (b_1, b_2, \cdots, b_n)$.

Linear combination of vectors: The linear combination of set of k, n-component vectors $\mathbf{a}_1, \mathbf{a}_2, \cdots, \mathbf{a}_k$ is a vector $\mathbf{a}$ with n-components given by the reiation

$$\mathbf{a} = \lambda_1\mathbf{a}_1 + \lambda_2\mathbf{a}_2 + \cdots + \lambda_k\mathbf{a}_k = \sum_{i=1}^{k} \lambda_i\mathbf{a}_i, \tag{1.3.1}$$

where λ_i are all scalar quantities.

Linear dependence of vectors: *A* set of k, n-component vectors $\mathbf{a}_1, \mathbf{a}_2, \cdots, \mathbf{a}_k$ are said to be *linearly dependent* if there exist scalars $\lambda_i (i = 1, 2, \cdots, k)$ with at least one $\lambda_i \neq 0$, such that

$$\lambda_1 \mathbf{a}_1 + \lambda_2 \mathbf{a}_2 + \cdots + \lambda_k \mathbf{a}_k = \mathbf{0}, \quad \text{or,} \quad \sum_{i=1}^{k} \lambda_i \mathbf{a}_i = \mathbf{0} \tag{1.3.2}$$

is satisfied, where $\mathbf{0}$ is an n-component null vector.

For example, in three dimensional vector space $\mathbf{a} = (1, 2, 3)$ and $\mathbf{b} = (4, 8, 12)$ are linearly dependent because $4\mathbf{a} - \mathbf{b} = \mathbf{0}$, where the scalar quantities are 4 and -1, at least one of which is $\neq 0$. Similarly the three vectors

$$\begin{aligned} \mathbf{a} &= (1, 3, -2), \\ \mathbf{b} &= (2, 1, -1) \\ \text{and} \quad \mathbf{c} &= (0, -5, 3) \end{aligned}$$

are linearly dependent because

$$2\mathbf{a} - \mathbf{b} + \mathbf{c} = 2(1, 3, -2) - (2, 1, -1) + (0, -5, 3) = (0, 0, 0) = \mathbf{0},$$

where the scalar quantities are $2, -1$ and 1 and at least one of which is non-zero. It is evident that if two non-null vectors be linearly dependent, one can be expressed as a scalar multiple of the other.

Linearly independent vectors: A set of k, n-component vectors $(k \leq n)$ $\mathbf{a}_1, \mathbf{a}_2, \cdots, \mathbf{a}_k$ are said to be linearly independent if the linear combination of all the vectors be equal to an n-component null vector only, when all the scalars are equal to zero. Mathematically, the set of vectors $\mathbf{a}_1, \mathbf{a}_2, \cdots, \mathbf{a}_k$ are said to be *linearly independent* if the condition

$$\sum_{i=1}^{k} \lambda_i \mathbf{a}_i = \mathbf{0} \tag{1.3.3}$$

is satisfied only when all the scalar quantities λ_i are zero.

The two vectors $(1, 3, -2)$ and $(4, 1, 3)$ are linearly independent because the condition $\lambda_1(1, 3, -2) + \lambda_2(4, 1, 3) = (0, 0, 0)$ will be satisfied if and only if $\lambda_1 = 0$ and $\lambda_2 = 0$. There are no other values of λ_1 and λ_2 which will satisfy the condition.

Similarly the three vectors $\mathbf{a} = (1, -1, 1)$, $\mathbf{b} = (1, 2, -1)$ and $\mathbf{c} = (3, 2, -5)$ are linearly independent. Because the condition

$$\begin{aligned} &\lambda_1 \mathbf{a} + \lambda_2 \mathbf{b} + \lambda_3 \mathbf{c} = \mathbf{0} \quad [\mathbf{0} \text{ is a 3-component null vector}] \\ \text{or,} \quad &\lambda_1(1, -1, 1) + \lambda_2(1, 2, -1) + \lambda_3(3, 2, -5) = (0, 0, 0) \end{aligned}$$

will be satisfied only if $\lambda_1 = \lambda_2 = \lambda_3 = 0$. There are no other values of λ_1, λ_2 and λ_3 which will satisfy the condition. Hence the vectors $\mathbf{a}, \mathbf{b}$ and $\mathbf{c}$ are linearly independent. From the discussions it is clear that any single non-null vector is always linearly independent.

Generating set of the vectors: A set of k, n-component vectors $\mathbf{a}_1, \mathbf{a}_2, \cdots, \mathbf{a}_k$ in a vector space E^n is said to be a generating set, if all vectors in the space can be expressed as a linear combination of the vectors $\mathbf{a}_1, \mathbf{a}_2, \cdots, \mathbf{a}_k$. The set is also known as *spanning set.* All vectors of the generating set may be linearly dependent or independent.

Basis set: A generating set of the vectors, all of which are linearly independent, is known as a *basis set.* The set of all n, n-component unit vectors $[\mathbf{e}_1, \mathbf{e}_2, \cdots, \mathbf{e}_i, \cdots, \mathbf{e}_n]$ are said to form a basis set in E^n and this basis set is known as the *standard basis set.*

The vectors $\mathbf{e}_i$ $(i = 1, 2, \cdots, n)$ are *linearly independent* and all the vectors in the space can be generated by the set of vectors $\mathbf{e}_i$ $(i = 1, 2, \cdots, n)$. For example, in E^3, any vector $\mathbf{a} = (a_1, a_2, a_3) \in E^3$ can be expressed as $\mathbf{a} = a_1\mathbf{e}_1 + a_2\mathbf{e}_2 + a_3\mathbf{e}_3$. Thus in E^3, $\mathbf{e}_1 = (1, 0, 0)$, $\mathbf{e}_2 = (0, 1, 0)$ and $\mathbf{e}_3 = (0, 0, 1)$, form a basis which is the standard basis in E^3. For a particular vector space there may have more than one basis set.

Some well known properties of vectors: We state all the properties without proof.

1. A set of vectors is either linearly independent or dependent, i.e., the terms linearly dependent and linearly independent are mutually exclusive.
2. A set of vectors with a null vector are always linearly dependent.
3. A non-empty subset of linearly independent set of vectors are always linearly independent and a superset of a linearly dependent set of vectors are always linearly dependent.
4. For a vector space, there must have at least one basis set.
5. The representation of a vector by a basis set is unique.
6. In E^n, $(n + 1)$ or more vectors must be linearly dependent.
7. The exact number of vectors in a basis set is n, in E^n.
8. If there are two or more basis sets in a vector space then each has the same number of vectors.
9. Any set of n linearly independent vectors in E^n, is a basis set.
10. The set of n vectors in E^n are linearly independent, if and only if the rank of the $n \times n$ matrix, formed with the components of n vectors as the elements of the matrix, be equal to n. If it is less than that, the vectors are linearly dependent, i.e., if the square matrix formed with the n vectors be non-singular, the vectors are L.I.; otherwise vectors are L.D.

The proofs of some important theorems on vectors.

Theorem 1.3.1 *The set of vectors containing a null vector are always linearly dependent.*

Proof: Let a set A of vectors be

$$A = \{\mathbf{a}_1, \mathbf{a}_2, \cdots, \mathbf{a}_k, \mathbf{0}\} \in E^n.$$

We requrie to prove that the vectors in A are linearly dependent.

The condition that

$$\lambda_1\mathbf{a}_1 + \lambda_2\mathbf{a}_2 + \cdots + \lambda_k\mathbf{a}_k + \lambda\mathbf{0} = \mathbf{0}$$

will be satisfied if $\lambda_1 = \lambda_2 = \cdots = \lambda_k = 0$ and $\lambda \neq 0$. [λ_j, $(j = 1, 2, \cdots, k)$, are all scalar quantities].

Therefore, according to the definition, the set of vectors are linearly dependent.

Theorem 1.3.2 (i) *The superset of any linear dependent set of vectors are linearly dependent and* (ii) *The non-empty subset of any linearly independent set of vectors are linearly independent.*

Proof: (i) Let A be a set of vectors given by

$$A = \{\mathbf{a}_1, \mathbf{a}_2, \cdots, \mathbf{a}_k\} \in E^n$$

and the vectors are linearly dependent. Any superset B of A is given by

$$B = \{\mathbf{a}_1, \mathbf{a}_2, \cdots, \mathbf{a}_k, \mathbf{a}_m, \cdots, \mathbf{a}_l\} \in E^n.$$

We require to prove that the vectors in the set B are linearly dependent.

As $\mathbf{a}_1, \mathbf{a}_2, \cdots, \mathbf{a}_k$ are linearly dependent, therefore

$$\sum_{i=1}^{k} \lambda_i\mathbf{a}_i = \mathbf{0}$$

will be astisfied with at least one $\lambda_i \neq 0$. [λ_i are all scalar quantities].

Hence the condition

$$\sum_{i=1}^{k} \lambda_i\mathbf{a}_i + 0\cdot\mathbf{a}_m + \cdots + 0\cdot\mathbf{a}_l = \mathbf{0}$$

will be satisfied with at least one $\lambda_i \neq 0$. Therefore, the vectors $\mathbf{a}_1, \mathbf{a}_2, \cdots, \mathbf{a}_k, \mathbf{a}_m, \cdots, \mathbf{a}_l$ are linearly dependent.

(ii) Let A be a set of vectors given by

$$A = \{\mathbf{a}_1, \mathbf{a}_2, \cdots, \mathbf{a}_k, \cdots, \mathbf{a}_n\} \in E^n$$

and the vectors are linearly independent.

Any non-empty subset B of A is given by

$$B = \{\mathbf{a}_1, \mathbf{a}_2, \cdots, \mathbf{a}_k\}.$$

We require to prove that the vectors in B are linearly independent.

As $\mathbf{a}_1, \mathbf{a}_2, \cdots, \mathbf{a}_k, \cdots, \mathbf{a}_n$ are linearly independent, therefore the condition that

$$\sum_{i=1}^{n} \lambda_i\mathbf{a}_i = \mathbf{0}$$

will be satisfied if and only if

$$\lambda_1 = \lambda_2 = \cdots = \lambda_k = \cdots = \lambda_n = 0.$$

Hence the condition that

$$\sum_{i=1}^{k} \lambda_i \mathbf{a}_i = \mathbf{0}$$

will be satisfied if and only if

$$\lambda_1 = \lambda_2 = \cdots = \lambda_k = 0.$$

Thus the vectors in the set B are linearly independent.

Theorem 1.3.3 *The representation of a vector by a basis set is unique.*

Proof: Let the set A of vectors be given by

$$A = \{\mathbf{a}_1, \mathbf{a}_2, \cdots, \mathbf{a}_n\}$$

is a basis set in E_n and let $\mathbf{a}$ be any vector in the same vector space E_n.

Let us make an assumption that the representation is not unique. Therefore, $\mathbf{a}$ can be expressed as the linear combination of n independent vectors $\mathbf{a}_1, \mathbf{a}_2, \cdots, \mathbf{a}_n$ given below:

$$\mathbf{a} = \lambda_1\mathbf{a}_1 + \lambda_2\mathbf{a}_2 + \cdots + \lambda_n\mathbf{a}_n \qquad (1.3.4)$$

$$\text{and} \quad \mathbf{a} = \mu_1\mathbf{a}_1 + \mu_2\mathbf{a}_2 + \cdots + \mu_n\mathbf{a}_n. \qquad (1.3.5)$$

(1.3.4)–(1.3.5) gives

$$\mathbf{0} = (\lambda_1 - \mu_1)\mathbf{a}_1 + (\lambda_2 - \mu_2)\mathbf{a}_2 + \cdots + (\lambda_n - \mu_n)\mathbf{a}_n. \qquad (1.3.6)$$

As the vectors $\mathbf{a}_1, \mathbf{a}_2, \cdots, \mathbf{a}_n$ are linearly independent therefore the condition (1.3.6) will be satisfied if and only if

$$\lambda_1 - \mu_1 = \lambda_2 - \mu_2 = \cdots = \lambda_n - \mu_n = 0$$

which indicates that $\lambda_1 = \mu_1$, $\lambda_2 = \mu_2, \cdots$ and $\lambda_n = \mu_n$.

Hence the representation is unique.

Theorem 1.3.4 Replacement theorem: *In E^n, if the set of vectors $\mathbf{a}_1, \mathbf{a}_2, \cdots, \mathbf{a}_r$ be a basis set and $\mathbf{b}$ is a vector in the same space, $\mathbf{b} \neq 0$, $\mathbf{b} \neq \mathbf{a}_i$ $(\mathbf{i} = 1, 2, \cdots, r)$ and if $\mathbf{b} = \sum_{i=1}^{r} \lambda_i \mathbf{a}_i$, where $\lambda_i \neq 0$ for some i, then $\mathbf{b}$ can replace $\mathbf{a}_l$ to form a new basis for $\lambda_l \neq 0$, i.e., if $\lambda_l \neq 0$, then $\mathbf{a}_1, \mathbf{a}_2, \cdots, \mathbf{a}_{l-1}, \mathbf{b}, \mathbf{a}_{l+1}, \cdots, \mathbf{a}_r$ form a new basis set.*

Proof: The set of vectors $\mathbf{a}_1, \mathbf{a}_2, \cdots, \mathbf{a}_r$ is a basis set in E^n. $\mathbf{b}$ is a vector given by

$$\mathbf{b} = \sum_{i=1}^{r} \lambda_i \mathbf{a}_i \text{ with } \lambda_l \neq 0. \qquad (1.3.7)$$

We are to prove that the set of vectors $\mathbf{a}_1, \mathbf{a}_2, \mathbf{a}_3, \cdots, \mathbf{a}_{l-1}, \mathbf{b}, \mathbf{a}_{l+1}, \cdots, \mathbf{a}_r$ is also a basis set in E^n, i.e., we are to prove that the vectors $\mathbf{a}_1, \mathbf{a}_2, \cdots, \mathbf{a}_{l-1}, \mathbf{b}, \mathbf{a}_{l+1}, \cdots, \mathbf{a}_r$ are linearly independent and each vector in E^n can be expressed as a linear combination of the set of vectors $\mathbf{a}_1, \mathbf{a}_2, \cdots, \mathbf{a}_{l-1}, \mathbf{b}, \mathbf{a}_{l+1}, \cdots, \mathbf{a}_r$.

Let us make an assumption that the vectors are not linearly independent. Therefore, the condition that

$$\sum_{\substack{i=1\\i\neq l}}^{r} \delta_i \mathbf{a}_i + \delta \mathbf{b} = \mathbf{0} \tag{1.3.8}$$

will be satisfied with at least one of the scalars δ_i $[i = 1, 2, \cdots, r, i \neq l]$ and δ, non-zero.

Now $\delta \neq 0$; because if $\delta = 0$, then from condition (1.3.8) we get

$$\sum_{\substack{i=1\\i\neq l}}^{r} \delta_i \mathbf{a}_i = \mathbf{0}$$

with at least one scalar δ_i not equal to zero which is impossible. [as the vectors $\mathbf{a}_1, \mathbf{a}_2, \cdots, \mathbf{a}_{l-1}, \mathbf{a}_{l+1}, \cdots, \mathbf{a}_r$ are linearly independent]

Now putting the value of $\mathbf{b}$ from (1.3.7) in (1.3.8) we get

$$\sum_{\substack{i=1,\\i\neq l}}^{r} \delta_i \mathbf{a}_i + \delta \sum_{i=1}^{r} \lambda_i \mathbf{a}_i = \mathbf{0}, \quad \text{or,} \quad \sum_{\substack{i=1\\i\neq l}}^{r} (\delta_i + \delta\lambda_i)\mathbf{a}_i + \delta\lambda_l \mathbf{a}_l = \mathbf{0} \tag{1.3.9}$$

Since $\delta \neq 0, \lambda_l \neq 0$, therefore $\delta\lambda_l \neq 0$, and hence the vectors $\mathbf{a}_1, \mathbf{a}_2, \cdots, \mathbf{a}_l, \cdots, \mathbf{a}_r$ are linearly dependent. This contradicts the fact. Therefore, the vectors $\mathbf{a}_1, \mathbf{a}_2, \cdots, \mathbf{a}_{l-1}, \mathbf{b}, \mathbf{a}_{l+1}, \cdots, \mathbf{a}_r$ are linearly independent.

Next, let $\mathbf{x}$ be any vector in E^n. Therefore,

$$\mathbf{x} = \sum_{i=1}^{r} \mu_i \mathbf{a}_i \quad [\mu_i \text{ are all scalars}] \tag{1.3.10}$$

From (1.3.7) we get

$$\mathbf{a}_l = \frac{\mathbf{b}}{\lambda_l} - \frac{1}{\lambda_l} \sum_{\substack{i=1\\i\neq l}}^{r} \lambda_i \mathbf{a}_i. \tag{1.3.11}$$

Putting the value of $\mathbf{a}_l$ in (1.3.10) we find that $\mathbf{x}$ can be expressed as a linear combination of the vectors $\mathbf{a}_1, \mathbf{a}_2, \cdots, \mathbf{a}_{l-1}, \mathbf{b}, \mathbf{a}_{l+1}, \cdots, \mathbf{a}_r$. Hence the theorem is proved.

For example, the three vectors $\mathbf{a} = (1, 2, 4)$, $\mathbf{b} = (1, -1, 1)$ and $\mathbf{c} = (-1, 3 - 1)$ from a basis set E^3.

$\mathbf{d} = (3, 0, 6)$ is another vector in the same space; $\mathbf{d} \neq 0$, $\mathbf{d} \neq \mathbf{a}$ or $\mathbf{b}$ or $\mathbf{c}$ and $\mathbf{d}$ can be expressed as $\mathbf{d} = \lambda_1\ (1, 2, 4) + \lambda_2(1, -1, 1) + \lambda_3(-1, 3, -1)$, where $\lambda_1 = 1$, $\lambda_2 = 2$ and $\lambda_3 = 0$ [at least one of which is not zero].

Hence

$$\mathbf{d} = \lambda_1 \mathbf{a} + \lambda_2 \mathbf{b} + \lambda_3 \mathbf{c}$$

when $\lambda_1 \neq 0$, $\lambda_2 \neq 0$ and $\lambda_3 = 0$.

Therefore, according to the above theorem $\mathbf{d}$ can replace $\mathbf{a}$ and $\mathbf{b}$ to form a new basis, i.e., $(\mathbf{d}, \mathbf{b}, \mathbf{c})$ and $(\mathbf{a}, \mathbf{d}, \mathbf{c})$ form new basis but as $\lambda_3 = 0$, $\mathbf{d}$ cannot replace $\mathbf{c}$ to form a new basis, i.e., $(\mathbf{a}, \mathbf{b}, \mathbf{d})$ is not a basis in E^3.

Theorem 1.3.5 *In E^n, the exact number of vectors in a basis is n.*

Proof: This can be proved by applying the above theorem.

We know that $(\mathbf{e}_1, \mathbf{e}_2, \cdots, \mathbf{e}_n)$ form a basis (standard) in E^n, the number of vectors of which is n. Now by using replacement theorem, by suitable replacement of one of the vectors by a vector, a new basis can be formed, the number of vectors of which is exactly n. In the same way it can be shown that in all cases the number of vectors in a basis is n.

Worked out examples

1. Find out the rank of matrices:

(a) $A = \begin{bmatrix} 1 & 2 & 1 \\ 3 & 1 & 5 \\ 3 & 4 & 7 \end{bmatrix}$ (b) $A = \begin{bmatrix} 2 & -3 & -1 \\ 4 & 1 & 5 \\ 6 & -2 & 4 \end{bmatrix}$.

Solution: (a) The determinant of the matrix A, given by

$$|A| = \begin{vmatrix} 1 & 2 & 1 \\ 3 & 1 & 5 \\ 3 & 4 & 7 \end{vmatrix} = -16 \neq 0.$$

Hence A, the square matrix of order three, is non-singular. Therefore, the rank of the matrix is 3.

(b) $|B| = 0$. Hence the rank of the matrix B is less than 3. Now we take any square sub-matrix B_1 of order two and $B_1 = \begin{bmatrix} 2 & -3 \\ 4 & 1 \end{bmatrix}$; $|B_1| = 14 \neq 0$. Hence the rank of the matrix B is 2.

2. Prove that the vectors (a) $\mathbf{a}_1 = (2, 3)$, $\mathbf{a}_2 = (-1, 2)$, $\mathbf{a}_3 = (3, 5)$ are linearly dependent and (b) $\mathbf{b}_1 = (4, 5)$, $\mathbf{b}_2 = (6, 2)$, $\mathbf{b}_3 = (8, 10)$ are linearly dependent.

Solution: In each case there are three vectors in E^3. Hence the vectors are linearly dependent.

Now in (i) let there be three constants λ_1, λ_2 and λ_3 [with at least one of them non-zero] such that the condition

$$\lambda_1 \mathbf{a}_1 + \lambda_2 \mathbf{a}_2 + \lambda_3 \mathbf{a}_3 = 0 \text{ be satisfied}$$

$$\text{or, } \lambda_1(2, 3) + \lambda_2(-1, 2) + \lambda_3(3, 5) = (0, 0).$$

Equating we get

$$2\lambda_1 - \lambda_2 + 3\lambda_3 = 0 \quad \text{(i)}$$

$$3\lambda_1 + 2\lambda_2 + 5\lambda_3 = 0. \quad \text{(ii)}$$

From (i) and (ii)

$$\frac{\lambda_1}{-5-6} = \frac{\lambda_2}{9-10} = \frac{\lambda_3}{4+3} = k = 1 \text{ (say)}.$$

Therefore, $\lambda_1 = -11$, $\lambda_2 = -1$ and $\lambda_3 = 7$. Hence the vectors are linearly dependent. [Putting $k = 0$ of course $\lambda_1 = \lambda_2 = \lambda_3 = 0$ is also a solution which is called a *trivial solution*.]

(b) As in (a) we have,

$$\lambda_1\mathbf{b}_1 + \lambda_2\mathbf{b}_2 + \lambda_3\mathbf{b}_3 = \mathbf{0} \text{ [with at least one of } \lambda_1, \lambda_2, \lambda_3 \text{ non-zero]}$$
$$\text{or, } \lambda_1(4,5) + \lambda_2(6,2) + \lambda_3(8,10) = (0,0).$$

Equating we get

$$4\lambda_1 + 6\lambda_2 + 8\lambda_3 = 0 \quad \text{(i)}$$
$$5\lambda_1 + 2\lambda_2 + 10\lambda_3 = 0. \quad \text{(ii)}$$

From (i) and (ii)

$$\frac{\lambda_1}{60-16} = \frac{\lambda_2}{40-40} = \frac{\lambda_3}{8-30} = k = 1 \text{ (say)}.$$

Therefore, $\lambda_1 = 44$, $\lambda_2 = 0$, $\lambda_3 = -22$ [Hence the vectors are L.D.]

3. Prove that the vectors (4, 3, 2), (2, 1, 4) and (2, 3, −8) are linearly dependent.

Solution: Square matrix A, from with the components of the three vectors is

$$A = \begin{bmatrix} 4 & 3 & 2 \\ 2 & 1 & 4 \\ 2 & 3 & -8 \end{bmatrix} \quad \text{and} \quad |A| = \begin{vmatrix} 4 & 3 & 2 \\ 2 & 1 & 4 \\ 2 & 3 & -8 \end{vmatrix} = 0.$$

Hence the rank of the matrix A is less than 3.

Therefore, the vectors are linearly dependent. [by prop. (10)]

2nd method: Let the vectors be linearly dependent and it is possible to find three scalar quantities $\lambda_1, \lambda_2, \lambda_3$ [with at least one of them non-zero] such that the relation $\lambda_1(4,3,2) + \lambda_2(2,1,4) + \lambda_3(2,3,-8) = (0,0,0)$ holds.

Equating we get,

$$4\lambda_1 + 2\lambda_2 + 2\lambda_3 = 0 \quad \text{(i)}$$
$$3\lambda_1 + \lambda_2 + 3\lambda_3 = 0 \quad \text{(ii)}$$
$$2\lambda_1 + 4\lambda_2 - 8\lambda_3 = 0. \quad \text{(iii)}$$

From (i) and (ii)

$$\frac{\lambda_1}{6-2} = \frac{\lambda_2}{6-12} = \frac{\lambda_3}{4-6} = k = 1 \text{ (say)}.$$
$$\text{or, } \lambda_1 = 4, \lambda_2 = -6, \lambda_3 = -2.$$

With these values of λ_1, λ_2 and λ_3 equation (iii) is also satisfied.

Hence the three vectors are L.D. and we may take $\lambda_1 = 2$, $\lambda_2 = -3$, $\lambda_3 = -1$ [dividing by 2].

4. Prove that the vectors (1, 2, 0), (0, 3, 1), $(-1, 0, 0)$ are linearly independent.

Solution: Square matrix A formed with the components of the three vectors is

$$A = \begin{bmatrix} 1 & 2 & 0 \\ 0 & 3 & 1 \\ -1 & 0 & 0 \end{bmatrix} \quad \text{and} \quad \det A = |A| = \begin{vmatrix} 1 & 2 & 0 \\ 0 & 3 & 1 \\ -1 & 0 & 0 \end{vmatrix} = -2 \neq 0.$$

Hence the rank of the square matrix A is 3. Therefore, the vectors are linearly independent.

2nd method: Let the vectors be linearly dependent and it is possible to find three scalar quantities λ_1, λ_2, λ_3 [at least one of them non-zero] such that $\lambda_1(1,2,0) + \lambda_2(0,3,1) + \lambda_3(-1,0,0) = (0,0,0)$ be satisfied.

Equating we get

$$\lambda_1 + 0 \cdot \lambda_2 - \lambda_3 = 0 \qquad \text{(i)}$$
$$2\lambda_1 + 3\lambda_2 + 0\lambda_3 = 0 \qquad \text{(ii)}$$
$$0 \cdot \lambda_1 + \lambda_2 - 0 \cdot \lambda_3 = 0. \qquad \text{(iii)}$$

From (i) and (ii)

$$\frac{\lambda_1}{3} = \frac{\lambda_2}{-2} = \frac{\lambda_3}{3} = k = 1 \text{ (say)}.$$

Therefore, $\lambda_1 = 3$, $\lambda_2 = -2$, $\lambda_3 = 3$. With these set of values, equation (iii) will not be satisfied. Therefore, it is not possible to find three scalar quantities λ_1, λ_2, λ_3 [with at least one of them non-zero] such that the condition $\lambda_1(1,2,0) + \lambda_2(0,3,1) + \lambda_3(-1,0,0) = (0,0,0)$ is satisfied. The conditions will be satisfied only for $\lambda_1 = \lambda_2 = \lambda_3 = 0$. Hence the vectors are linearly independent.

3rd method: Let us assume that the vectors be linearly dependent and it is possible to find three scalar quantities λ_1, λ_2, λ_3 [at least one of them non-zero] such that the relation $\lambda_1(1,2,0) + \lambda_2(0,3,1) + \lambda_3(-1,0,0) = (0,0,0)$ be satisfied.

Equating we get

$$\lambda_1 + 0\lambda_2 - \lambda_3 = 0$$
$$2\lambda_1 + 3\lambda_2 + 0\lambda_3 = 0$$
$$0\lambda_1 + \lambda_2 + 0\lambda_3 = 0.$$

Now the coefficient matrix $= \begin{bmatrix} 1 & 0 & -1 \\ 2 & 3 & 0 \\ 0 & 1 & 0 \end{bmatrix} = A^t$ [as shown in the first method]

and $\det A^t = \det A = -2 \neq 0$. Then by Cramer's rule or by matrix inverse method it can be said that there exists unique solution for λ_1, λ_2, λ_3 and only solution is $\lambda_1 = \lambda_2 = \lambda_3 = 0$. Hence our assumption is wrong and thus the vector and not L.D. and therefore they are L.I.

5. (a) Prove that $\mathbf{a} = (2,3)$ and $\mathbf{b} = (4,-1)$ from a basis in E^2 and express $(10,8)$ as the linear combination of the basis vectors.

(b) Prove that the vectors (1, 2, 3), (2, 4, 1), (3, 2, 9) form a basis in three dimensional vector space E^3 and also prove that $(2,16/3,1)$ is a vector which may replace any one the three vectors to form a new basis set.

(c) Prove that the vectors (1, 1, 0) and (3, 0, 1) are L.I. and express (5, 2, 1) as the L.C. of the vectors (1, 1, 0) and (3, 0, 1). [C.U. 1983]

(d) Prove that in E^3, the vectors $(1,0,-1)$ and (0, 2, 3) are L.I. but the vector (1, 2, 3) cannot be expressed as the L.C. of the above vectors.

Solution: (a) There are two vectors in E^2 and $B = \begin{bmatrix} 2 & 3 \\ 4 & -1 \end{bmatrix}$ and $|B| = -14 \neq 0$. Hence the vectors are linearly independent and vector $(a_1, a_2) \in E^2$ can be expressed as the L.C. of the vector (2, 3) and $(4, -1)$ as $(a_1, a_2) = c_1(2,3) + c_2(4,-1)$, c_1, c_2 being scalars $\Rightarrow a_1 = 2c_1 + 4c_2$, $a_2 = 3c_1 - c_2$ has unique solution for c_1 and c_2 whatever may be the value of a_1, a_2 $[\because \begin{vmatrix} 2 & 4 \\ 3 & -1 \end{vmatrix} = -14 \neq 0]$ and evidently they form a basis in E^2.

Now $(10, 8) = \lambda_1 \mathbf{a} + \lambda_2 \mathbf{b}$ (say)

$= \lambda_1(2,3) + \lambda_2(4,-1)$.

Equating we get

$$10 = 2\lambda_1 + 4\lambda_2$$

$$\text{and} \quad 8 = 3\lambda_1 - \lambda_2.$$

Solving we get

$$\lambda_1 = 3, \lambda_2 = 1.$$

Thus $(10, 8) = 3\mathbf{a} + \mathbf{b}$.

(b) $A = \begin{bmatrix} 1 & 2 & 3 \\ 2 & 4 & 1 \\ 3 & 2 & 9 \end{bmatrix}$ and $|A| = \begin{vmatrix} 1 & 2 & 3 \\ 2 & 4 & 1 \\ 3 & 2 & 9 \end{vmatrix} = -20 \neq 0$, thus $r(A) = 3$.

Hence the three vectors are linearly independent.

Now let any vector $\mathbf{a} = (a_1, a_2, a_3) \in E^3$. We shall prove that $\mathbf{a}$ can be expressed as the L.C. of the vectors (1, 2, 3), (2, 4, 1), (3, 2, 9).

Let $(a_1, a_2, a_3) = c_1(1,2,3) + c_2(2,4,1) + c_3(3,2,9)$, c_1, c_2, c_3 being three scalars.

Equating we get

$$\left.\begin{aligned} a_1 &= c_1 + 2c_2 + 3c_3 \\ a_2 &= 2c_1 + 4c_2 + 2c_3 \\ a_3 &= 3c_1 + c_2 + 9c_3 \end{aligned}\right\}. \quad \text{(i)}$$

Now the coefficient matrix is

$$\begin{bmatrix} 1 & 2 & 3 \\ 2 & 4 & 2 \\ 3 & 1 & 9 \end{bmatrix} = A^t \text{ and } \det A^t = \det A = -20 \neq 0.$$

Therefore, there exists unique solution for c_1, c_2 and c_3 whatever may be the value of a_1, a_2 and a_3 and the vector $\mathbf{a} = (a_1, a_2, a_3)$ can be expressed as the L.C. of the vectors (1, 2, 3), (2, 4, 1) and (3, 2, 9). Hence the vectors from a basis for E^3.

Important remark: In E^3, any three vectors which are linearly independent from a basis for E^3. This property may be used in answering short answer type questions.

Let

$$(2, 16/3, 1) = \lambda_1(1,2,3) + \lambda_2(2,4,1) + \lambda_3(3,2,9).$$

Then

$$\left.\begin{aligned} \lambda_1 + 2\lambda_2 + 3\lambda_3 &= 2 \\ 2\lambda_1 + 4\lambda_2 + 2\lambda_3 &= 16/3 \\ 3\lambda_1 + \lambda_2 + 9\lambda_3 &= 1 \end{aligned}\right\}.$$

Solving these three equations we get

$$\lambda_1 = 1, \lambda_2 = 1, \lambda_3 = -1/3.$$

Hence

$$(2, 16/3, 1) = 1(1,2,3) + 1(2,4,1) - 1/3(3,2,9).$$

As (2, 16/3, 1) is not a null vector, $\lambda_1 \neq 0$, $\lambda_2 \neq 0$ and $\lambda_3 \neq 0$ therefore the vector (2, 16/3, 1) and any other two vectors will form a new basis set for E^3.

(c) The two vectors are L.I. as the vector (1, 1, 0) cannot be expressed in the manner (1, 1, 0) $= c(3,0,1)$ for any scalar value of c and evidently the vectors are not L.D. and thus they are L.I.

Let (5, 2, 1) $= c_1(1,1,0) + c_2(3,0,1)$, c_1, c_2 being scalars.

Equating we get

$$5 = c_1 + 3c_2, 2 = c_1 + 0\cdot c_2, 1 = 0\cdot c_1 + c_2.$$

For $c_1 = 2$ and $c_2 = 1$ all three conditions are satisfied and then (5, 2, 1) = 2 (1, 1, 0) + 1 (3, 0, 1).

(d) The vectors $(1,0,-1)$ and (0, 2, 3) are L.I. as $(1,0,-1)$ cannot be expressed by the relation $(1,0,-1) = c(0,2,3)$, c being any scalar. Thus the vectors are not L.D. and hence they are L.I.

Let (1, 2, 3)$= c_1(1,0,-1) + c_2(0,2,3)$, c_1, c_2 being scalars.

Equating we get $1 = c_1 + 0\cdot c_2$, $2 = 0\cdot c_1 + 2c_2$ and $3 = -c_1 + 3c_2$.

From the first two conditions we get $c_1 = 1, c_2 = 1$. But with these values the third condition will not be satisfied. Hence we cannot find two scalar quantities c_1 and c_2 such that the relation (1, 2, 3) $= c_1(1,0,-1) + c_2(0,2,3)$ holds. Hence $(1,2,3)$ cannot be expressed as the L.C. of the vectors.

Remark: Any vector in E^3 can be expressed as the L.C. of the set of three vectors in E^3 provided they form a basis for E^3. Here the two vectors do not form a basis for E^3.

Objective and Short Answer Type Questions with Answers

Vector

1. Define: (a) vector, (b) linear combination of a set of vectors, (c) Linearly dependent set of vectors, (d) Linearly independent set of vectors, (e) Basis set, (f) Standard basis or unit basis. [*Ans:* See sec. 1.3.]

2. Which of the statements given below are true?

   (a) A set of vectors are either linearly dependent or linearly independent.
   (b) A set of vectors containing a null vector are always linearly dependent.
   (c) The number of vectors in a basis set is $2n$ in E^n.
   (d) The representation of a vector by a basis is unique.
   (e) Any superset of a linearly independent of set of vector is linearly independent.
   (f) Any superset of a linearly dependent set of vectors is linearly dependent.
   (g) All basis sets in E^n are generating sets.
   (h) The transpose of a row vector is a column vector.
   (i) All generating sets are basis sets.

   [*Ans:* (a) true, (b) true, (c) false, (d) true, (e) false, (f) true, (g) true, (h) true, (i) false.]

3. Define the inner product or scalar product of two vectors. State the nature of the vectors associated with the inner product.

   [*Ans:* In the case of inner product of two vectors **a** and **b** (having same number of components) the first vector must be a row vector and the second one is a column vector and $\mathbf{a} \cdot \mathbf{b} \equiv \mathbf{b}'\mathbf{a}'$, where $\mathbf{a}'$ is is the transpose of **a** etc.]

4. A vector **a** in E^3 is given by $\mathbf{a} = a_1\mathbf{e}_1 + a_2\mathbf{e}_2 + a_3\mathbf{e}_3$, where $\mathbf{e}_1, \mathbf{e}_2$ etc. have their usual meanings. What are the components of **a**? [*Ans:* $\mathbf{a} = (a_1, a_2, a_3)$.]

5. Define a unit vector in E^n.

6. Vectors $\mathbf{a}_1, \mathbf{a}_2, \mathbf{a}_3$ form a basis in E^3. **a** is also a non-null vector in E^3. **a** is expressed as the linear combination given by $\mathbf{a} = 2\,\mathbf{a}_1 + 0 \cdot \mathbf{a}_2 - 5\mathbf{a}_3$.

   (a) Do the vectors $\mathbf{a}, \mathbf{a}_2, \mathbf{a}_3$ form a new basis in E^3?
   (b) Do the vectors $\mathbf{a}, \mathbf{a}_1, \mathbf{a}_2$ form a new basis in E^3?
   (c) Do the vectors $\mathbf{a}, \mathbf{a}_1, \mathbf{a}_3$ form a new basis in E^3?

   Give reasons for your answer in each case.

   [*Ans:* As the coefficient of $\mathbf{a}_2$ in the above linear combination is zero, **a** cannot replace $\mathbf{a}_2$ to form a new basis.

   Hence the vectors $\mathbf{a}_1, \mathbf{a}, \mathbf{a}_3$ cannot form a new basis. In other two cases they form a new basis.]

7. Does a basis contain a null vector? Give reasons for your answer.

   [*Ans:* No; because a set of vectors containing a null vector are always linearly dependent. Hence the vectors do not form a basis.]

8. (a) Give a suitable example of generating set as well as basis in E^3.

   (b) Prove that vectors $(1, 4, -1), (2, 5, -2)$ are linearly independent in E^3 but the vector $(1, 7, -1)$ cannot be expressed as the linear combination of the above two vectors.

   (c) Prove that the vector $(5, 1, -3)$ and $(2, 1, 5)$ are linearly independent in E^3 but the vector (7, 2, 4) cannot be expressed as the linear combination of the above vectors. Give reasons for your answer.

   (d) Prove that the following vectors in E^4 are L.D. (i) $(2, 4, -1, 2)$ and (ii) $(8, 16, -4, 8)$.

   (e) Prove that vectors $(1, 1, 0, -2)$ and $(0, 2, 4, -1)$ in E^4 are L.I.

   (f) Prove that the vectors (1, 2, 3) and (3, 2, 1) are linearly independent in E^3 but they do not form a basis set of vectors. Give reasons for your answer.

[*Ans:* (a) The standard basis set (1, 0, 0), (0, 1, 0) and (0, 0, 1) is a suitable example of basis set in E^3. As the set of vector form a basis set, evidently the set is a generating set.

(b) $(1, 4, -1)$ cannot be expressed as $(1, 4, -1) = c(2, 5, -2)$ [c being any scalar.] Hence they are L.I. $(1, 7, -1) = 3(1, 4, -1) - (2, 5, -2)$.

(c) The vectors are L.I. but they are two in numbers. Hence they cannot form a basis in E^3. Thus any vector in E^3 cannot be expressed as the linear combination of the above vectors.

(d) $(2, 4, -1, 2) = 1/2(8, 16, -4, 8)$. Hence the vectors are L.D.

(e) (1, 1, 0, 2) cannot be expressed as $(1, 1, 0, -2) = c(0, 2, 4, -1)$ [c being any scalar.] Thus the vectors are L.I.

(f) The reason is same as given in (c).]

9. (a) Do the vectors (4, 2, 2), (2, 1, 4) and $(2, 3, -8)$ form a basis for E^3? Give reasons for your answer.

   (b) Do the vectors (1, 2, 3), (2, 4, 1) and (3, 2, 9) constitute a basis for E^3? Give reasons.

   (c) Prove that the vectors (1, 1, 0), $(1, -1, 0)$, (0, 0, 1) form a basis for E^3. [C.U.(P) 1983]

   (d) Do the vectors (1, 2, 3), (4, 5, 6) and (3, 6, 9) form a basis for E^3? [C.U.(P) 1981, 1995]

   (e) Prove that the vectors $\mathbf{a} = (1, 2, 1), \mathbf{b} = (2, 3, 0)$ and $\mathbf{c} = (1, 2, 2)$ form a basis for E^3. [C.U.(P) 1987]

   (f) Prove that the vectors (1, 1, 0), (0, 1, 1) and (1, 2, 1) form a basis of E^3. [C.U.(P) 1990]

[*Ans:* (a) Vectors are linearly dependent; thus they do not form a basis. (b) There are three vectors in E^3 and they are linearly independent. Hence they constitute a basis in E^3. (c) Reason is same as (b).]

10. (a) Prove that a set of vector containing only one non-null vector is always linearly independent.

    (b) If a set of vectors are linearly dependent at least one of them can be expressed as a linear combination of the others.

(c) If two non-null vectors are linearly dependent then one of them is a scalar multiple of the other.

[*Ans:* (a) Let **a** be a non-null vector; therefore $\lambda \mathbf{a} = \mathbf{0}$ will be satisfied if and only if $\lambda = 0$ which indicates that **a** is linearly independent.

(b) Let $\mathbf{a}_1, \mathbf{a}_2, \cdots, \mathbf{a}_n$ be linearly dependent. They $\lambda_1 \mathbf{a}_1 + \lambda_2 \mathbf{a}_2 + \cdots + \lambda_n \mathbf{a}_n = 0$ will be satisfied, with at least one of $\lambda_1, \cdots \lambda_n$ not equal to zero. Let $\lambda_1 \neq 0$ then

$$\mathbf{a}_1 = -\frac{\lambda_2}{\lambda_1}\mathbf{a}_2 - \frac{\lambda_3}{\lambda_1}\mathbf{a}_3 - \cdots - \frac{\lambda_n}{\lambda_1}\mathbf{a}_n.$$

Hence $\mathbf{a}_1$ can be expressed as a linear combination of the others.

(c) If two non-null vectors $\mathbf{a}_1$ and $\mathbf{a}_2$ are linearly dependent then $\lambda_1 \mathbf{a}_1 + \lambda_2 \mathbf{a}_2 = \mathbf{0}$ [with at least one of λ_1 and λ_2 not zero] will be satisfied if and only if both $\lambda_1 \neq 0$, $\lambda_2 \neq 0$.

Therefore,

$$\mathbf{a}_1 = -\frac{\lambda_2}{\lambda_1}\mathbf{a}_2 \quad \text{and} \quad \mathbf{a}_2 = -\frac{\lambda_1}{\lambda_2}\mathbf{a}_1 \text{ etc.}$$

11. (a) Express $\mathbf{x} = [4, 5]$ as a linear combination of $\mathbf{a} = [1, 3], \mathbf{b} = [2, 2]$.

[C.U.(P) 1983]

(b) Prove that the vectors (4, 5), (6, 2), (8, 10) are linearly dependent and express null vector as a linear combination of these vectors.

[*Ans:* (a) $\mathbf{x} = \frac{1}{2}\mathbf{a} + \frac{7}{4}\mathbf{b}$. (b) The vectors are linearly dependent as there are (2+ 1) vectors in E^2.]

Exercise 1

1. Prove that if two non-null vectors are linearly dependent then one of them is a scalar multiple of the other.

2. Prove that the following sets of vectors are linearly independent:

   (a) $\mathbf{a} = (2, 1, 4), \mathbf{b} = (1, -1, -2), \mathbf{c} = (3, 1, -2)$.

   (b) $\mathbf{a} = (1, 2, 1), \mathbf{b} = (3, 1, 5), \mathbf{c} = (3, 4, 7)$.

   (c) $\mathbf{a} = (1, 2, -1), \mathbf{b} = (3, -1, 2), \mathbf{c} = (2, -2, 3)$.

   (d) $\mathbf{a} = (1, 2, 3), \mathbf{b} = (3, -2, 1), \mathbf{c} = (4, 2, 1)$.

   (e) $(1, -1, 1, 2), (1, 2, -1, 4)$ and $(1, 1, -1, -5)$.

3. Prove that the following sets of vectors are linearly dependent:

   (a) (2, 3), (3, 4), (10, 15).

   (b) $(1, 2, -4), (3, 2-1), (6, 8, -13)$.

   (c) $\mathbf{a} = (-1, 3, -5), \mathbf{b} = (2, 0, 4), \mathbf{c} = (1, 3, -1)$.

   (d) $(2, 2, 8), (1, 0, 4), (1, 2, 4)$. [C.U.(P) 1989]

   (e) $\mathbf{a} = (1, 1, 2, 4), \mathbf{b} = (2, -1, -5, 2), \mathbf{c} = (2, 1, 1, 6)$.

4. (a) Prove that the vectors $\mathbf{a} = (2, 3)$ and $\mathbf{b} = (1, 4)$ form a basis for E^2 and express the vector $\mathbf{c} = (5, 15)$ as the linear combination of **a** and **b**.

   (b) Prove that the vectors $\mathbf{a} = (1, 0, 1)$, $\mathbf{b} = (1, 2, 3)$ and $\mathbf{c} = (2, 3, 4)$ form a basis in E^3 and express $\mathbf{d} = (5, 5, 9)$ as the linear combination of the vectors.

   (c) Express (5, 0) as the linear combination of the vectors (4, 1) and (3, 2).

(d) Prove that the vectors $\mathbf{a} = (-1, 2, 2)$, $\mathbf{b} = (2, -3, 4)$ and $\mathbf{c} = (4, 6, 3)$ form a basis in E^3 and express $\mathbf{d} = (4, 7, 11)$ as the linear combination of the basis vectors.

(e) Do the vectors $\mathbf{a} = (4, -1)$ and $\mathbf{b} = (-2, 1)$ form a basis in E^2? If so, express $\mathbf{c} = (10, -5)$ as the linear combination of the basis vectors.

5. (a) Prove the vectors: (i) $(1, 0, 0)$, $(0, 1, 0)$, $(0, 0, 1)$ and (ii) $(1, 0, 0)$, $(1, 1, 0)$, $(1, 1, 1)$ are two basis sets in the vector space E^3.

(b) Prove that the vectors (1, 1, 0), (0, 1, 1) and (1, 2, 1) form a basis in E^3. [C.U.(P) 1990]

(c) Prove that the vectors (1, 1, 0), $(1, -1, 0)$ and (0, 0, 1) forms a basis for E^3. [C.U.(P) 1983]

6. Prove that vectors (1, 2, 3), (2, 4, 1) and (3, 2, 9) is a basis set in $V_3(F)$ and prove that (2, 6, 1) is a vector which may replace any one of the three vectors to form a new basis.

7. Prove that the vectors (2, 4, 7), (4, 8, 9) are linarly independent.

[*Hints:* One is not a scalar multiple of the other.]

8. If a set of vectors are linearly dependent then prove that at least one of them can be expressed as the linear combination of the remaining others.

9. (a) Prove that the vectors $\mathbf{a} = (2, -3)$ and $\mathbf{b} = (1, 4)$ form a basis in E^2 and express $\mathbf{c} = (3, 1)$ as the linear combination of the basis vectors. Hence prove that $(\mathbf{a}, \mathbf{c})$ and $(\mathbf{b}, \mathbf{c})$ form two basis in E^2.

(b) Express $\mathbf{d} = (2, 0, -3)$ as the linear combination of the basis vectors $\mathbf{a} = (1, 0, 0)$, $\mathbf{b} = (0, 1, 0)$ and $\mathbf{c} = (0, 0, 1)$ and hence prove that the vectors $\mathbf{a}, \mathbf{c}, \mathbf{d}$ will not be a basis, but the set of vectors $\mathbf{a}, \mathbf{b}, \mathbf{d}$ and $\mathbf{c}, \mathbf{b}, \mathbf{d}$ will form basis sets.

[*Hints:* $\mathbf{a}, \mathbf{b}, \mathbf{c}$ form a basis in E^3 and $\mathbf{d} = 2\mathbf{a} + 0\mathbf{b} - 3\mathbf{c}$, $\mathbf{d}$ is non-null and thus $\mathbf{d}, \mathbf{a}, \mathbf{c}$ are not L.I. but $\mathbf{d}, \mathbf{a}, \mathbf{b}$ and $\mathbf{d}, \mathbf{c}, \mathbf{b}$ are L.I. (by replacement theorem).]

10. (a) Prove that in E^3, the set of vectors (2, 1, 4), (1, 0, 2) are linearly independent are express (4, 1, 8) as the linear combination of the above vectors.

(b) Prove that in E^3, though the vectors $(0, -1, 2)$ and (1, 2, 0) are linearly independent, but the vectors $(0, -2, 1)$ cannot be expressed as the linear combination of the above vectors. Give reasons.

(c) Given the basis vectors $\mathbf{a}_1 = (1, 0, 0)$, $\mathbf{a}_2 = (0, 1, 1)$ and $\mathbf{a}_3 = (0, 0, 1)$ for E^3. Indicate one vector which can be removed from the basis and be replaced by $\mathbf{b} = (4, 3, 6)$ while still maintaining basis. [C.U.(H) 1990]

Answers

4. (a) $\mathbf{c} = \mathbf{a} + 3\mathbf{b}$, (b) $\mathbf{d} = 2\mathbf{a} + \mathbf{b} + \mathbf{c}$, (c) $2\mathbf{a} - \mathbf{b}$, (d) $\mathbf{d} = \mathbf{a} + \mathbf{b} + \mathbf{c}$. 9. (a) $\mathbf{c} = \mathbf{a} + \mathbf{b}$.
10. (a) (2,1,4)+2(1,0,2), (c) Any one of the three vectors.

Chapter 2

Motivation and Mathematical Formulation of Linear Programming Problems

2.1 Introduction

With the limited available resources (such as raw materials, manpower, capital, power and technical appliance etc.) the main object of an industry is to produce different products in such a way that maximum profit may be earned by selling them at market prices. Similarly, the main aim of a housewife is to buy the goodgrains, vegetables, fruits and other food materials at a minimum cost which will satisfy the minimum need (regarding food values, calories, proteins, vitamins etc.) of the members of her family. All these (the nature of production of different commodities etc.) can be done mathematically by formulating a problem which is known as a *programming problem.* Some *restrictions* or *constraints* are to be adopted to formulate the problem. The function which is to be optimized (such as profit, cost etc. either maximized or minimized) is known as the *objective function.* Almost in all types of problems, the objective function and the constraints are of linear type, and these problems are known as the *Linear programming problems.*

There are different types of problems which can be put in the form of linear programming problems. The problems discussed above are known as *production allocation problem* and *diet problem* respectively.

Consider the problem of transporting various amounts of a single product from different warehouses to different destinations with the help of different transport systems such that the cost of transport be minimum. This type of problem is known as *transportation problem.*

2.2 Motivation of the subject

Motivation of the subject is to make overall maximum profit or minimum loss satisfying the constraints. Initially Mathematicians try to solve by using Calculus, but they failed to do so. In the last part of the fifth decade Mr. Danzig and other Mathematicians try to solve the problem and they together discovered the simplex method to solve the problems. Very recently, Dr. Nagendra Karmakar discovered a method to solve the problems more easily and that method is known as *Karmakar Algorithm.* But this method has not been discussed in the book.

2.3 Mathematical formulation of the problems

Almost in each and every sphere of life, all practical problems can be put in the form of programming problem and if the objective function be a linear function and if the constraints be all linear constraints, then the programming problems are called *linear programming problems.* Collecting all informations or data about the problem initially these are expressed in ordinary language in a descriptive manner and from this, the problem are expressed strictly in mathematical form before solving them by any method. Below given are some illustrative examples from which we can get some idea, how the problems can be put in strictly mathematical form.

2.3.1 Production allocation problem

▶ **Example 2.3.1** *Four different metals namely, iron, copper, zinc and manganese are required to produce three commodities* A, B *and* C*. To produce one unit of* A*,* 40 kg *iron,* 30 kg *copper,* 7 kg *zinc and* 4 kg *manganese are needed. Similarly to produce one unit of* B*,* 70 kg *iron,* 14 kg *copper and* 9 kg *manganese are needed and for producing one unit of* C*,* 50 kg *iron,* 18 kg *copper and* 8 kg *zinc are required. The total available quantities of metals are* 1 metric ton *iron,* 5 quintals *of copper,* 2 quintals *of zinc and manganese each. The profits are* Rs. 300, Rs. 200 *and* Rs. 100 *in selling per one unit of A, B and C respectively. Formulate the problem mathematically.*

Solution: Let z be the total profit and the problem is to maximize z.

z is known as the *objective function.*

All the available quantities and the quantities required to produce different commodities are given below in the tabular form:

	Iron	Copper	Zinc	Manganese
Total	1000 kg	500 kg	200 kg	200 kg
A	40 kg	30 kg	7 kg	4 kg
B	70 kg	14 kg	0 kg	9 kg
C	50 kg	18 kg	8 kg	0 kg

To get the maximum profit let x_1 units of A, x_2 units of B and x_3 units of C are to be produced.

Then total quantity of iron needed is $(40x_1 + 70x_2 + 50x_3)$ kg.

Similarly total quantity of copper needed is $(30x_1 + 14x_2 + 18x_3)$ kg.

Zinc $(7x_1 + 0x_2 + 8x_3)$ kg.

Manganese

$$(4x_1 + 9x_2 + 0x_3) \text{ kg.}$$

and from the condition of the problem

$$\begin{aligned} 40x_1 + 70x_2 + 50x_3 &\leq 1000 \\ 30x_1 + 14x_2 + 18x_3 &\leq 500 \\ 7x_1 + 0x_2 + 8x_3 &\leq 200 \\ 4x_1 + 9x_2 + 0x_3 &\leq 200 \end{aligned}$$

The objective function

$$z = 300x_1 + 200x_2 + 100x_3$$

which is to be maximized. Hence the problem can be formulated as,

$$\text{Maximize,} \quad z = 300x_1 + 200x_2 + 100x_3$$

subject to

$$\begin{aligned} 40x_1 + 70x_2 + 50x_3 &\leq 1000 \\ 30x_1 + 14x_2 + 18x_3 &\leq 500 \\ 7x_1 + 0x_2 + 8x_3 &\leq 200 \\ 4x_1 + 9x_2 + 0x_3 &\leq 200 \end{aligned}$$

As none of the commodity produced be negative in number then $x_1 \geq 0$, $x_2 \geq 0$, $x_3 \geq 0$.

All these inequations are known as *constraints* or *restrictions.*

2.3.2 Diet problem

► **Example 2.3.2** *A patient needs daily* 5 mg, 20 mg *and* 15 mg *of vitamins A, B and C respectively. The vitamins available from a mango, an orange and an apple, are* 0.5 mg *of A,* 1 mg *of B,* 1 mg *of C;* 2 mg *of B,* 3 mg *of C;* 0.5 mg *of A,* 3 mg *of B and* 1 mg *of C respectively. If the cost of a mango, an orange and an apple be* Rs. 0.50, Rs. 0.25 *and* Rs. 0.40 *respectively, find the minimum cost of collecting the fruits so that daily requirement of the patient be met. Formulate the problem mathematically.*

Solution: The problem is to find the minimum cost of buying the food materials. Let z be the objective function.

Let x_1 mangoes, x_2 oranges and x_3 apples be bought to get the minimum daily requirement of vitamins so that the cost be minimum. Objective function z is given by $z = 0.50x_1 + 0.25x_2 + 0.40x_3$.

The quantity of vitamin A, available from the fruits, is

$$(0.5x_1 + 0x_2 + 0.5x_3) \text{ mg.}$$

Similarly the quantity of vitamin B available, is

$$(1x_1 + 2x_2 + 3x_3) \text{ mg.}$$

and vitamin C is

$$(1x_1 + 3x_2 + 1x_3) \text{ mg.}$$

From the condition of the problem

$$\begin{aligned} 0.5x_1 + 0x_2 + 0.5x_3 &\geq 5 \\ x_1 + 2x_2 + 3x_3 &\geq 20 \\ x_1 + 3x_2 + x_3 &\geq 15 \end{aligned}$$

and $x_1 \geq 0$, $x_2 \geq 0$, $x_3 \geq 0$.

Hence the problem is, Minimize,

$$z = 0.50x_1 + 0.25x_2 + 0.40x_3$$

subject to

$$\begin{aligned} 0.5x_1 + 0x_2 + 0.5x_3 &\geq 5 \\ x_1 + 2x_2 + 3x_3 &\geq 20 \\ x_1 + 3x_2 + x_3 &\geq 15 \end{aligned}$$

and $x_1, x_2, x_3 \geq 0$.

2.3.3 Transportation problem

▶ **Example 2.3.3** *Three different types of lorries A, B and C have been used to transport* 60 tons *solid and* 35 tons *liquid substance. A type lorry can carry* 7 tons *solid and* 3 tons *liquid. B type lorry can carry* 6 tons *solid and* 2 tons *liquid and C type lorry can carry* 3 tons *solid and* 4 tons *liquid. The cost of transport are* Rs. 500, Rs. 400 *and* Rs. 450 *per lorry of A, B and C type respectively. Find the minimum transport cost. Formulate the problem mathematically.*

Solution: The problem is to minimize the transport cost. Let z be the objective function.

Let x_1 lorries of A type, x_2 lorries of B type and x_3 lorries of C types be used to transport the materials so that the cost of transport be minimum. The objective function z is given by

$$z = 500x_1 + 400x_2 + 450x_3$$

The quantity of solid, transported by the lorries is

$$(7x_1 + 6x_2 + 3x_3) \text{ tons.}$$

The quantity of liquid, transported by the lorries is

$$(3x_1 + 2x_2 + 4x_3) \text{ tons.}$$

From the condition of the problem

$$\begin{aligned} 7x_1 + 6x_2 + 3x_3 &\geq 60 \\ 3x_1 + 2x_2 + 4x_3 &\geq 35. \end{aligned}$$

Hence the problem is, Minimize,

$$z = 500x_1 + 400x_2 + 450x_3$$

subject to

$$7x_1 + 6x_2 + 3x_3 \geq 60$$
$$3x_1 + 2x_2 + 4x_3 \geq 35.$$

and $x_1, x_2, x_3 \geq 0$.

▶ **Example 2.3.4** *An electronic company manufactures two radio models each on a separate production line. The daily capacity of the first line is 60 radios and that of the second is 75 radios. Each unit of the first model uses 10 pieces of a certain electronic component, where as each unit of the second model requires 8 pieces of the same component. The maximum daily availability of the special components is 800 pieces. The profit per unit of models 1 and 2 are Rs. 500 and Rs. 400 respectively. Determine graphically or otherwise the optimal daily production of each model.*

[C.U.(H) '91,'94,'96]

Solution: The problem is a problem of maximization.

Let x_1 and x_2 be the number of two radio models each on a separate production line. Therefore the objective function is

$$z = 500x_1 + 400x_2$$

which is to be maximized. [Since profit per of unit of models 1 and 2 are Rs. 500 and Rs. 400 respectively.]

Since the daily capacity of the first line and the second line are 60 and 75 radios respectively, then we have

$$x_1 \leq 60$$
$$x_2 \leq 75.$$

The total electronic components required $= 10x_1 + 8x_2$.

From the condition of the problem $10x_1 + 8x_2 \leq 800$.

Hence the problem is to maximize,

$$z = 500x_1 + 400x_2$$

subject to

$$x_1 \leq 60$$
$$x_2 \leq 75$$
$$10x_1 + 8x_2 \leq 800$$

and $x_1, x_2 \geq 0$.

By solving it using the graphical method (Chapter 7) it can be shown that the maximum profit will Rs. 40,000 but there will be different types of production for which the profit will be maximum such as for $(x_1 = 60, x_2 = 25)$; $(x_1 = 20, x_2 = 75)$, etc. profit will be maximum.

▶ **Example 2.3.5** *An agricultural firm has* 180 tons *of Nitrogen fertilisers,* 50 tons *of Phosphate and* 220 tons *of Potash. It will be able to sell* 3 : 3 : 4 *mixtures of these substances at a profit of* Rs. 15 *per ton and* 2 : 4 : 2 *mixtures at a profit of* Rs. 12 per *ton respectively. Pose a linear programming problem to how many tons of these two mixtures should be prepared to obtain the maximum profit?*

[C.U.(P)'93, 97]

Solution: Let the 3 : 3 : 4 mixture be called A and that of 2 : 4 : 2 be called B.

Let x_1 and x_2 tons of these two mixtures be produced to get maximum overall profit. Thus the objective function is $z = 15x_1 + 12x_2$ which is to be maximized.

Let us denote nitrogen, phosphate and potash as N, Ph, P respectively, in the mixture A and,

$$\frac{N}{3} = \frac{Ph}{3} = \frac{P}{4} = k_1 \text{ (say)}.$$

Hence

$$N = 3k_1,\ Ph = 3k_1,\ P = 4k_1, \text{ so that } x_1 = 10k_1.$$

Similarly for the mixture B,

$$N = 2k_2,\ Ph = 4k_2,\ P = 2k_2 \text{ so that } x_2 = 8k_2.$$

Thus the constraints are

$$\frac{3}{10}x_1 + \frac{1}{4}x_2 \leq 180$$

[∵ In the mixture A, amount of nitrogen $= \frac{3k_1}{10k_1}x_1 = (3/10)x_1$]

Similarly,

$$\frac{3}{10}x_1 + \frac{1}{2}x_2 \leq 250$$

$$\text{and} \quad \frac{2}{5}x_1 + \frac{1}{4}x_2 \leq 220, \quad x_1, x_2 \geq 0.$$

Thus the constraints are

$$6x_1 + 5x_2 \leq 3600$$
$$3x_1 + 5x_2 \leq 2500$$
$$8x_1 + 5x_2 \leq 4400, \quad x_1, x_2 \geq 0$$

and the problem is, maximize $z = 15x_1 + 12x_2$.

▶ **Example 2.3.6** *A coin to be minted contains* 40% *silver,* 50% *copper,* 10% *nickel. The mint has available alloys A. B, C and D having the following composition and costs:*

	% Silver	*% Copper*	*% Nickel*	*Costs*
A	*30*	*60*	*10*	*Rs. 11.00*
B	*35*	*35*	*30*	*Rs. 12.00*
C	*50*	*50*	*0*	*Rs. 16.00*
D	*40*	*45*	*15*	*Rs. 14.00*

Present the problem of getting the alloys with specific composition at minimum cost in the form of a L.P.P. (Solution is not necessary.) [C.U.(P)'94, '99]

Solution: The question is incomplete since the amount of the metals are not given. We assume that the total quantity of metals is 1000 kg. and further assume that the prices of the alloys given in the table are per kg.

Let x_1, x_2, x_3, x_4 kgs. be the quantities of alloys A, B, C, D used for the purpose in kg. so that

$$x_1 + x_2 + x_3 + x_4 \leq 1000.$$

The objective function is

$$z = 11x_1 + 12x_2 + 16x_3 + 14x_4$$

and the constraints are

$$\begin{aligned} 0.3x_1 + 0.35x_2 + 0.5x_3 + 0.4x_4 &\leq 400 \text{ for silver} \\ 0.6x_1 + 0.35x_2 + 0.5x_3 + 0.45x_4 &\leq 500 \text{ for copper} \\ 0.1x_1 + 0.3x_2 \qquad\qquad + 0.15x_4 &\leq 100 \text{ for nickel}, \quad x_1, x_2, x_3, x_4 \geq 0. \end{aligned}$$

Thus the L.P.P. is, minimize

$$z = 11x_1 + 12x_2 + 16x_3 + 14x_4$$

subject to

$$\begin{aligned} 0.3x_1 + 0.35x_2 + 0.5x_3 + 0.4x_4 &\leq 400 \\ 0.6x_1 + 0.35x_2 + 0.5x_3 + 0.45x_4 &\leq 500 \\ 0.1x_1 + 0.3x_2 \qquad\qquad + 0.15x_4 &\leq 100 \\ x_1 + x_2 + x_3 + x_4 &\leq 1000 \quad x_1, x_2, x_3, x_4 \geq 0. \end{aligned}$$

▶ **Example 2.3.7** *A soft drink plant has two bottling machines A and B. It produces and sells* 500 ml. *and* 800 ml. *bottles. The following data are available.*

Machine	*500 ml*	*800 ml*
A	*100/minute*	*40/minute*
B	*60/minute*	*75/minute*

The machines can be run 8 hours per day and 5 days per week. Weekly productions of the drink cannot exceed 30,00,000 ml. *and the market can absorb* 25,000 *bottles of* 500 ml *and* 7,000 *bottles of* 800 ml. *per week. Profit on two types of bottles is* 15 *paise and* 25 *paise respectively. The planner wishes to maximize his profit to all the productions and marketing restrictions. Formulate it as a linear programming problem.* [C.U.(P)'98]

Solution: Let x_1 and x_2 be number of 500 ml and 800 ml bottles produced to get an over all maximum profit. Then the profit is

$$\left(x_1 \times \frac{15}{100} + x_2 \times \frac{25}{100}\right) \text{Rs.} = (0.15x_1 + 0.25x_2) \text{ Rs.} \quad \text{(say)}.$$

From the market condition, we get

$$x_1 \leq 25{,}000$$
$$x_2 \leq 7{,}000$$

Total production of 500 ml bottles $= (100 + 60) \times 60 \times 8 \times 5$ per week.

Total production of 800 ml bottles $= (40 + 75) \times 60 \times 8 \times 5$ per week.

Thus the production of 500 ml and 800 ml bottles per week are 3,84,000 and 2,76,000 respectively. As the number of bottles that can be produced are greater than 25,000 and 7,000 respectively then these conditions are redundant.

The amount of soft drinks is $500x_1 + 800x_2$ ml.

Then

$$500x_1 + 800x_2 \leq 30{,}00{,}000.$$

Thus the problem is,

$$\text{Maximize,} \quad z = 0.15x_1 + 0.25x_2$$

subject to

$$\begin{aligned} 500x_1 + 800x_2 &\leq 30{,}00{,}000 \\ x_1 \quad\quad &\leq \quad 25{,}000 \\ x_2 &\leq \quad 7{,}000, \quad x_1, x_2 \geq 0. \end{aligned}$$

▶ **Example 2.3.8** *A hospital has the following minimum requirement for nurses.*

Period	*Clock time (24 hours day)*	*Minimum number of nurses required*
1	6 A.M.–10 A.M.	60
2	10 A.M.– 2 P.M.	70
3	2 P.M.– 6 P.M.	60
4	6 P.M.–10 P.M.	50
5	10 P.M.– 2 A.M.	20
6	2 A.M.– 6 A.M.	30

Nurses report to the hospital wards at the beginning of each period and work for eight consecutive hours. The hospital wants to determine the minimum number of nurses so that there may be sufficient number of nurses available for each period. Formulate this as a L.P.P. [C.U.(H)'83, '97]

Solution: This is a minimization problem.

Let $x_1, x_2, \cdots, x_6$ be the number of nurses required for the period $1, 2, \cdots, 6$, etc.

Then the objective function is

$$\text{Minimize,} \quad z = x_1 + x_2 + \cdots + x_6.$$

Now the constraints can be written in the following manner.

x_1 nurses work for the period 1 and 2 and x_2 nurses work for the period 2 and 3 etc.

Thus for the period 2

$$x_1 + x_2 \geq 70.$$

Similarly,

$$x_2 + x_3 \geq 60$$
$$x_3 + x_4 \geq 50$$
$$x_4 + x_5 \geq 20$$
$$x_5 + x_6 \geq 30$$
$$x_6 + x_1 \geq 60$$

and $x_j \geq 0,\ j = 1, 2, \cdots, 6$.

2.4 Mathematical formulation of a L.P.P.

From the above discussions of the mathematical formulation of Linear Programming Problems (L.P.P.), general linear programming problem can be stated mathematically as follows.

Find out a set of values $x_1, x_2, \cdots, x_n$ which will optimize (either maximize or minimize) the linear function

$$z = c_1x_1 + c_2x_2 + c_3x_3 + \cdots + c_nx_n$$

subject to the restrictions

$$\left.\begin{array}{l} a_{11}x_1 + a_{12}x_2 + \cdots + a_{1n}x_n \ (\leq = \geq)\ b_1 \\ a_{21}x_1 + a_{22}x_2 + \cdots + a_{2n}x_n \ (\leq = \geq)\ b_2 \\ \cdots\cdots\cdots\cdots\cdots\cdots\cdots\cdots\cdots\cdots\cdots\cdots \\ a_{m1}x_1 + a_{m2}x_2 + \cdots + a_{mn}x_n \ (\leq = \geq)\ b_m \end{array}\right\} \qquad (2.4.1)$$

and the non-negative restrictions $x_j \geq 0,\ j = 1, 2, \cdots, n$,

where a_{ij}, c_j and b_i's $(i = 1, \cdots, m)$ are all constants and x_j's are variables. Each of the linear expressions on the left side connected to the corresponding constants on the right side, by only one of the signs "$\leq$", "$=$" and "$\geq$", is known as a *constraint*. A constraint is either an equation or an inequation.

The linear function

$$z = c_1x_1 + c_2x_2 + \cdots + c_nx_n$$

is known as the *objective function*.

By using the matrix and vector notations the problem can be expressed in a compact form given below.

Optimize $z = \mathbf{cx}$ subject to the restrictions

$$A\mathbf{x}(\leq = \geq)\mathbf{b}, \quad \mathbf{x} \geq \mathbf{0}, \tag{2.4.2}$$

where $A = [a_{ij}]$ is a $m \times n$ co-efficient matrix.

$\mathbf{c} = (c_1, c_2, \cdots, c_n)$ is a n-component row vector, which is known as a *cost* or *price* vector as it determines the cost of production or the price of a commodity.

$\mathbf{x} = [x_1, x_2, \cdots, x_n]$ is a n-component column vector which is known as *decision variable or legitimate variable vector.*

$\mathbf{b} = [b_1, b_2, \cdots, b_m]$ is a m-component column vector which is known as *requirement vector* and

$\mathbf{0} = [0, \cdots, 0]$ is a n-component *null column vector.*

If all constraints are equations then the L.P.P. is reduced to, Optimize $z = \mathbf{cx}$ or simply optimize, $\mathbf{cx}$ subject to

$$A\mathbf{x} = \mathbf{b}, \quad \mathbf{x} \geq \mathbf{0}. \tag{2.4.3}$$

This form is called the *standard form.*

[See (3–15) and (3–16) of page 76 [Linear Programming by Hadly] and page 48 of [Linear Programming by Gass]].

In all practical discussions $b_i \geq 0$, $i = 1, 2, \cdots, m$. If some of them be negative, then make them positive, multiplying both sides of the linear constraints, having negative values of b_i, by (-1).

Feasible solution to a L.P.P. : A set of values of the variables, which satisfy all the constraints and all the non-negative restrictions of variables, is known as the *feasible solution* (F.S.) to the L.P.P.

Optimal solution to a L.P.P. : A feasible solution to a L.P.P. which makes the objective function an optimum is known as the *optimal solution* to the L.P.P.

There are various methods of finding the optimal solution of a L.P.P. : (i) *Geometrical method or graphical method* and (ii) *Algebraical method (simplex method).*

Before solving a L.P.P. by simplex method, we shall discuss the method of finding out the solution set of a system of m linear equations with n variables $(n > m)$ in the Chapter 4.

2.5 Problem of maximization and minimization

A particular L.P.P. is either a problem of maximization or a problem of minimization. The problem of minimization of the objective function z is nothing but the problem of maximization of the function $(-z)$ and vice versa and

$$\min z = -\max(-z)$$

with same set of constraints and the same solution set etc.

We shall prove this theorem in some later chapter.

Short Answer Type Questions

1. What is a linear programming problem?
2. Write down the mathematical form of a general L.P.P.
   [*Ans.* See Art. 2.4.1.]
3. What is the relation between the optimum value of maximization and minimization problem ?
4. Write down the standard form of a general L.P.P.
   [*Ans.* See Art. 2.4.3]

2.6 Formulation of the problems mathematically

1. A particular company manufactures two products A and B. These products are processed on the same machine. It takes 25 minutes to process one unit of product A and 15 minutes for one unit of product B and the machine operates for a maximum of 35 hours in a week. Product A requires 1 kg. and product B, 2.5 kg. of raw-material per unit, the supply of which is 170 kgs per week. If the net profit from the product are Rs. 100 and Rs. 450 per unit respectively, find how much of each product should be produced per week, in order to get maximum profit. [Formulate it mathematically].

2. A manufacturer of lether belts makes three types of belts A, B and C which are processed on three machines M_1, M_2 and M_3. Belt A requires 2 hours on machine M_1 and 3 hours on machine M_3. Belt B requires 3 hours on machine M_1, 2 hours on machine M_2 and 2 hours on machine M_3 and Belt C requires 5 hours on machine M_2 and 4 hours on machine M_3. There are 8 hours of time per day available on machine M_1, 10 hours of time per day available on machine M_2 and 15 hours of time per day available on machine M_3. The profit per unit of A, B and C are Rs. 3.00, Rs. 5.00 and Rs. 4.00 respectively. Find out the daily production of each type of belts such that the profit be maximum.

3. A firm can produce three types of cloth say A, B and C. Three kinds of wool are required for it, say red wool, green wool and blue wool. One unit length of type A cloth needs 2 yards of red wool and 3 yards of blue wool; one unit length of type B cloth needs 3 yards of red wool, 2 yards of green wool and 2 yards of blue wool; and one unit length of type C cloth needs 5 yards of green wool and 4 yards of blue wool. The firm has only a stock of 8 yards of red wool, 10 yards of green wool and 15 yards of blue wool. It is assumed that the income obtained from one unit length of type A cloth is Rs. 3.00, of type B cloth is Rs. 5.00 and that of type C cloth is Rs. 4.00. Formulate the above problem as a linear programming problem.

4. A person requires 10, 12 and 12 units of chemical A, B and C respectively for his garden. A liquid product contains 3, 2 and 1 unit of A, B and C respectively per jar. A dry product contains 1, 2 and 4 units of A, B and C per packet. If the liquid product sells for Rs. 3.00 per jar and the dry product sells for Rs. 2 per packet then formulate the problem as a linear programming problem. [C.U.(P)'87]

5. Food X contains 6 units of vitamin A per gram and 7 units of vitamin B per gram and cost 12 paise per gram. Food Y contains 8 units of vitamin A per gram and 12 units of vitamin B and cost 20 paise per gram. The daily minimum requirement of vitamin A and B are 100 units and 120 units respectively. Find the minimum cost of product mix. (Formulate the problem). [C.U.(H)'93, (P)'95]
6. A manufacturer of furniture makes two products, chairs and tables. Processing of these products is done on two machines A and B. A chairs requires 2 hours on machine A and 6 hours on machine B. A table requires 5 hours on machine A and no time on machine B. There are 16 hours of time per day available on machine A and 20 hours on machine B. Profit gained by the manufacturer from a chair and table is Rs. 10 and Rs. 50 respectively. What should be the daily production of each of the two products to obtain maximum profit? (Formulate the problem).
7. A television company has three major departments for the manufacture of two models, A and B. Monthly capacities are given as follows.

	Per unit time requirement (hours)		Hours available this month
	Model A	Model B	
Department I	4.0	2.0	1600
Department II	2.5	1.0	1200
Department III	4.5	1.5	1600

The marginal profit of model A is Rs. 400 each and that of model B is Rs. 100 each. Assuming that the company can sell any quantity of either product due to favourable market conditions, determine the highest possible profit for this month. (Formulate it).
8. A company sells two different products A and B. The company makes a profit of Rs. 40 and Rs. 30 per unit of products A and B respectively. The two products are produced in a common production process and are sold in two different markets. The production process has a capacity of 30,000 man hours. It takes 3 hours to produce one unit of A and one hour to produce one unit of B. The market has been surveyed and the company officials feel that the maximum number of unit of A that can be sold is 8,000 and the maximum of B is 12,000 units. Subject to these limitations, formulate the problem as a L.P.P.
9. A furniture manufacturer wishes to determine the number of tables and chairs to be made by him in order to optimise the use of his available resources. These products utilize two different types of timber and he has in hand 1,500 board feet of the first type and 1,000 board feet of the second type. He has 800 man hours available for the total job. Each table and chair requires 5 and 1 board feet respectively of the first type timber and 2 and 3 board feet of the second type. 3 man hours are required to make a table and 2 man hours are needed to make a chair. The manufacturer makes a profit of Rs. 12 on a table and Rs. 5 on a chair. Write down the complete linear programming formulation of the problem in terms of maximising the profit. [C.U.(P)'86]
10. A tailor has 80 sq m of cotton material and 120 sq m of woollen material. A suit requires 1 sq m of cotton and 3 sq m of woollen material and a dress

requires 2 sq m of each. A suit sells for Rs. 500 and a dress sells for Rs. 400. Pose a L.P.P. in terms of maximizing the income. [B.U.(P)'96]

11. A farmer has a 100 acre farm. He can sell all tomatoes, lettuce or radishes he can raise. The priçe he can obtain is Rs. 1.00 per kg. for tomatoes Rs. 0.75 a head for lettuce and Rs. 2.00 per kg. for radishes. The average yield per acre is 2,000 kgs of tomatoes, 3,000 heads of lettuce and 1,000 kgs of radishes. Fertilizer is available at Rs. 0.50 per kg. and the amount per acre is 100 kgs each for tomatoes and lettuce and 50 kgs for radishes. Labour required for sowing, cultivating and harvesting per acre is 5 man days for tomatoes and radishes and 6 man-days for lettuce. A total of 400 man-days of labour are available at Rs. 20.00 per man-day. Formulate this problem as a linear programming model to maximize the total profit.

12. A factory is engaged in manufacturing three products, A, B and C which involve lathe work, grinding and assembling. The cutting, grinding and assembling times required for one unit of A are 2, 1 and 1 hours respectively. Similarly they are 3, 1, 3 hours for one unit of B and 1, 3, 1 hours for one unit of C. The profits on A, B and C are Rs. 2, Rs. 2 and Rs. 4 per unit respectively. Assuming that there are available 300 hours of lathe time, 300 hours of grinder time and 240 hours of assembly time, how many units of each product should be produced to maximize profit? (Formulate mathematically). [C.U.(H)'87,(P)'80]

13. A manufacturer makes red and blue pen. A red pen takes twice as much as time to make a blue pen one. If the manufacturer makes only blue pens, 500 can be made in a day. A red pen sells for Rs. 8 and at most 150 can be sold in a day. A blue pen sells for Rs. 5 and at most 250 can be sold in a day. The manufacturer desires to maximize his revenue. Formulate the manufacturer's problem as a linear programme. [C.U. M.Com.'89]

[*Hints:* Let there be 500 units of time available and 2 and 1 unit of time required to produce red and blue pen. Thus, if x_1 and x_2 be the number of red and blue pen then the problem is to maximize $8x_1 + 5x_2$ s.t. $2x_1 + x_2 \leq 500$, $x_1 \leq 150$, $x_2 \leq 250$, $x_1, x_2 \geq 0$.]

14. At a cattle breeding firm, it is prescribed that the food ration for one animal must contain at least 14, 22 and 11 units of nutrients A, B and C respectively. Two different kinds of fodder are available. Each unit weight of these two contains the following amounts of the three nutrients. [C.U.(H)'88, (P)'92]

	Fodder 1	Fodder 2
Nutrient A	2	1
Nutrient B	2	3
Nutrient C	1	1

It is given that the cost of fodder 1 and 2 are 3 and 2 monetary units respectively. Formulate the problem of finding the minimum cost of purchasing the fodders as a L.P.P.

15. ABC company wishes to plan its advertising strategy. There are two media under consideration, call them magazine I and magazine II. Magazine I has a reach of 2,000 potential customers and magazine II has a reach of 3,000 potential customers. The cost per page of advertising is Rs. 400 and Rs. 600 in magazines I and II respectively. The firm has a monthly budget of Rs. 6,000. There is an important requirement that the total reach for the income group

under Rs. 20,000 per annum, should not exceed 4,000 potential customers. (The reach in magazine I and II for this income group is 400 and 200 poential customers). The firm wants to buy as many pages for different advertisements in the two magazines so as to maximise the total reach. Formulate the above as L.P.P. [C.U.(H)'86]

16. A medicine firm manufacturing two types of medicines A and B, can make a profit of Rs. 20 per bottle of A and Rs. 30 per bottle of B. Both A and B need for their production two essential chemicals C and D. Each bottle of A requires 3 litres of C and 2 litres of D and each bottle of B requires 2 litres of C and 4 litres of D. The total supply of these chemicals are restricted to 210 litres of C and 300 litres of D. Type B medicine contains alcohol and so its manufacture is restricted to 65 bottles per day. Construct a linear programming problem in terms of maximizing the daily profit of the product. [C.U.(P)'91]

17. A manufacturer of furnitures makes two products; chairs and tables. Processing of the products is done on two machines A and B. A chair requires 2 hours an machine A and 6 hours on machine 13. A table requires 5 hours on machine A and 2 hours on machine B. There are 16 hours of time per day available on machine A and 22 hours on machine B. Profit gained by the manufacturer from a chair and a table is Re. 1 and Rs. 5 respectively. Formulate a linear programming problem to maximize profit per day. [C.U.(P)'90]

Answers

1. Maximize $z = 100x_1 + 450x_2$ subject to $5x_1 + 3x_2 \leq 420$, $2x_1 + 5x_2 \leq 340$, $x_1 \geq 0$, $x_2 \geq 0$.
2. Maximize $z = 3x_1 + 5x_2 + 4x_3$ subject to $2x_1 + 3x_2 \leq 8$, $2x_2 + 5x_3 \leq 10$, $3x_1 + 2x_2 + 4x_3 \leq 15$, $x_1, x_2, x_3 \geq 0$.
3. Same as Example 2.
4. Minimize $z = 3x_1 + 2x_2$ subject to $3x_1 + x_2 \geq 10$, $2x_1 + 2x_2 \geq 12$, $x_1 + 4x_2 \geq 12$, $x_1 \geq 0$, $x_2 \geq 0$.
5. Minimize $z = 12x_1 + 20x_2$ subject to $6x_1 + 8x_2 \geq 100$, $7x_1 + 12x_2 \geq 120$, $x_1 \geq 0$, $x_2 \geq 0$.
6. Maximize $z = 10x_1 + 50x_2$ sibject to $2x_1 + 5x_2 \leq 16$, $6x_1 \leq 20$, $x_1 \geq 0$, $x_2 \geq 0$.
7. Maximize $z = 400x_1 + 100x_2$ subject to $4x_1 + 2x_2 \leq 1,600$, $2.5x_1 + x_2 \leq 1,200$, $4.5x_1 + 1.5x_2 \leq 1,600$, $x_1 \geq 0$, $x_2 \geq 0$.
8. Maximize $z = 40x_1 + 30x_2$ subject to $3x_1 + x_2 \leq 3,0000$, $x_1 \leq 8,000$, $x_2 \leq 12,000$, $x_1 \geq 0$, $x_2 \geq 0$.
9. Maximize $z = 12x_1 + 5x_2$ subject to $5x_1 + x_2 \leq 1,500$, $2x_1 + 3x_2 \leq 1,000$, $3x_1 + 2x_2 \leq 800$, $x_1, x_2 \geq 0$.
10. Maximize $z = 500x_1 + 400x_2$ subject to $x_1 + 2x_2 \leq 80$, $3x_1 + 2x_2 \leq 120$, $x_1, x_2 \geq 0$.
11. Maximize $z = 1,850x_1 + 2,080x_2 + 1,875x_3$ subject to $x_1 + x_2 + x_3 \leq 100$, $5x_1 + 6x_2 + 5x_3 \leq 400$, $x_1 \geq 0$, $x_2 \geq 0$, $x_3 \geq 0$.
12. Maximize $z = 2x_1 + 2x_2 + 4x_3$ subject to $2x_1 + 3x_2 + x_3 \leq 300$, $x_1 + x_2 + 3x_3 \leq 300$, $x_1 + 3x_2 + x_3 \leq 240$, $x_1 \geq 0$, $x_2 \geq 0$, $x_3 \geq 0$.
14. Minimize $z = 3x_1 + 2x_2$ subject to $2x_1 + x_2 \geq 14$, $2x_1 + 3x_2 \geq 22$, $x_1 + x_2 \geq 11$, $x_1, x_2 \geq 0$.
15. Maximize $z = 2,000x_1 + 3,000x_2$ subject to $400x_1 + 600x_2 \leq 6,000$, $400x_1 + 200x_2 \leq 4,000$, $x_1, x_2 \geq 0$.
16. Maximize $z = 20x_1 + 30x_2$ subject to $3x_1 + 2x_2 \leq 210$, $2x_1 + 4x_2 \leq 300$, $x_2 \leq 65$, $x_1, x_2 \geq 0$.
17. Maximize $z = x_1 + 5x_2$ subject to $2x_1 + 5x_2 \leq 16$, $6x_1 + 2x_2 \leq 22$, $x_1, x_2 \geq 0$.

Chapter 3

Slack and Surplus Variables

3.1 Introduction

There are two methods of solving a L.P. Problem. (1) *Geometrical method* and (2) *Algebraical method.* In algebraical method, the problem can be solved only when all constraints are equations. We now show how the constraints can be converted into equations. The algebraical method is called the *simplex method* or *simplex algorithm.*

3.2 Slack variables

When the constraints are inequations connected by the sign "≤", in each inequation, a variable is added to the left hand side of it to convert it into an equation. These variables are known as *slack variables.* For example,

$$x_1 - 3x_2 + 4x_3 \leq 4$$

is an inequation connected with the sign "≤". Then a variable x_4 is added to the left hand side of the constraint inequation and then it is converted into an equation

$$x_1 - 3x_2 + 4x_3 + x_4 = 4.$$

The variable x_4 is known as a *slack variable.* Similarly, a slack variable x_5 is added to the left hand side of the constraint inequation

$$3x_1 + 2x_2 - x_3 \leq 10$$

to convert it into an equation

$$3x_1 + 2x_2 - x_3 + x_5 = 10.$$

From the discussion it is clear that the *slack variables are non-negative quantities.*

3.3 Surplus variables

When the constraints are inequations connected by the sign "≥" then in each inequation a variable is subtracted from the left hand side to convert it into an equation. These variables are known as the *surplus variables.* For example,

$$x_1 + 3x_2 + x_3 \geq 12$$

is an inequation connected with the sign "$\geq$" and subtracting a variable x_6 from the left hand side of the inequation, it is being converted into an equation

$$x_1 + 3x_2 + x_3 - x_6 = 12.$$

The variable x_6 is the surplus variable.

$$x_1 - 3x_2 + 7x_3 \geq -20$$

is being converted into an equation

$$x_1 - 3x_2 + 7x_3 - x_7 = -20$$

subtracting the surplus variable x_7 from the left hand side of the inequation. The surplus variables are also *non-negative quantities.*

But if we multiply both sides of the equation

$$x_1 - 3x_2 + 7x_3 - x_7 = -20$$

by (-1) to make the constant at the right hand side positive, the equation will be

$$-x_1 + 3x_2 - 7x_3 + x_7 = 20$$

and in that case the surplus variable will change into a slack variable.

Let a general L.P.P., containing r variables and m constraints be

$$\text{Optimize,} \quad z = c_1x_1 + c_2x_2 + \cdots + c_rx_r$$

subject to

$$a_{i1}x_1 + a_{i2}x_2 + a_{ij}x_j + \cdots + a_{ir}x_r \leq = \geq b_i, \; i = 1, 2, \cdots, m;$$

$$x_j \geq 0, \; j = 1, 2, \cdots, r, \tag{3.3.1}$$

where one and only one of the signs $\leq = \geq$ holds for each constraint; but the sign may vary from one constraint to another. Let out of m constraints $k(0 \leq k \leq m)$ constraints be inequations. Then introducing k variables (slack or surplus) $x_{r+1}, x_{r+2}, \cdots, x_{r+k}$ one to each of the inequations by using the technique discussed above all constraints can be converted into equations containing $r + k = n$ (say) variables. We further assume that $n \geq m$. Now the objective function

$$z = c_1x_1 + c_2x_2 + \cdots + c_rx_r$$

is similarly accommodated with k slack and surplus variables $x_{r+1}, x_{r+2}, \cdots, x_n$ assuming the cost components of all slack and surplus variables, zero. Then the adjusted z is equivalent to

$$z_{ad} = c_1x_1 + c_2x_2 + \cdots + c_rx_r + 0x_{r+1} + 0x_{r+2} + \cdots + 0x_n$$

and the problem (3.3.1) can be written in matrix notation as, Optimize (maximize or minimize) $z_{ad} = \mathbf{cx}$ subject to

$$A\mathbf{x} = \mathbf{b}, \quad \mathbf{x} \geq \mathbf{0}, \tag{3.3.2}$$

where A is a $m \times n$ matrix, known as *coefficient matrix* given by

$$A = (\mathbf{a}_1, \mathbf{a}_2, \cdots, \mathbf{a}_n),$$

where $\mathbf{a}_j = [a_{1j}, a_{2j}, \cdots, a_{mj}]$ is a column vector associated with the variables $x_j[j = 1, 2, \cdots, n]$

$\mathbf{c} = (c_1, c_2, \cdots, c_r, \overbrace{0, 0, \cdots, 0}^{k\text{-components}})$ an n-component row vector.

$\mathbf{x} = [x_1, x_2, \cdots, x_r, x_{r+1}, \cdots, x_n]$ an n-component column vector.

$\mathbf{b} = [b_1, b_2, \cdots, b_m]$ an m-component column vector.

The components of $\mathbf{b}$ can be made positive by proper adjustment.

It is interesting to note that the column vectors associated with all slack variables are unit vectors.

As the cost components associated with the slack and surplus variables are zero, then it can be easily verified that the solution set which will optimize z_{ad} with constraints given in (3.3.2) will also optimize z with constraints given in (3.3.1) and the optimum value of z and z_{ad} will be same. The proof is left to the readers. Hence in order to solve the L.P.P. given in (3.3.1) we shall solve the L.P.P. given in (3.3.2). For further discussion hence we shall use the same notation for z and z_{ad}.

► **Example 3.3.1** *Transform the following L.P.P. accordingly such that the simplex method may be applicable.*

$$\textit{Maximize } z = 2x_1 + 3x_2 - 4x_3$$

subject to

$$\begin{aligned} 4x_1 + 2x_2 - x_3 &\leq 4 \\ -3x_1 + 2x_2 + 3x_3 &\geq 6 \\ x_1 + x_2 - 3x_3 &= 8, \quad x_1 \geq 0, x_2 \geq 0, x_3 \geq 0. \end{aligned}$$

Solution: First and second constraints are inequations. Hence we shall have to add two variables. First one can be converted into an equation by adding slack variable x_4 and second one can be converted into an equation by subtracting surplus variable x_5 from the left hand side of the constraints. Hence the transformed problem can be written as

$$\text{Maximize } z = 2x_1 + 3x_2 - 4x_3 + 0x_4 + 0x_5$$

subject to

$$\begin{aligned} 4x_1 + 2x_2 - x_3 + x_4 \quad &= 4 \\ -3x_1 + 2x_2 + 3x_3 \quad - x_5 &= 6 \\ x_1 + x_2 - 3x_3 \quad &= 8, \quad x_j \geq 0, [j = 1, 2, \cdots, 5] \end{aligned}$$

It can be easily verified that the column vector associated with the slack variable x_4 is a unit vector $\mathbf{e}_1 = [1, 0, 0]$

▶ **Example 3.3.2**

$$\textit{Maximize } z = x_1 - x_2 + x_3$$

subject to

$$\begin{aligned} x_1 + \ x_2 - 3x_3 &\geq \ 4 \\ 2x_1 - 4x_2 + \ x_3 &\geq -5 \\ x_1 + 2x_2 - 2x_3 &\leq \ 3 \ , \quad x_1 \geq 0, \ x_2 \geq 0, \ x_3 \geq 0. \end{aligned}$$

Solution: The problem can be transformed as given below

$$\text{Maximize } z = x_1 - x_2 + x_3 + 0x_4 + 0x_5 + 0x_6$$

subject to

$$\begin{aligned} x_1 + \ x_2 - 3x_3 - x_4 \qquad\qquad &= \ 4 \\ 2x_1 - 4x_2 + \ x_3 \qquad - x_5 \qquad &= -5 \\ x_1 + 2x_2 - 2x_3 \qquad\qquad + x_6 &= \ 3 \ , \quad x_j \geq 0, \ [j = 1, 2, \cdots, 6] \end{aligned}$$

x_4 and x_5 are surplus variables and x_6 is a slack variable. Changing the second component of **b** vector, -5 to 5, the second equation can be written as

$$-2x_1 + 4x_2 - x_3 + x_5 = 5$$

and in that case the surplus variable x_5 is converted into a slack variable.

▶ **Example 3.3.3**

$$\textit{Maximize } z = 2x_1 + 3x_2$$

subject to

$$\begin{aligned} x_1 + \ x_2 \qquad &\leq 10 \\ 2x_1 + 3x_2 + x_3 &\geq 28 \ , \quad x_1 \geq 0, \ x_2 \geq 0, \ x_3 \geq 0. \end{aligned}$$

Solution:

$$\text{Maximize } z = 2x_1 + 3x_2 + 0x_3 + 0x_4 + 0x_5$$

subject to

$$\begin{aligned} x_1 + \ x_2 + 0x_3 + x_4 \qquad &= 10 \\ 2x_1 + 3x_2 + \ x_3 \qquad - x_5 &= 28 \ , \quad x_j \geq 0 \ [j = 1, 2, \cdots, 5] \end{aligned}$$

Note. Here the coefficient of x_3 in the objective function is zero. But x_3 is not a slack or surplus variable.

▶ **Example 3.3.4** *Express the following minimization problem as standard maximization problem by introducing slack and surplus variables.*

$$\textit{Minimize } z = 4x_1 - x_2 + 2x_3$$

subject to

$$\begin{aligned} 4x_1 + \ x_2 - x_3 &\leq \ 7 \\ 2x_1 - 3x_2 + x_3 &\leq 12 \\ x_1 + \ x_2 + x_3 &= \ 8 \\ 4x_1 + 7x_2 - x_3 &\geq 16 \ , \quad x_1, \ x_2, \ x_3 \geq 0. \end{aligned}$$

Solution: After the introduction of the slack variables in the first two constraints and surplus variable in the fourth constraint, the converted problem is

$$\text{Minimize } z = 4x_1 - x_2 + 2x_3 + 0x_4 + 0x_5 + 0x_6$$

subject to

$$\begin{aligned} 4x_1 + x_2 - x_3 + x_4 \qquad\qquad &= 7 \\ 2x_1 - 3x_2 + x_3 \qquad + x_5 \qquad &= 12 \\ x_1 + x_2 + x_3 \qquad\qquad &= 8 \\ 4x_1 + 7x_2 - x_3 \qquad\qquad - x_6 &= 16\ , \quad x_1, x_2, \cdots, x_6 \geq 0; \end{aligned}$$

x_1, x_2, x_3 legitimate variables, x_4, x_5 slack variables and x_6 is a surplus variable. Now this problem is to be written in standard maximization problem

$$\text{Maximize } (-z) = z^* = -4x_1 + x_2 - 2x_3 + 0x_4 + 0x_5 + 0x_6$$

subject to

$$\begin{aligned} 4x_1 + x_2 - x_3 + x_4 \qquad\qquad &= 7 \\ 2x_1 - 3x_2 + x_3 \qquad + x_5 \qquad &= 12 \\ x_1 + x_2 + x_3 \qquad\qquad &= 8 \\ 4x_1 + 7x_2 - x_3 \qquad\qquad - x_6 &= 16\ , \quad x_1, x_2, \cdots, x_6 \geq 0 \end{aligned}$$

and Minimum value of z = negative value of the maximum of $(-z)$ with the same solution set [Proof is given at some later chapter].

Remark: Note that the two problems will have the same optimal solution set (if it exists at all) but the optimal values of the objective functions will be quite different which has been stated in the last two lines of the answer. Thus without mentioning it in the second problem cannot be considered as the standard maximising form of the first problem.

▶ **Example 3.3.5** *Rewrite the L.P.P. in standard maximization form by supplying slack and surplus variables.* *[C.U.(P)'86]*

$$\textit{Minimize } z = 3x_1 - 2x_2 + 4x_3$$

subject to

$$\begin{aligned} x_1 - x_2 + 3x_3 &\geq 1 \\ 2x_1 + 3x_2 - 5x_3 &\geq -3 \\ 4x_1 + 2x_2 \qquad &\geq 2, \quad x_1, x_2, x_3 \geq 0. \end{aligned}$$

State which are the slack and which are surplus variables.

Solution: For the general convention, we shall make $\mathbf{b} \geq 0$, thus the second constraint is

$$-2x_1 - 3x_2 + 5x_3 \leq 3.$$

Now the standard maximization form is

$$\text{Maximize } (-z) = -3x_1 + 2x_2 - 4x_3 + 0x_4 + 0x_5 + 0x_6$$

subject to

$$
\begin{aligned}
x_1 - x_2 + 3x_3 - x_4 \qquad\qquad &= 1 \\
-2x_1 - 3x_2 + 5x_3 \qquad + x_5 \qquad &= 3 \\
4x_1 + 2x_2 \qquad\qquad\qquad + x_6 &= 2\,, \quad x_j \geq 0,\ j = 1, 2, \cdots, 6
\end{aligned}
$$

where the minimum value of z = negative of the maximum value of $(-z)$ with the same solution set or simply

$$\min z = -\max(-z).$$

x_4 is a surplus and x_5, x_6 are two slack variables.

3.4 Variable unrestricted in sign

The difference of two non-negative variables is a variable unrestricted in sign. Let x_1 and x_2 be two non-negative variables. The difference of these two variables is a variable x_3 which is unrestricted in sign. If $x_1 > x_2$, then $x_3 > 0$; if $x_1 < x_2$, then $x_3 < 0$ and if $x_1 = x_2$, then $x_3 = 0$.

▶ **Example 3.4.1** *Write down the following L.P.P., where the variables are non-negative and in standard form.*

$$\textit{Maximize} \quad z = 2x_1 + 3x_2 - x_3$$

subject to

$$
\begin{aligned}
4x_1 + x_2 + x_3 &\geq 4 \\
7x_1 + 4x_2 - x_3 &\leq 25\,, \quad x_1, x_3 \geq 0,
\end{aligned}
$$

x_2 is unrestricted in sign.

Solution: Introducing slack and surplus variables etc. and writing $x_2 = x_2' - x_2''$, where $x_2' \geq 0$, $x_2'' \geq 0$, the problem in the standard form

$$\text{maximize} \quad z = 2x_1 + 3x_2' - 3x_2'' - x_3 + 0x_4 + 0x_5$$

subject to

$$
\begin{aligned}
4x_1 + x_2' - x_2'' + x_3 - x_4 \qquad &= 4 \\
7x_1 + 4x_2' - 4x_2'' - x_3 \qquad + x_5 &= 25\,, \quad x_1, x_1', x_2'', x_3 \geq 0,
\end{aligned}
$$

x_4, surplus and x_5, slack variable and all variables are non-negative.

Exercise 3

1. Transform the following constraints into equations by using slack or surplus variables.

$$
\begin{aligned}
x_1 + 2x_2 - 3x_3 &\geq 2 \\
x_1 + x_2 + 4x_3 &\leq 16 \\
x_1 + x_2 - x_3 &= 4\,, \quad x_1 \geq 0, x_2 \geq 0, x_3 \geq 0.
\end{aligned}
$$

2. Transform the following L.P.P. into the form where all constraints are of equality type. [C.U.(H)'80]

(a) Maximize $z = 2x_1 + x_2 - 6x_3 - x_4$

$$\text{subject to} \quad \begin{aligned} 3x_1 \quad\quad\quad\quad + x_4 &\le 25 \\ x_1 + x_2 + x_3 + x_4 &= 20 \\ 4x_1 + 6x_2 \quad\quad\quad &\ge 5 \\ 2 \le 2x_1 + 3x_3 + 2x_4 &\le 30, \quad x_j \ge 0, j = 1, 2, 3, 4. \end{aligned}$$

(b) Minimize $z = x_1 + x_2 - x_3$

$$\text{subject to} \quad \begin{aligned} x_1 + x_2 - x_3 &\ge 4 \\ 2x_1 + x_2 - 3x_3 &\le 24 \\ 2x_1 - x_2 + x_3 &\ge 10, \quad x_1 \ge 0, x_2 \ge 0, x_3 \ge 0. \end{aligned}$$

(c) Maximize $z = 4x_1 + x_2 - x_3$

$$\text{subject to} \quad \begin{aligned} x_1 - x_2 + x_3 &\ge 16 \\ 2x_1 + x_2 + x_3 &\ge 18, \quad x_1 \ge 0, x_2 \ge 0, x_3 \ge 0. \end{aligned}$$

3. Express the following minimization problems into standard maximization problems.

(a) Minimize $z = 2x_1 - x_2 + x_3$

$$\text{subject to} \quad \begin{aligned} 4x_1 + x_2 + x_3 &= 6 \\ 7x_1 + 3x_2 + 2x_3 &\ge 20 \\ 4x_1 + 7x_2 - 3x_3 &\le 10, \quad x_1, x_2, x_3 \ge 0. \end{aligned}$$

(b) Minimize $z = 3x_1 + 2x_2 - x_3$

$$\text{subject to} \quad \begin{aligned} 2x_1 + x_2 + x_3 &\ge 6 \\ 7x_1 - 13x_2 - 2x_3 &\le 2 \\ 9x_1 + 2x_2 + x_3 &= 4, \quad x_1, x_2, x_3 \ge 0. \end{aligned}$$

Answers

1.

$$\begin{aligned} x_1 + 2x_2 - 3x_3 - x_4 \quad\quad &= 2 \\ x_1 + x_2 + 4x_3 \quad\quad + x_5 &= 16 \\ x_1 + x_2 - x_3 \quad\quad\quad\quad &= 4, \quad x_j \ge 0, j = 1, 2, \cdots, 5 \end{aligned}$$

2. (a) Maximize $z = 2x_1 + x_2 - 6x_3 - x_4 + 0x_5 + 0x_6 + 0x_7 + 0x_8$

$$\text{subject to} \quad \begin{aligned} 3x_1 \quad\quad\quad\quad + x_4 + x_5 \quad\quad\quad\quad &= 25 \\ x_1 + x_2 + x_3 + x_4 \quad\quad\quad\quad\quad &= 20 \\ 4x_1 + 6x_2 \quad\quad\quad\quad - x_6 \quad\quad &= 5 \\ 2x_1 + 3x_3 + 2x_4 \quad\quad\quad - x_7 \quad &= 2 \\ 2x_1 + 3x_3 + 2x_4 \quad\quad\quad\quad + x_8 &= 30, \end{aligned}$$

$$x_j \ge 0, \; j = (1, 2, \cdots, 8)$$

(b) Minimize $z = x_1 + x_2 - x_3 + 0x_4 + 0x_5 + 0x_6$

$$\begin{aligned} \text{subject to} \quad x_1 + x_2 - x_3 - x_4 \qquad\qquad &= 4 \\ 2x_1 + x_2 - 3x_3 \qquad + x_5 \qquad &= 4 \\ 2x_1 - x_2 + x_3 \qquad\qquad - x_6 &= 10\ , \quad x_1, x_2, x_3 \geq 0, \end{aligned}$$

x_4 and x_6 surplus variables and x_5 slack variable, all of them ≥ 0.

(c) Maximize $z = 4x_1 + x_2 - x_3 + 0x_4 + 0x_5$

$$\begin{aligned} \text{subject to} \quad x_1 - x_2 + x_3 - x_4 \qquad &= 16 \\ 2x_1 + x_2 + x_3 \qquad - x_5 &= 18\ , \quad x_1, x_2, x_3 \geq 0, \end{aligned}$$

x_4 and x_5 are surplus variables ≥ 0.

3. (a) Maximize $(-z) = -2x_1 + x_2 - x_3 + 0x_4 + 0x_5$

$$\begin{aligned} \text{subject to} \quad 4x_1 + x_2 + x_3 \qquad\qquad &= 6 \\ 7x_1 + 3x_2 + 2x_3 - x_4 \qquad &= 20 \\ 4x_1 + 7x_2 - 3x_3 \qquad + x_5 &= 10\ , \quad x_1, x_2, x_3 \geq 0, \end{aligned}$$

x_4 surplus and x_5 slack variable all ≥ 0, where

minimum value of $z = -$ maximum value of $(-z)$

with the same solution set or simply min $z = -\max(-z)$.

(b) Maximize $(-z) = -3x_1 - 2x_2 + x_3 + 0x_4 + 0x_5$

$$\begin{aligned} \text{subject to} \quad 2x_1 + x_2 + x_3 - x_4 \qquad &= 6 \\ 7x_1 - 13x_2 - 2x_3 \qquad + x_5 &= 2 \\ 9x_1 + 2x_2 + x_3 \qquad\qquad &= 4\ , \quad x_1, x_2, x_3 \geq 0 \end{aligned}$$

x_4 surplus and x_5 slack variable all ≥ 0, where

minimum value of $z = -$ maximum value of $(-z)$

with the same solution set or simply min $z = -\max(-z)$.

Chapter 4

Basic Solutions of a Set of Simultaneous Linear Equations

4.1 Introduction

Let us consider m linear equations with n variables $(n > m)$ and let the set of equations be

$$\begin{aligned}
a_{11}x_1 + a_{12}x_2 + \cdots + a_{1j}x_j + \cdots + a_{1n}x_n &= b_1 \\
a_{21}x_1 + a_{22}x_2 + \cdots + a_{2j}x_j + \cdots + a_{2n}x_n &= b_2 \\
\cdots\cdots\cdots\cdots\cdots\cdots\cdots\cdots\cdots\cdots\cdots\cdots \\
\cdots\cdots\cdots\cdots\cdots\cdots\cdots\cdots\cdots\cdots\cdots\cdots \\
a_{i1}x_1 + a_{i2}x_2 + \cdots + a_{ij}x_j + \cdots + a_{in}x_n &= b_i \\
\cdots\cdots\cdots\cdots\cdots\cdots\cdots\cdots\cdots\cdots\cdots\cdots \\
\cdots\cdots\cdots\cdots\cdots\cdots\cdots\cdots\cdots\cdots\cdots\cdots \\
a_{m1}x_1 + a_{m2}x_2 + \cdots + a_{mj}x_j + \cdots + a_{mn}x_n &= b_m
\end{aligned}$$

These set of equations can be written in compact form

$$A\mathbf{x} = \mathbf{b}, \quad \text{where} \tag{4.1.1}$$

$A = \text{coefficient matrix} = [a_{ij}]_{\substack{i=1,2,\cdots,m \\ j=1,2,\cdots,n}}$

$\mathbf{x} = [x_1, x_2, \cdots, x_j, \cdots, x_n]$

$\mathbf{b} = [b_1, b_2, \cdots, b_i, \cdots, b_m]$

We further assume that $R(A) = m$, which indicates that all equations are linearly independent and none of them are redundant.

These set of equations can also be written in the form

$$\mathbf{a}_1x_1 + \mathbf{a}_2x_2 + \cdots + \mathbf{a}_jx_j + \cdots + \mathbf{a}_nx_n = \mathbf{b}, \tag{4.1.2}$$

where $\mathbf{a}_j = [a_{1j}, a_{2j}, \cdots, a_{ij}, \cdots, a_{mj}]$ an m-component column vector and all $\mathbf{a}_j$ vectors are non-null vectors. These vectors are called the *activity vectors.* From the theory of linear algebra involving matrix, we can say that here infinitely many solutions exist. Now we find a particular type solutions of the set of equations, which are finite in number, the discussions of which are given below.

4.2 Basic solutions

From the set of n column vectors $\mathbf{a}_j (j = 1, 2, \cdots, n)$ we select arbitrarily a set of m vectors $(m < n)$ which are linearly independent [there exists at least one such set of vectors since $R(A) = m$]. These set of vectors constitute a basis B. The vectors which are not included in the basis are called *non-basis vectors.* Assuming that all variables attached with non-basis vectors, zero, we get a set of m equations with m variables and since the matrix is a basis matrix, then B is non-singular, i.e., $\det B \neq 0$ and there exists a unique solution for the set of m equations containing m variables. This solution is called a *basic solution.* The variables attached with the linearly independent set of vectors $\mathbf{a}_j$ are called *basic variables* and the number of basic variables will be always m and $(n - m)$ variables whose values are assumed to be zero, attached with non-basis vectors are called *non-basic variables.* With the assumption that all non-basic variables be zero, the set of equations reduces to

$$B\mathbf{x}_B = \mathbf{b}, \tag{4.2.1}$$

where B is the basic matrix, $\mathbf{x}_B$, the column vectors having m components (which are called the *basic variables*). Now using the matrix inverse method of finding the solution of the set of equations

$$\begin{aligned} (B^{-1}B)\mathbf{x}_B &= B^{-1}\mathbf{b} \\ \text{or,} \quad I_m\mathbf{x}_B = \mathbf{x}_B &= B^{-1}\mathbf{b} \end{aligned} \tag{4.2.2}$$

$\mathbf{x}_B$, the m component column vector is written $\mathbf{x}_B = [x_{B1}, x_{B2}, \cdots, x_{Bi}, \cdots, x_{Bm}]$. Practically the general solution $\mathbf{x}_B$ is written in the manner.

$\mathbf{x}_B = [B^{-1}\mathbf{b}, \mathbf{0}]$, where $\mathbf{0}$ vector is a $(n - m)$ component null column vector which contains the values of all non-basic variables and the basic and non-basic variables must be posted in their respective positions.

Since out of n vectors, m vectors constitute a basis, then theoretically maximum number of basis matrices be ${}^nC_m$ and hence the maximum number of basic solutions will be ${}^nC_m$. It may be less than ${}^nC_m$ in a particular problem and for finite n, ${}^nC_m$ is finite and thus basic solutions are finite in number. Now we formally define the basic solution in the manner.

Basic solution: Given a system of m simultaneous linear equations containing n variables $(n > m)$ and the set of equations be $A\mathbf{x} = \mathbf{b}$, $R(A) = m$. If any $m \times m$, non-singular matrix be arbitrarily selected from A and if we assume all $(n - m)$ variables zero, which are not associated with the column vectors of the matrix, the solution so obtained is called a *basic solution.* The variables which are attached with m-vectors of the non-singular matrix are called *basic variables* and remaining $(n-m)$ variables whose values are assumed to be zero, are called *non-basic variables.* The values of m-components of the basic solution may be positive, negative or zero. From this we can conclude that a solution is said to a basic solution if the vectors $\mathbf{a}_j$ associated with non-zero variables be linearly independent. This condition is necessary and sufficient. The following solved problems will give us a clear idea about basic solutions. We write down the inverse of the (2×2) basis matrix which will be immensely helpful in solving the problems. if the matrix be $A = \begin{bmatrix} a_{11} & a_{12} \\ a_{21} & a_{22} \end{bmatrix}$ and the matrix be non-singular, i.e., $\det A = a_{11}a_{22} - a_{12}a_{21} \neq 0$, then

$$A^{-1} = \frac{1}{\det A}\begin{bmatrix} a_{22} & -a_{12} \\ -a_{21} & a_{11} \end{bmatrix}.$$

This is only true for 2×2 non-singular matrix. For example,

$$A = \begin{bmatrix} 5 & 1 \\ 4 & 2 \end{bmatrix}, \quad \det A = 6 \neq 0.$$

$$A^{-1} = \frac{1}{6}\begin{bmatrix} 2 & -1 \\ -4 & 5 \end{bmatrix} = \begin{bmatrix} 1/3 & -1/6 \\ -2/3 & 5/6 \end{bmatrix} \text{ etc.}$$

4.3 Definitions

4.3.1 Non-degenerate Basic Solution

If all components of a solution set corresponding to the basic variables are non-zero quantities then the basic solution is known as the *non-degenerate basic solution.* In the worked out Example 4.3.1 all basic solutions are non-degenerate as in each basic solution the components of solution set corresponding to the basic variables are non-zero quantities.

4.3.2 Degenerate Basic Solution

If some components of the solution set corresponding to the basic variables are zero, the basic solution is known as a *degenerate basic solution.*

In a linear programming problem, the variables are essentially non-negative quantities, i.e., variable vector $\mathbf{x} \geq \mathbf{0}$.

4.3.3 Feasible solution to a Linear Programming Problem

If all components of a solution set are non-negative quantities then the solution is known as a *feasible solution* (F.S.) *of the* L.P.P.

4.3.4 All Feasible Solution to a L.P.P.

If all components of a solution set of L.P.P. are positive integers including zero then the solution is known as *all feasible solution* (A.F.S.) of the L.P.P. In the worked out Example 4.3.1, $\mathbf{x}_1$ and $\mathbf{x}_3$ are the feasible solutions but $\mathbf{x}_2$ is not a feasible solution. The solution set (2, 3, 4) is not only a feasbile solution but also an *all feasible solution.*

4.3.5 Basic Feasible Solution (B.F.S.) of a L.P.P.

The solution set of a L.P.P. which is feasbile as well as basic is known as the *basic feasible solution* of the problem. A solution will be B.F.S. if all components of the solution set corresponding to the basic variables are non-negative quantities. $\mathbf{x}_1$ and $\mathbf{x}_3$ are B.F.S. of the worked out Example 4.3.1.

4.3.6 Non-degenerate B.F.S.

The solution of a L.P.P. of which all components corresponding to the basic variables are positive quantities, is known as *non-degenerate* B.F.S. In the worked out example 4.3.1 $\mathbf{x}_1$ and $\mathbf{x}_3$ are non-degenerate B.F.S. also.

4.3.7 Degenerate B.F.S.

The feasible solution set of L.P.P. of which some components corresponding to the basic variables are zero, is known as the *degenerate* B.F.S.

From the defintion, it is clear that all components corresponding to the basic variables are non-negative quantities in case of a B.F.S. If all of them are positive, then it is a non-degenerate B.F.S. and if at least one of them is zero then it is a degenerate B.F.S.

► **Example 4.3.1** *Find the basic solution or solutions, if there be any, of the set of equations*

$$2x_1 + 4x_2 - 2x_3 = 10$$
$$10x_1 + 3x_2 + 7x_3 = 33$$

Solution: The set of equations can be written in the manner

$$\mathbf{a}_1 x_1 + \mathbf{a}_2 x_2 + \mathbf{a}_3 x_3 = \mathbf{b}, \qquad \text{(i)}$$

where $\mathbf{a}_1 = \begin{bmatrix} 2 \\ 10 \end{bmatrix}$, $\mathbf{a}_2 = \begin{bmatrix} 4 \\ 3 \end{bmatrix}$, $\mathbf{a}_3 = \begin{bmatrix} -2 \\ 7 \end{bmatrix}$, and $\mathbf{b} = \begin{bmatrix} 10 \\ 33 \end{bmatrix}$;

$A = \begin{bmatrix} 2 & 4 & -7 \\ 10 & 3 & 7 \end{bmatrix}$, $R(A) = 2$.

Thus the equations are linearly independent.

The three square sub-matrices from A are

$$B_1 = (\mathbf{a}_1, \mathbf{a}_2) = \begin{bmatrix} 2 & 4 \\ 10 & 3 \end{bmatrix},$$

$$B_2 = (\mathbf{a}_1, \mathbf{a}_3) = \begin{bmatrix} 2 & -2 \\ 10 & 7 \end{bmatrix}$$

$$\text{and} \quad B_3 = (\mathbf{a}_2, \mathbf{a}_3) = \begin{bmatrix} 4 & -2 \\ 3 & 7 \end{bmatrix}.$$

$$\det B_1 = -34 \neq 0, \quad \det B_2 = 34 \neq 0, \quad \det B_3 = 34 \neq 0.$$

All three square sub-matrices are non-singular and therefore all of them are basis matrices and thus there exist three basic solutions. For basis B_1,

$$B_1^{-1} = \frac{1}{-34}\begin{bmatrix} 3 & -4 \\ -10 & 2 \end{bmatrix}$$

and the basic solution

$$\mathbf{x}_{B_1} = B_1^{-1}\mathbf{b} = \frac{1}{-34}\begin{bmatrix} 3 & -4 \\ -10 & 2 \end{bmatrix}\begin{bmatrix} 10 \\ 33 \end{bmatrix} = \begin{bmatrix} 3 \\ 1 \end{bmatrix}.$$

Hence $\mathbf{x}_{B_1} = [x_{B_1}, x_{B_2}, 0] = [x_1 = 3, x_2 = 1, x_3 = 0]$, x_1 and x_2 are basic variables and $x_3 = 0$ is non-basic. For basis, B_2, $B_2^{-1} = \frac{1}{34}\begin{bmatrix} 7 & 2 \\ -10 & 2 \end{bmatrix}$.

Basic solution $\mathbf{x}_{B_2} = B_2^{-1}\mathbf{b} = \frac{1}{34}\begin{bmatrix} 7 & 2 \\ -10 & 2 \end{bmatrix}\begin{bmatrix} 10 \\ 33 \end{bmatrix} = \begin{bmatrix} 4 \\ -1 \end{bmatrix}$.

Hence $\mathbf{x}_{B_2} = [x_{B_1}, 0, x_{B_2}] = [x_1 = 4, x_2 = 0, x_3 = -1]$, x_1 and x_3 are basic variables and $x_2 = 0$, non-basic. For basis B_3, $B_3^{-1} = \frac{1}{34}\begin{bmatrix} 7 & 2 \\ -3 & 4 \end{bmatrix}$.

Basic solution $\mathbf{x}_{B_3} = B_3^{-1}\mathbf{b} = \frac{1}{34}\begin{bmatrix} 7 & 2 \\ -3 & 4 \end{bmatrix}\begin{bmatrix} 10 \\ 33 \end{bmatrix} = \begin{bmatrix} 4 \\ 3 \end{bmatrix}$.

Thus $\mathbf{x}_{B_3} = [0, x_{B_1}, x_{B_2}] = [x_1 = 0, x_2 = 4, x_3 = 3]$. Here x_2 and x_3 are basic variables and $x_1 = 0$ is non-basic.

Note 1. $\mathbf{x}_{B_1}$ and $\mathbf{x}_{B_3}$ are basic feasible solutions and $\mathbf{x}_{B_2}$ is basic but non-feasible.

2. For $\mathbf{x}_{B_1}$, if we take $x_3 = 0$, the equations reduce to

$$\begin{aligned} 2x_1 + 4x_2 &= 10 \\ \text{and} \quad 10x_1 + 3x_2 &= 33 \end{aligned}$$

and using the method of solving the two linear simultaneous equations with two unknowns, which we have done in school algebra, we get $x_1 = 3$, $x_2 = 1$ and the basic solution is $\mathbf{x}_{B_1} = [x_1 = 3, x_2 = 1, x_3 = 0]$, x_3 is non-basic. But before testing, that x_1, x_2 be basic variables, i.e., x_3 to be non-basic variable, we cannot at random assume $x_3 = 0$ and in that case we may put into a confusion which will be discussed in problems given afterwards. Using this procedure, we can find all basic solutions, but the students are advised not to use this method and insisted them to use the matrix inverse method to find the solutions, since only this method will truely help the students to understand the procedure of solving L.P.P. by simplex method.

▶ **Example 4.3.2** *Two linear simultaneous equations (linearly independent) with four unknowns (variables) are given below.*

$$\begin{aligned} 4x_1 + 2x_2 + 3x_3 - 8x_4 &= 6 \\ 3x_1 + 5x_2 + 4x_3 - 6x_4 &= 8 \end{aligned}$$

(a) *How many basic solutions are there?*

(b) *Find all of them.*

(c) *Discuss the nature of each and every basic solution.*

Solution: Here the equations can be written in the manner

$$\mathbf{a}_1 x_1 + \mathbf{a}_2 x_2 + \mathbf{a}_3 x_3 + \mathbf{a}_4 x_4 = \mathbf{b},$$

where the column vectors associated with the variables are

$$\mathbf{a}_1 = \begin{bmatrix} 4 \\ 3 \end{bmatrix}, \ \mathbf{a}_2 = \begin{bmatrix} 2 \\ 5 \end{bmatrix}, \ \mathbf{a}_3 = \begin{bmatrix} 3 \\ 4 \end{bmatrix}, \ \mathbf{a}_4 = \begin{bmatrix} -8 \\ -6 \end{bmatrix} \text{ and } \mathbf{b} = \begin{bmatrix} 6 \\ 8 \end{bmatrix}.$$

(a) From the four vectors $\mathbf{a}_1, \cdots, \mathbf{a}_4$, taking two at a time, ${}^4C_2 = 6$ square sub-matrices are

$$B_1 = (\mathbf{a}_1, \mathbf{a}_2) = \begin{bmatrix} 4 & 2 \\ 3 & 5 \end{bmatrix}, \ B_2 = (\mathbf{a}_1, \mathbf{a}_3) = \begin{bmatrix} 4 & 3 \\ 3 & 4 \end{bmatrix}, \ B_3 = (\mathbf{a}_1, \mathbf{a}_4) = \begin{bmatrix} 4 & -8 \\ 3 & -6 \end{bmatrix}.$$

$$B_4 = (\mathbf{a}_2, \mathbf{a}_3) = \begin{bmatrix} 2 & 3 \\ 5 & 4 \end{bmatrix}, \ B_5 = (\mathbf{a}_2, \mathbf{a}_4) = \begin{bmatrix} 2 & -8 \\ 5 & -6 \end{bmatrix}, \ B_6 = (\mathbf{a}_3, \mathbf{a}_4) = \begin{bmatrix} 3 & -8 \\ 4 & -6 \end{bmatrix}.$$

$$\det B_1 = 14 \neq 0, \quad \det B_2 = 7 \neq 0, \quad \det B_3 = 0,$$
$$\det B_4 = -7 \neq 0, \quad \det B_5 = 28 \neq 0, \quad \det B_6 = 14 \neq 0.$$

Here five square sub-matrices are non-singular and hence there exist five basis matrices and corresponding to which there exist exactly five basic solutions.

(b) Now, for basis matrix B_1, $B_1^{-1} = \dfrac{1}{14}\begin{bmatrix} 5 & -2 \\ -3 & 4 \end{bmatrix}$ and the basic solution

$$\mathbf{x}_1 = B_1^{-1}\mathbf{b} = \frac{1}{14}\begin{bmatrix} 5 & -2 \\ -3 & 4 \end{bmatrix}\begin{bmatrix} 6 \\ 8 \end{bmatrix} = \begin{bmatrix} 1 \\ 1 \end{bmatrix}.$$

Hence the basic solution $\mathbf{x}_1 = [x_1 = 1,\ x_2 = 1,\ x_3 = x_4 = 0]$, x_1, x_2 basic and x_3 and x_4 are non-basic variables.

For basis matrix B_2, $B_2^{-1} = \dfrac{1}{7}\begin{bmatrix} 4 & -3 \\ -3 & 4 \end{bmatrix}$,

$$\mathbf{x}_2 = \frac{1}{7}\begin{bmatrix} 4 & -3 \\ -3 & 4 \end{bmatrix}\begin{bmatrix} 6 \\ 8 \end{bmatrix} = \begin{bmatrix} 0 \\ 2 \end{bmatrix}.$$

Hence the basic solution

$\mathbf{x}_2 = [x_1 = 0 \text{ (basic)},\ x_2 = 0 \text{ (non-basic)}\ x_3 = 2 \text{ (basic)},\ \mathbf{x}_4 = 0 \text{ (non-basic)}]$.

$\mathbf{x}_2$ is a degenerate basic solution and for a degenerate basic solution, basic and non-basic variables must be stated clearly.

Similarly, proceeding the same way, we can say for basis B_4,

$\mathbf{x}_4 = [x_1 = 0$ (non-basic), $x_2 = 0$ (basic), $x_3 = 2$ (basic), $x_4 = 0$ (non-basic)].

For basis B_5, $\mathbf{x}_5 = [x_1 = 0, x_2 = 1, x_3 = 0, x_4 = -1/2]$.

For basis B_6, $\mathbf{x}_6 = [x_1 = x_2 = 0$ (non-basic), $x_3 = 2$ (basic), $x_4 = 0$ (basic)].

Since B_3 is not a basis which contains the vectors $\mathbf{a}_1$ and $\mathbf{a}_4$, then there exists no basic solution corresponding to the matrix B_3 which indicates that the two variables x_1 and x_4 together cannot be considered as basic variables; on the other hand the set of two variables x_2 and x_3 simultaneously cannot be treated as non-basic.

(c) $\mathbf{x}_1$ is a non-degenerate B.F.S.

$\mathbf{x}_2$ is a degenerate B.F.S.

$\mathbf{x}_4$ is a degenerate B.F.S.

$\mathbf{x}_5$ is a non-degenerate B.S. but not feasible.

$\mathbf{x}_6$ is a degenerate B.F.S.

▶ **Example 4.3.3** *Show that* $x_1 = 5$, $x_2 = 0$, $x_3 = -1$ *is a basic solution of the system of equations*

$$\begin{aligned} x_1 + 2x_2 + x_3 &= 4 \\ 2x_1 + x_2 + 5x_3 &= 5 \end{aligned}$$

Find the other basic solutions, if there be any. [C.U.(H) '85; (P) '91, '99]

Solution: The set of values $x_1 = 5, x_2 = 0, x_3 = -1$ satisfy the two equations and the vectors $\mathbf{a}_1 = \begin{bmatrix} 1 \\ 2 \end{bmatrix}$, $\mathbf{a}_3 = \begin{bmatrix} 1 \\ 5 \end{bmatrix}$ associated with non-zero variables x_1 and x_3 are linearly independent since $B = (\mathbf{a}_1, \mathbf{a}_3) = \begin{bmatrix} 1 & 1 \\ 2 & 5 \end{bmatrix}$ and $\det B = 3 \neq 0$. Hence the solution is basic but not feasible. Now, two other square matrices are (taking two at a time from $\mathbf{a}_1, \mathbf{a}_2$ and $\mathbf{a}_3$). $B_1 = (\mathbf{a}_1, \mathbf{a}_2) = \begin{bmatrix} 1 & 2 \\ 2 & 1 \end{bmatrix}$ $\det B_1 = -3 \neq 0$.

$B_2 = (\mathbf{a}_2, \mathbf{a}_3) = \begin{bmatrix} 2 & 1 \\ 1 & 5 \end{bmatrix}$ and $\det B_2 = 9 \neq 0$. Thus B_1 and B_2 are both basis and there are other two B.S. corresponding to B_1 and B_2. Now, B.S. $\mathbf{x}_1$ corresponding to B_1 is

$$B_1^{-1}\mathbf{b} = -\frac{1}{3}\begin{bmatrix} 1 & -2 \\ -2 & 1 \end{bmatrix}\begin{bmatrix} 4 \\ 5 \end{bmatrix} = \begin{bmatrix} 2 \\ 1 \end{bmatrix}, \quad \text{i.e.,} \quad \mathbf{x}_1 = [2, \ 1, \ 0]$$

and B.S. corresponding to B_2 is

$$B_2^{-1}\mathbf{b} = \frac{1}{9}\begin{bmatrix} 5 & -1 \\ -1 & 2 \end{bmatrix}\begin{bmatrix} 4 \\ 5 \end{bmatrix} = \begin{bmatrix} 5/3 \\ 2/3 \end{bmatrix}, \quad \text{i.e.,} \quad \mathbf{x}_2 = [0, \ 5/3, \ 2/3].$$

[**Note:** The last two B.S. are feasible and non-degenerate.]

▸ **Example 4.3.4** *The two linearly independent equations with three variables are given below.*

$$\begin{aligned} 2x_1 - 3x_2 + \ \ 5x_3 &= 10 \\ 4x_1 + \ \ x_2 + 10x_3 &= 20 \end{aligned}$$

Find, if possible, a basic solution with x_2, a non-basic variable.

Solution: The equations are linearly independent. But the vectors

$$\mathbf{a}_1 = \begin{bmatrix} 2 \\ 4 \end{bmatrix}, \quad \mathbf{a}_3 = \begin{bmatrix} 5 \\ 10 \end{bmatrix}$$

are not linearly independent since

$$B = (\mathbf{a}_1, \mathbf{a}_3) = \begin{bmatrix} 2 & 5 \\ 4 & 10 \end{bmatrix}, \quad \det B = 0.$$

Thus, if we initially assume x_2 as non-basic whose value is zero the two equations reduce to

$$\begin{aligned} 2x_1 + \ 5x_3 &= 10 \\ 4x_1 + 10x_3 &= 20. \end{aligned}$$

Here the two equations are consistent and they are the same equation. Infinite number of solutions satisfy the equations. For example, $x_1 = 5/2$, $x_3 = 1$ is a solution but $[x_1 = 5/2,\ x_2 = 0,\ x_3 = 1]$ is not a basic solution. Similarly, $x_1 = -5$, $x_3 = 4$ is a solution. But $[x_1 = -5,\ x_2 = 0,\ x_3 = 4]$ is not a basic solution.

▸ **Example 4.3.5** *Find all basic solutions of the set of equations given below. (Problem is taken from Linear Programming, G. Hadley. page* 70, 2–14)

$$\begin{aligned} x_1 + 2x_2 + 3x_3 + 4x_4 &= 7 \\ 2x_1 + \ \ x_2 + \ \ x_3 + 2x_4 &= 3 \end{aligned}$$

Solution: Here $R(A) = 2$, the equations are linearly independent. The set of equations can be written in the manner

$$\mathbf{a}_1 x_1 + \mathbf{a}_2 x_2 + \mathbf{a}_3 x_3 + \mathbf{a}_4 x_4 = \mathbf{b},$$

where $\mathbf{a}_1 = \begin{bmatrix} 1 \\ 2 \end{bmatrix}$, $\mathbf{a}_2 = \begin{bmatrix} 2 \\ 1 \end{bmatrix}$, $\mathbf{a}_3 = \begin{bmatrix} 3 \\ 1 \end{bmatrix}$, $\mathbf{a}_4 = \begin{bmatrix} 4 \\ 2 \end{bmatrix}$ and $\mathbf{b} = \begin{bmatrix} 7 \\ 3 \end{bmatrix}$.

$$B_1 = (\mathbf{a}_1, \mathbf{a}_2) = \begin{bmatrix} 1 & 2 \\ 2 & 1 \end{bmatrix}, \quad \det B_1 = -3 \neq 0,$$

$$B_2 = (\mathbf{a}_1, \mathbf{a}_3) = \begin{bmatrix} 1 & 3 \\ 2 & 1 \end{bmatrix}, \quad \det B_2 = -5 \neq 0,$$

$$B_3 = (\mathbf{a}_1, \mathbf{a}_4) = \begin{bmatrix} 1 & 4 \\ 2 & 2 \end{bmatrix}, \quad \det B_3 = -6 \neq 0,$$

$$B_4 = (\mathbf{a}_2, \mathbf{a}_3) = \begin{bmatrix} 2 & 3 \\ 1 & 1 \end{bmatrix}, \quad \det B_4 = -1 \neq 0,$$

$$B_5 = (\mathbf{a}_2, \mathbf{a}_4) = \begin{bmatrix} 2 & 4 \\ 1 & 2 \end{bmatrix}, \quad \det B_5 = 0,$$

$$B_6 = (\mathbf{a}_3, \mathbf{a}_4) = \begin{bmatrix} 3 & 4 \\ 1 & 2 \end{bmatrix}, \quad \det B_6 = 2 \neq 0.$$

Five matrices are non-singular except the matrix B_5. Thus there exist five basic solutions of the above problem.

For B_1, the basic solution is

$$\mathbf{x}_1 = \frac{1}{-3}\begin{bmatrix} 1 & -2 \\ -2 & 1 \end{bmatrix}\begin{bmatrix} 7 \\ 3 \end{bmatrix} = \begin{bmatrix} -1/3 \\ 11/3 \end{bmatrix}.$$

$$\text{Then } \mathbf{x}_1 = \left[-\frac{1}{3}, \frac{11}{3}, 0, 0\right].$$

For B_2, the basic solution is

$$\mathbf{x}_2 = -\frac{1}{5}\begin{bmatrix} 1 & -3 \\ -2 & 1 \end{bmatrix}\begin{bmatrix} 7 \\ 3 \end{bmatrix} = \begin{bmatrix} 2/5 \\ 11/5 \end{bmatrix}.$$

$$\text{Then } \mathbf{x}_2 = \left[\frac{7}{5}, 0, \frac{11}{5}, 0\right]$$

For B_3, the basic solution is

$$\mathbf{x}_3 = -\frac{1}{6}\begin{bmatrix} 2 & -4 \\ -2 & 1 \end{bmatrix}\begin{bmatrix} 7 \\ 3 \end{bmatrix} = \begin{bmatrix} -1/3 \\ 11/6 \end{bmatrix}.$$

$$\text{Then } \mathbf{x}_3 = \left[-\frac{7}{3}, 0, 0, \frac{11}{6}\right]$$

For B_4, the basic solution is

$$\mathbf{x}_4 = \frac{1}{-1}\begin{bmatrix} 1 & -3 \\ -1 & 2 \end{bmatrix}\begin{bmatrix} 7 \\ 3 \end{bmatrix} = \begin{bmatrix} 2 \\ 1 \end{bmatrix}.$$

$$\text{Then } \mathbf{x}_4 = [0, 2, 1, 0]$$

For B_6, the basic solution is

$$\mathbf{x}_6 = \frac{1}{2}\begin{bmatrix} 2 & -4 \\ -1 & 3 \end{bmatrix}\begin{bmatrix} 7 \\ 3 \end{bmatrix} = \begin{bmatrix} 1 \\ 1 \end{bmatrix}.$$

$$\text{Then } \mathbf{x}_6 = [0, 0, 1, 1]$$

▶ **Example 4.3.6** *How many basic solutions are there in the following linearly independent set of equations? Find all of them.*

$$\begin{aligned} 2x_1 - x_2 + 3x_3 + x_4 &= 6 \\ 4x_1 - 2x_2 - x_3 + 2x_4 &= 10 \end{aligned}$$

Solution: The equations are linearly independent. There are two equations with four variables. Then there are at most ${}^4C_2 = 6$ basic solutions, with two basic variables and $4 - 2 = 2$ non-basic variables.

The column vectors

$$\mathbf{a}_1 = \begin{bmatrix} 2 \\ 4 \end{bmatrix}, \mathbf{a}_2 = \begin{bmatrix} -1 \\ -2 \end{bmatrix}, \mathbf{a}_3 = \begin{bmatrix} 3 \\ -1 \end{bmatrix}, \mathbf{a}_4 = \begin{bmatrix} 1 \\ 2 \end{bmatrix} \text{ and } \mathbf{b} = \begin{bmatrix} 6 \\ 10 \end{bmatrix}.$$

The six squares matrices, taking two at a time from $\mathbf{a}_1, \mathbf{a}_2, \cdots, \mathbf{a}_4$ are

$$B_1 = (\mathbf{a}_1, \mathbf{a}_2) = \begin{bmatrix} 2 & -1 \\ 4 & -2 \end{bmatrix}, \quad B_2 = (\mathbf{a}_1, \mathbf{a}_3) = \begin{bmatrix} 2 & 3 \\ 4 & -1 \end{bmatrix},$$

$$B_3 = (\mathbf{a}_1, \mathbf{a}_4) = \begin{bmatrix} 2 & 1 \\ 4 & 2 \end{bmatrix}, \quad B_4 = (\mathbf{a}_2, \mathbf{a}_3) = \begin{bmatrix} -1 & 3 \\ -2 & -1 \end{bmatrix},$$

$$B_5 = (\mathbf{a}_2, \mathbf{a}_4) = \begin{bmatrix} -1 & 1 \\ -2 & 2 \end{bmatrix}, \quad B_6 = (\mathbf{a}_3, \mathbf{a}_4) = \begin{bmatrix} 3 & 1 \\ -1 & 2 \end{bmatrix},$$

$$\det B_1 = 0, \quad \det B_2 = -14 \neq 0, \quad \det B_3 = 0,$$
$$\det B_4 = 7 \neq 0, \quad \det B_5 = 0, \quad \det B_6 = 7 \neq 0.$$

There are only three non-singular matrices and corresponding to these matrices (which are considered to be basis matrices) there exist only three basic solutions.

For basis matrix B_2, $B_2^{-1} = -\dfrac{1}{14}\begin{bmatrix} -1 & -3 \\ -4 & 2 \end{bmatrix}$ and the basic solution

$$\mathbf{x}_{B_2} = B_2^{-1}\mathbf{b} = -\frac{1}{14}\begin{bmatrix} -1 & -3 \\ -4 & 2 \end{bmatrix}\begin{bmatrix} 6 \\ 10 \end{bmatrix} = \begin{bmatrix} 18/7 \\ 2/7 \end{bmatrix}.$$

Hence B.S. = [18/7, 0, 2/7, 0].

For basis matrix B_4,

$$B_4^{-1} = \frac{1}{7}\begin{bmatrix} -1 & -3 \\ 2 & -1 \end{bmatrix} \text{ and } \mathbf{x}_{B_4} = \frac{1}{7}\begin{bmatrix} -1 & -3 \\ 2 & -1 \end{bmatrix}\begin{bmatrix} 6 \\ 10 \end{bmatrix} = \begin{bmatrix} -36/7 \\ 2/7 \end{bmatrix}.$$

Hence the B.S. corresponding to the basis B_4 is $[0, -36/7, 2/7, 0]$.

For basis matrix B_6,

$$B_6^{-1} = \frac{1}{7}\begin{bmatrix} 2 & -1 \\ 1 & 3 \end{bmatrix} \text{ and } \mathbf{x}_{B_6} = \frac{1}{7}\begin{bmatrix} 2 & -1 \\ 1 & 3 \end{bmatrix}\begin{bmatrix} 6 \\ 10 \end{bmatrix} = \begin{bmatrix} 2/7 \\ 36/7 \end{bmatrix}.$$

Hence the B.S. corresponding to the basis B_6 is [0, 0, 2/7, 36/7].

Note: Of the three B.S., two are feasible and one is non-feasible and all the three B.S. are non-degenerate.

▶ **Example 4.3.7** *Two linearly independent equations with three variables are given below*

$$\begin{aligned} 3x_1 + x_2 - 6x_3 &= 1 \\ 2x_1 - 3x_2 - 4x_3 &= 5 \end{aligned}$$

Find, if possible, the basic solution with x_2 as a non-basic variable.

Solution: There are two equations (linearly independent) with three variables. Then in each basic solution, there will have two basic variables and one non-basic variable. If x_2 is taken as non-basic, then the remaining x_1 and x_3 will be considered basic provided the vectors $\mathbf{a}_1 = \begin{bmatrix} 3 \\ 2 \end{bmatrix}$ and $\mathbf{a}_3 = \begin{bmatrix} -6 \\ -4 \end{bmatrix}$ associated with the variables x_1 and x_3 respectively be linearly independent. Now $B = (\mathbf{a}_1, \mathbf{a}_3) = \begin{bmatrix} 3 & -6 \\ 2 & -2 \end{bmatrix}$ and $\det B = 0$. Thus the two vectors are not L.I. and hence x_1 and x_3 simultaneously cannot be considered as basic variables and hence there exists no B.S. with x_2 as non-basic variable.

▶ **Example 4.3.8** (a) *Prove that $x_1 = 2$, $x_2 = 0$ and $x_3 = 1$ is a basic solution to set of equations,*

$$\begin{aligned} 2x_1 + x_2 - x_3 &= 3 \\ x_1 + 2x_2 + 3x_3 &= 5 \end{aligned}$$

(b) *Prove that $x_1 = 2$, $x_2 = -1$ and $x_3 = 0$ is a solution set but not a basic solution to the set of equations,*

$$\begin{aligned} 3x_1 - 2x_2 + x_3 &= 8 \\ 9x_1 - 6x_2 + 4x_3 &= 24 \end{aligned}$$

Solution: (a) $A = \begin{bmatrix} 2 & 1 & -1 \\ 1 & 2 & 3 \end{bmatrix}$, $\mathbf{b} = \begin{bmatrix} 3 \\ 5 \end{bmatrix}$, $R(A) = 2$.

Hence the two equations are linearly independent. The set of values satisfy the equations. Hence the set is a solution set.

Now the vectors $\mathbf{a}_1 = \begin{bmatrix} 2 \\ 1 \end{bmatrix}$ and $\mathbf{a}_3 = \begin{bmatrix} -1 \\ 3 \end{bmatrix}$ associated with non-zero variables x_1 and x_3 are L.I. Hence the solution is a B.S.

To test that $\mathbf{a}_1$ and $\mathbf{a}_3$ are L.I., we have $B = (\mathbf{a}_1, \mathbf{a}_3) = \begin{bmatrix} 2 & -1 \\ 1 & 3 \end{bmatrix}$ and $|B| = 7 \neq 0$. Hence the vectors are L.I.

(b) $A = \begin{bmatrix} 3 & -2 & 1 \\ 9 & -6 & 4 \end{bmatrix}$ and $\mathbf{b} = \begin{bmatrix} 8 \\ 24 \end{bmatrix}$, $R(A) = 2$.

Hence the equations are linearly independent. The set of values satisfy the equations. Hence the set is a solution set. But the vectors $\mathbf{a}_1$ and $\mathbf{a}_2$ associated with non-zero variables are not L.I. Hence the solution set is not basic.

▶ **Example 4.3.9** *Prove that x_3 will be a non-basic variable in a basic solution of the set of equations,*

$$\begin{aligned} x_1 + 4x_2 - \quad x_3 &= 3 \\ 5x_1 + 2x_2 + 3x_3 &= 4 \end{aligned}$$

and find that B.S. [C.U.(P) '93,'86]

Solution: Here, there are two linearly independent equations with three variables. Then for x_3 non-basic, x_1 and x_2 must be basic, the condition for which is $\mathbf{a}_1 = \begin{bmatrix} 1 \\ 5 \end{bmatrix}$, $\mathbf{a}_2 \begin{bmatrix} 4 \\ 2 \end{bmatrix}$ must be L.I. Here the vectors $\mathbf{a}_1$ and $\mathbf{a}_2$ are L.I. As the matrix $B = (\mathbf{a}_1, \mathbf{a}_2) = \begin{bmatrix} 1 & 4 \\ 5 & 2 \end{bmatrix}$ and $\det B = -18 \neq 0$. Hence x_1 and x_2 together can be considered as basic variables and the corresponding B.S.

$$\mathbf{x} = B^{-1}\mathbf{b} = \frac{1}{-18}\begin{bmatrix} 2 & -4 \\ -5 & 1 \end{bmatrix}\begin{bmatrix} 3 \\ 4 \end{bmatrix} = \begin{bmatrix} 5/9 \\ 11/18 \end{bmatrix}.$$

Hence the B.S. is $\mathbf{x} = [5/8, 11/18, 0]$ with x_3 non-basic.

Exercise 4

1. (a) Define a basic solution of a system of m linearly independent equations with n unknowns $(n > m)$.

   (b) How many basic solutions are there in the given set of linearly independent equations

$$\begin{aligned} 2x_1 - 5x_2 + \quad 6x_3 &= 9 \\ 6x_1 + \quad x_2 + 18x_3 &= 11 \end{aligned}$$

   Find all of them.

2. Find out all basic solutions of the set of equations

   (*a*) $\begin{aligned} x_1 + 2x_2 - x_3 &= 4 \\ 2x_1 + \quad x_2 + x_3 &= 10 \end{aligned}$ (*b*) $\begin{aligned} 3x_1 + \quad x_2 - \quad x_3 &= 5 \\ 6x_1 + 4x_2 - 2x_3 &= 2 \end{aligned}$

   (*c*) How many basic solutions are there in the following set of equations. Find all of them.

$$\begin{aligned} x_1 + 3x_2 + 2x_3 + 3x_4 &= 10 \\ 2x_1 - \quad x_2 + 4x_3 + 6x_4 &= 16 \end{aligned}$$

   (*d*) $\begin{aligned} 2x_1 + x_2 + 4x_3 &= 11 \\ 3x_1 + x_2 + 5x_3 &= 14\ ? \end{aligned}$ [C.U.(P)'82]

3. Find out all basic solutions to the following equations identifying in each case the basis vectors

$$\begin{aligned} x_1 + x_2 + x_3 &= 4 \\ 2x_1 + 5x_2 - 2x_3 &= 3 \end{aligned}$$

[C.U.(H)'81]

4. Find all the basic solutions of the system of equations,

$$\begin{aligned} x_1 + x_2 + 2x_3 &= 9 \\ 3x_1 + 2x_2 + 5x_3 &= 22 \end{aligned}$$

[C.U.(P)'80]

5. Determine all basic feasible solutions of the set of equations

$$\begin{aligned} 2x_1 + 6x_2 + 2x_3 + x_4 &= 3 \\ 6x_1 + 4x_2 + 4x_3 + 6x_4 &= 2 \end{aligned}$$

Identify the basic and non-basic variables in each case. [C.U.(P)'88]

6. Find out all basic feasible solutions of the set of equations

$$\begin{aligned} 2x_1 + 3x_2 - x_3 + 4x_4 &= 8 \\ -x_1 + 2x_2 - 6x_3 + 7x_4 &= 3 \end{aligned}$$

[J.U.M.Sc.'81]

7. Prove that x_1 and x_3 cannot be simultaneously considered as the basic variables of the set of equations

$$\begin{aligned} 2x_1 + x_2 - x_3 + 4x_4 &= 6 \\ 6x_1 + 3x_2 - 3x_3 + x_4 &= 10 \end{aligned}$$

Find all the basic solutions with x_1 as one basic variable.

8. How many basic solutions are there in the following set of equations

$$\begin{aligned} 2x_1 - 5x_2 + x_3 + 3x_4 &= 4 \\ 3x_1 - 10x_2 + 2x_3 + 6x_4 &= 12 \ ? \end{aligned}$$

Find all basic solution or solutions with x_1 as one of the basic variables.

9. Prove that $x_1 = 2$, $x_2 = 1$, $x_3 = 0$, $x_4 = 2$ is a solution of the set of equations but not basic.

$$\begin{aligned} 2x_1 - 3x_2 + x_3 &= 1 \\ 4x_1 - 6x_2 - x_3 &= 2 \\ 3x_1 + 4x_2 + 2x_3 + x_4 &= 12 \end{aligned}$$

Find one basic solution, where x_3 and x_4 are two basic variables.

10. For the set of equations

$$\begin{aligned} x_1 + x_2 + 2x_3 &= 9 \\ 3x_1 + 2x_2 + 5x_3 &= 22 \end{aligned}$$

Show that the feasible solution (2, 3, 2) is not basic. Find all the basic feasible solutions. [C.U.(P)'87]

11. Find the basic feasible solutions of the following equation

$$2x_1 + 3x_3 - x_3 = 6.$$

Answers

1. 2, [2, −1, 0], [0, −1, 2/3]. **2.** (a) [16/3, −2/3, 0], [14/3, 0, 2/3], [0, 14/3, 16/3]; (b) [3, −4, 0], [0, −4, −9]; (c) 3, [58/7, 4/7, 0, 0], [0, 4/7, 23/7, 0]; (d) [3, 5, 0], [1/2, 0, 5/2], [0, −1, 3]. **3.** [17/3, −5/3, 0], [0, 11/7, 17/7]. **4.** [4, 5, 0], [−1, 0, 5], [0, 1, 4]. **5.** [$x_1 = 0, x_2 = 1/2, x_3 = x_4 = 0$ (non-basic)]. [$x_2 = 1/2$, $x_3 = 0$, $x_1 = x_4 = 0$ (non-basic)]; [$x_2 = 1/2$, $x_4 = 0$, $x_1 = x_3 = 0$ (non-basic)]. All the three solutions are degenerate. **6.** [1, 2, 0, 0], [22/9, 0, 0, 7/9], [0, 45/16, 7/16, 0], [0, 0, 44/17, 45/17]. **7.** Det $\begin{bmatrix} 2 & -1 \\ 6 & -3 \end{bmatrix} = 0$. So x_1 and x_3 cannot be considered simultaneously as the basic variables. **8.** 3, [−4, −12/5, 0, 0], [−4, 0, 12, 0], [−4, 0, 0, 4]. **9.** [$x_1 = 0$, $x_2 = -1/3$, $x_3 = 0$ (basic), $x_4 = 40/3$]. **10.** [4, 5, 0], [0, 1, 4]. **11.** [3, 0, 0], [0, 2, 0].

Chapter 5

N-Dimensional Euclidean Space and Convex Set

5.1 Introduction

There are two types of vectors, row and column vectors. All n-component row vectors may be considered as points in an n-dimensional space which is known as *n-dimensional vector space* or *n-dimensional Euclidean space.* Similarly all n-component column vectors may be considered as points in n-dimensional vector space. The vector space is generally denoted by V_n or E^n or R^n etc. The points in a particular vector space are either all row or column vectors.

Point set: Point sets are sets whose elements are all points in E^n. Point set is usually denoted by capital letter X or S. If the set X contains m points $\mathbf{x}_1, \mathbf{x}_2, \cdots, \mathbf{x}_m$, then X is given by

$$X = \{\mathbf{x} : \mathbf{x} = \mathbf{x}_j, \quad [j = 1, 2, \cdots, m]\}.$$

Line: If $\mathbf{x}_1 = (x_{11}, x_{12}, \cdots, x_{1n})$ and $\mathbf{x}_2 = (x_{21}, x_{22}, \cdots, x_{2n})$ be two points in E^n, then the line joining the points $\mathbf{x}_1$ and $\mathbf{x}_2$, $[\mathbf{x}_2 \neq \mathbf{x}_1]$ is a set X of points given by

$$X = \{\mathbf{x} : \mathbf{x} = \lambda\mathbf{x}_1 + (1 - \lambda)\mathbf{x}_2, \quad \text{for all real } \lambda\}.$$

For different values of λ, different points will be obtained which all lie in a line which is a line in E^n.

Line segment: If $\mathbf{x}_1$ and $\mathbf{x}_2$ be two distinct points in E^n, then the line segment joining these two points is a set X of points given by

$$X = \{\mathbf{x} : \mathbf{x} = \lambda\mathbf{x}_1 + (1 - \lambda)\mathbf{x}_2, 0 \leq \lambda \leq 1\}.$$

Hyperplane: A hyperplane in E^n is a set X of points given by $X = \{\mathbf{x} : \mathbf{c}\mathbf{x} = k\}$, where $\mathbf{c}$ is a row vector, given by $\mathbf{c} = (c_1, c_2, \cdots, c_n)$ and not all $c_j = 0$ and $\mathbf{x} = [x_1, x_2, \cdots, x_n]$ is an n-component column vector.

A hyperplane can be defined as a set of points which will satisfy

$$c_1x_1 + c_2x_2 + \cdots + c_nx_n = k.$$

A hyperplane divides the space E^n into three mutually exclusive disjoint sets given by

$$\begin{aligned} X_1 &= \{\mathbf{x} : \mathbf{cx} > k\} \\ X_2 &= \{\mathbf{x} : \mathbf{cx} = k\} \\ X_3 &= \{\mathbf{x} : \mathbf{cx} < k\}. \end{aligned}$$

The sets X_1 and X_2 are called as *open half spaces.*

The objective function $z = \mathbf{cx}$ and the constraints with equality sign $A\mathbf{x} = \mathbf{b}$ are all hyperplanes in a L.P.P.

Hypersphere: A hypersphere in E^n with centre at $\mathbf{a} = (a_1, a_2, \cdots, a_n)$ and radius $\epsilon > 0$ is defined to be a set X of points given by

$$X = \{\mathbf{x} : |\mathbf{x} - \mathbf{a}| = \epsilon\}, \text{ where } \mathbf{x} = (x_1, x_2, \cdots, x_n).$$

The equation can be written as

$$(x_1 - a_1)^2 + (x_2 - a_2)^2 + \cdots + (x_n - a_n)^2 = \epsilon^2.$$

The hypersphere in two dimension is a circle and in three dimension a sphere.

An ϵ-neighbourhood: An ϵ-neighbourhood about a point $\mathbf{a}$ is defined as the set X of points lying inside the hypersphere with centre at $\mathbf{a}$ and radius $\epsilon > 0$, i.e., the ϵ-neighbourhood about the point $\mathbf{a}$ is a set of points given by

$$X = \{\mathbf{x} : |\mathbf{x} - \mathbf{a}| < \epsilon\}.$$

An interior point of a set: A point $\mathbf{a}$ is said to be an interior point of the set S if there exists an ϵ-neighbourhood about the point $\mathbf{a}$ which contains only points of the same set. From the definition it is clear that an interior point of a set S must be an element of the set S.

Boundary point of a set: A point $\mathbf{a}$ is said to be a boundary point of a set S if every ϵ-neighbourhood about $\mathbf{a}$ ($\epsilon > 0$, however, small it may be) containts points which are in the set S and the points which are not in the set S. A boundary point may or may not be an element of the set.

Open set: A set S is said to be open if it contains only interior points.

Closed set: A set S is said to be closed set if it contains all its boundary points. [For more precise definitions, consult any Text book of Real Analysis.]

Bounded set: A set S is said to be strictly bounded set if there exists a positive number r such that for any point $\mathbf{x}$ belonging to S, $|\mathbf{x}| \leq r$. If for every $\mathbf{x}$, all its components have lower limits only then the set is bounded from below.

5.2 Convex Combination and Convex Sets

Convex combination of a set of points: The convex combination of a set of k points $\mathbf{x}_1, \mathbf{x}_2, \cdots, \mathbf{x}_k$ in a space E^n is also a point $\mathbf{x}$ in the same space, given by

$$\mathbf{x} = \lambda_1 \mathbf{x}_1 + \lambda_2 \mathbf{x}_2 + \cdots + \lambda_k \mathbf{x}_k,$$

where λ_i is real scalar and ≥ 0 for all i and $\sum_{i=1}^{k} \lambda_i = 1$.

For different values of the scalar quantities λ_i $[i = 1, 2, \cdots, k]$ satisfying the conditions $\sum_{i=1}^{k} \lambda_i = 1$ and $\lambda_i \geq 0$, $\forall i$, a set of points will be obtained from the convex combinations of the set of k finite points which is a point set in E^n. This point set is known as the *convex polyhedron.*

The point set, convex polyhedron X is given by

$$X = \left\{ \mathbf{x} : \mathbf{x} = \sum_{i=1}^{k} \lambda_i \mathbf{x}_i, \sum_{i=1}^{k} \lambda_i = 1 \text{ and } \lambda_i \geq 0, \ \forall i \right\}.$$

From the definition it is also clear that line segment is a convex combination of the two distinct points in the same vector space.

Convex set: A point set is said to be a convex set if the convex combination of any two points of the set is in the set. In other words if the line segment joining any two distinct points of the set is in the set then the set is known as a *convex set;* otherwise the set is a non-convex set. Convex set is usually denoted by C or X.

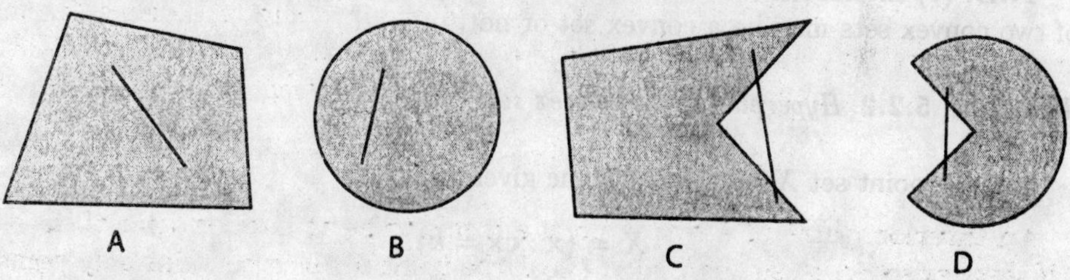

Fig. 5.1: (i) A and B : Convex Set **(ii) C and D : Non-convex Set**

Extreme points of the convex set: A points $\mathbf{x}$ is said to be an extreme point of the convex set C if it cannot be expressed as the convex combination of any other two distinct points $\mathbf{x}_1$ and $\mathbf{x}_2$ of the set C, i.e., $\mathbf{x}$ cannot be expressed in the form $\mathbf{x} = \lambda \mathbf{x}_1 + (1 - \lambda)\mathbf{x}_2$, $0 < \lambda < 1$. From the definition, it is clear that all extreme points of the convex set C are the boundary points but all boundary points may not be necessarily extreme points. Every point of the boundary of a circle is an extreme point of the convex set which includes the boundary and the interior of the circle. The extreme points of a rectangle are its four vertices.

Convex hull: If X be a point set, then convex hull of X which is denoted by $C(X)$, is the set of all convex combinations of sets of points from X. If the set X consists of a finite number of points then the convex hull $C(X)$ is called as a *convex polyhedron.* We shall prove that convex polyhedron is a convex set. If X is a point set, consists of only eight vertices of a cube then $C(X)$ is the whole cube which is also a convex polyhedron. A circle with all its interior points is a convex hull because all boundary points are extreme points which are not finite. Thus a circle with all its interior points is a convex hull but not a convex polyhedron. Hence the conclusion is that all convex polyhedrons are convex hull but all convex hulls may not be convex polyhedron.

For a convex polyhedron, any point in the set can be expressed as a convex combination of its extreme points.

Simplex: A: simplex is an n-dimensional convex polyhedron having exactly $(n+1)$ vertices. A simplex in zero dimension is a point; in one dimension, a line segment; in two dimension, a triangle and in three dimension, a tetrahedron.

Theorem 5.2.1 *Intersection of two convex sets is also a convex set:*

Let X_1 and X_2 be two convex sets and $X = X_1 \cap X_2$. It is required to prove that X is also a convex set.

Let $\mathbf{x}_1$ and $\mathbf{x}_2$ be any two distinct points of X. Then $\mathbf{x}_1$ and $\mathbf{x}_2$ are also points of the sets X_1 and X_2. Let $\mathbf{x}_3$ be a point given by

$$\mathbf{x}_3 = \lambda \mathbf{x}_1 + (1-\lambda)\mathbf{x}_2, \quad 0 \le \lambda \le 1.$$

As $\mathbf{x}_3$ is a convex combination of the two points $\mathbf{x}_1$ and $\mathbf{x}_2$ then $\mathbf{x}_3$ is a point of X_1 as well as a point of X_2 which indicates that $\mathbf{x}_3$ is a point on $X_1 \cap X_2 = X$. But $\mathbf{x}_3$ is the convex combination of two distinct points of X. Hence X is a convex set.

N.B. (1) Intersection of finite number of convex sets is a convex set. (2) Union of two convex sets may be a convex set or not.

Theorem 5.2.2 *Hyperplane is a convex set.*

Let the point set X be a hyperplane given by

$$X = \{\mathbf{x} : \mathbf{cx} = k\}.$$

Let $\mathbf{x}_1$ and $\mathbf{x}_2$ be any two distinct points of $\mathbf{cx} = k$. Then $\mathbf{cx}_1 = k$ and $\mathbf{cx}_2 = k$.

Let $\mathbf{x}_3$ be a point given by

$$\mathbf{x}_3 = \lambda \mathbf{x}_1 + (1-\lambda)\mathbf{x}_2, \quad 0 \le \lambda \le 1.$$

Therefore,

$$\mathbf{cx}_3 = \lambda \mathbf{cx}_1 + (1-\lambda)\mathbf{cx}_2 = \lambda k + (1-\lambda)k = k$$

which indicates that $\mathbf{x}_3$ is also a point of $\mathbf{cx} = k$.

But $\mathbf{x}_3$ is the convex combination of two distinct points $\mathbf{x}_1$ and $\mathbf{x}_2$ of X. Hence X is a convex set. Set X is closed also.

Note: Similarly it can be proved that $X = \{\mathbf{x} : \mathbf{cx} >< k\}$ are also convex sets.

Theorem 5.2.3 *Convex polyhedron is a convex set.*

Let S be a point set consisting of finite number of points $\mathbf{x}_1, \mathbf{x}_2, \cdots, \mathbf{x}_k$ in R^n.

Convex polyhedron $\mathbf{C}(S) = X = \left\{\mathbf{x} : \mathbf{x} = \sum_{i=1}^{k} \lambda_i \mathbf{x}_i, \lambda_i \ge 0, \sum_{i=1}^{k} \lambda_i = 1\right\}.$

It is required to prove that X is a convex set.

Let $\mathbf{u}$ and $\mathbf{v}$ be any two distinct points of X given by

$$\mathbf{u} = \sum_{i=1}^{k} a_i \mathbf{x}_i, \quad a_i \geq 0, \quad \sum_{i=1}^{k} a_i = 1.$$

$$\mathbf{v} = \sum_{i=1}^{k} b_i \mathbf{x}_i, \quad b_i \geq 0, \quad \sum_{i=1}^{k} b_i = 1.$$

Consider $\mathbf{w}$, where

$$\mathbf{w} = \lambda \mathbf{u} + (1 - \lambda)\mathbf{v}, \quad 0 \leq \lambda \leq 1$$

then

$$\mathbf{w} = \lambda \sum_{i=1}^{k} a_i \mathbf{x}_i + (1 - \lambda) \sum_{i=1}^{k} b_i \mathbf{x}_i = \sum_{i=1}^{k} \{\lambda a_i + (1 - \lambda) b_i\} \mathbf{x}_i = \sum_{i=1}^{k} c_i \mathbf{x}_i,$$

where $c_i = \lambda a_i + (1 - \lambda) b_i$.

Now

$$\sum_{i=1}^{k} c_i = \lambda \sum_{i=1}^{k} a_i + (1 - \lambda) \sum_{i=1}^{k} b_i = 1$$

and $c_i \geq 0$ as $a_i \geq 0$, $b_i \geq 0$ and $0 \leq \lambda \leq 1$.

Hence $\mathbf{w}$ is also a point of X, which is a convex combination of two distinct points $\mathbf{u}$ and $\mathbf{v}$ of X. Hence X is a convex set.

Theorem 5.2.4 *The set of all feasible solutions to a L.P.P.* $A\mathbf{x} = \mathbf{b}$, $\mathbf{x} \geq \mathbf{0}$ *is a closed convex set.*

Let X be the point set of all the feasible solutions of the problem $A\mathbf{x} = \mathbf{b}$, $\mathbf{x} \geq \mathbf{0}$.

If the set X has only one point then there is nothing to prove.

If X has at least two distinct points $\mathbf{x}_1$ and $\mathbf{x}_2$ then $A\mathbf{x}_1 = \mathbf{b}$, $\mathbf{x}_1 \geq \mathbf{0}$ and $A\mathbf{x}_2 = \mathbf{b}$, $\mathbf{x}_2 \geq \mathbf{0}$.

Consider a point $\mathbf{x}_3$ such that $\mathbf{x}_3 = \lambda \mathbf{x}_1 + (1 - \lambda)\mathbf{x}_2$, $\quad 0 \leq \lambda \leq 1$.

Thus,

$$A\mathbf{x}_3 = \lambda A\mathbf{x}_1 + (1 - \lambda) A\mathbf{x}_2 = \lambda \mathbf{b} + (1 - \lambda)\mathbf{b} = \mathbf{b}.$$

Again $\mathbf{x}_3 \geq \mathbf{0}$ as $\mathbf{x}_1 \geq \mathbf{0}$, $\mathbf{x}_2 \geq \mathbf{0}$ and $0 \leq \lambda \leq 1$.

Then $\mathbf{x}_3$ is also a feasible solution to the problem $A\mathbf{x} = \mathbf{b}$, $\mathbf{x} \geq \mathbf{0}$.

But $\mathbf{x}_3$ is the convex combination of two distinct points $\mathbf{x}_1$ and $\mathbf{x}_2$ of the set X. Thus X is a convex set.

Now the finite number of constraints represented by $A\mathbf{x} = \mathbf{b}$ are closed sets and also the set of inequations (finite) represented by $\mathbf{x} \geq \mathbf{0}$ are closed sets and therefore the intersection of finite number of closed sets which is the set of all feasible solutions is a closed set.

Similarly we can establish that the set of feasible solutions to a L.P.P. $A\mathbf{x}(\leq=\geq)\mathbf{b}$, $\mathbf{x} \geq \mathbf{0}$ is a closed convex set.

Note: (1) For different values of λ, satisfying the condition $0 \leq \lambda \leq 1$, an infinite set of points will be obtained. Therefore, if a L.P.P. has at least two feasible solutions then it has infinite number of feasible solutions.

(2) There are two types of convex sets of F.S. of a L.P.P. (i) strictly bounded or convex polyhedron and (ii) bounded from below only or convex polytope.

Theorem 5.2.5 *All B.F.S. of the set of equations* $A\mathbf{x} = \mathbf{b}$, $\mathbf{x} \geq \mathbf{0}$ *are extreme points of the convex set of feasible solutions of the equations and conversely.*

Proof. Let $A = (\mathbf{a}_1, \mathbf{a}_2, \cdots, \mathbf{a}_n)$ be the coefficient matrix of order $m \times n$ $(n > m)$ and let us assume that B be a basis matrix $B = (\mathbf{a}_1, \mathbf{a}_2, \cdots, \mathbf{a}_m)$, where $\mathbf{a}_1, \mathbf{a}_2, \cdots, \mathbf{a}_m$ are the column vectors corresponding to first m variables $x_1, x_2, \cdots, x_m$.

Let $\mathbf{x}$ be a B.F.S. and $\mathbf{x}$ is given by

$$\mathbf{x} = [\mathbf{x}_B, \mathbf{0}], \tag{5.2.1}$$

where $\mathbf{x}_B = B^{-1}\mathbf{b}$ and $\mathbf{0}$ is a $(n-m)$ component null column vector.

It is required to prove that $\mathbf{x}$ is an extreme point of the convex set X of the feasible solution of the equations $A\mathbf{x} = \mathbf{b}$, $\mathbf{x} \geq \mathbf{0}$.

Let $\mathbf{x}$ be not an extreme point of the convex set X.

Thus there exist two points $\mathbf{x}_1$ and $\mathbf{x}_2[\mathbf{x}_1 \neq \mathbf{x}_2]$ in X such that it is possible to express $\mathbf{x}$, as given by

$$\mathbf{x} = \lambda\mathbf{x}_1 + (1-\lambda)\mathbf{x}_2, \quad 0 < \lambda < 1 \quad \text{[note the limit of } \lambda],$$

where $\mathbf{x}_1$ and $\mathbf{x}_2$ are given by

$$\mathbf{x}_1 = [\mathbf{u}_1, \mathbf{v}_1] \text{ and } \mathbf{x}_2 = [\mathbf{u}_2, \mathbf{v}_2], \tag{5.2.2}$$

where $\mathbf{u}_1$ contains m components of $\mathbf{x}_1$, corresponding to the variables $x_1, x_2, \cdots, x_m$ and $\mathbf{v}_1$ contains the remaining $(n-m)$ components of $\mathbf{x}_1$. Similarly $\mathbf{u}_2$ and $\mathbf{v}_2$ contain the first m and the remaining $(n-m)$ components of $\mathbf{x}_2$ respectively.

Thus

$$\begin{aligned}\mathbf{x} &= \lambda[\mathbf{u}_1, \mathbf{v}_1] + (1-\lambda)[\mathbf{u}_2, \mathbf{v}_2] \\ &= [\lambda\mathbf{u}_1 + (1-\lambda)\mathbf{u}_2, \lambda\mathbf{v}_1 + (1-\lambda)\mathbf{v}_2].\end{aligned}$$

As $\mathbf{x} = [\mathbf{x}_B, \mathbf{0}]$, then equating the components corresponding to the last $(n-m)$ variables, we get

$$\lambda\mathbf{v}_1 + (1-\lambda)\mathbf{v}_2 = \mathbf{0}$$

which is possible only when $\mathbf{v}_1 = \mathbf{0}$ and $\mathbf{v}_2 = \mathbf{0}$ [as $\mathbf{v}_1 \geq \mathbf{0}, \mathbf{v}_2 \geq \mathbf{0}$ and $0 < \lambda < 1$]. Thus

$$\mathbf{x}_1 = [\mathbf{u}_1, \mathbf{0}] \text{ and } \mathbf{x}_2 = [\mathbf{u}_2, \mathbf{0}].$$

Hence $\mathbf{u}_1$ and $\mathbf{u}_2$ are the m components of the solution set corresponding to the basic variables $x_1, x_2, \cdots, x_m$ for which the basis matrix is B. Then $\mathbf{u}_1 = B^{-1}\mathbf{b}$ and $\mathbf{u}_2 = B^{-1}\mathbf{b}$. Hence $\mathbf{x}_B = \mathbf{u}_1 = \mathbf{u}_2$. Therefore, three points $\mathbf{x}, \mathbf{x}_1$ and $\mathbf{x}_2$ are not different and thus $\mathbf{x}$ cannot be expressed as the convex combination of two distinct points. So a B.F.S. $\mathbf{x}$ is also an extreme point. Similar is the case for all B.F.S.

Part II: Let $\mathbf{x}$ be an extreme point of the convex set X of all feasible solutions of the equations

$$A\mathbf{x} = \mathbf{b}, \quad \mathbf{x} \geq \mathbf{0}.$$

It is required to prove that $\mathbf{x}$ is a B.F.S.

Let

$$\mathbf{x} = [x_1, x_2, \cdots, x_k, \overbrace{0, \cdots, 0}^{n-k}]$$

where $x_j \geq 0$ for $j = 1, 2, \cdots, k$.

If the column vectors $\mathbf{a}_1, \mathbf{a}_2, \cdots, \mathbf{a}_k$ associated with the variables $x_1, x_2, \cdots, x_k$ respectively are L.I. (which is possible only for $k \leq m$) then $\mathbf{x}$, the extreme point of the convex set, is a B.F.S. and we have nothing to prove.

If $\mathbf{a}_1, \mathbf{a}_2, \cdots, \mathbf{a}_k$ are not L.I. then

$$\sum_{j=1}^{k} \mathbf{a}_j x_j = \mathbf{b} \tag{5.2.3}$$

$$\text{and} \quad \sum_{j=1}^{k} \mathbf{a}_j \lambda_j = \mathbf{0} \tag{5.2.4}$$

with at least one $\lambda_j \neq 0$.

Let $\delta > 0$, then from (1) and (2) we get

$$\sum_{j=1}^{k} (x_j \pm \delta\lambda_j)\mathbf{a}_j = \mathbf{b}. \tag{5.2.5}$$

Consider δ in the interval $0 < \delta < l$, where

$$l = \min_j \left(\frac{x_j}{|\lambda_j|}\right), \quad \lambda_j \neq 0.$$

Then $x_j \pm \delta\lambda_j \geq 0$ for $j = 1, 2, \cdots, k$.

Hence the two points

$$\mathbf{x}_1 = [x_1 + \delta\lambda_1, x_2 + \delta\lambda_2, \cdots, x_k + \delta\lambda_k, \overbrace{0, \cdots, 0}^{n-k}]$$

$$\text{and} \quad \mathbf{x}_2 = [x_1 - \delta\lambda_1, x_2 - \delta\lambda_2, \cdots, x_k - \delta\lambda_k, \overbrace{0, \cdots, 0}^{n-k}]$$

are the points of the convex set X.

Now

$$\frac{1}{2}\mathbf{x}_1 + \frac{1}{2}\mathbf{x}_2 = \mathbf{x}.$$

Then $\mathbf{x}$ can be expressed as

$$\mathbf{x} = \lambda\mathbf{x}_1 + (1-\lambda)\mathbf{x}_2, \quad \text{where } \lambda = 1/2.$$

Thus $\mathbf{x}$ is being expressed as the convex combination of two distinct points $\mathbf{x}_1$ and $\mathbf{x}_2$ in X which is against the assumption that $\mathbf{x}$ is an extreme point. Thus the column vectors $\mathbf{a}_1, \mathbf{a}_2, \cdots, \mathbf{a}_k$ are L.I. and hence $\mathbf{x}$ is a B.F.S.

Note: There is one to one correspondence between the extreme points and B.F.S. in the case of non-degenerate B.F.S.

► **Example 5.2.1** *Extreme points are finite in number.*

There are at most ${}^nC_m$ B.S. to a set of m equations with n unknowns. Hence the maximum number of B.F.S. to a L.P.P. is ${}^nC_m$ which is finite provided n is finite. This proves the theorem.

Theorem 5.2.6 *Any point of a convex polyhedorn can be expressed as a convex combination of its extreme points.*

Proof. Let the finite number of extreme points of the convex polyhedron be $\mathbf{x}_1, \mathbf{x}_2, \cdots, \mathbf{x}_n$. Let $\mathbf{x}$ be any point of the convex polyhedorn. We take any extreme point (say) $\mathbf{x}_1$ and join it with $\mathbf{x}$ by a line segment and extend it to meet either of a boundary line or a boundary surface (section of a hyperplane). Let the point of intersection be $\mathbf{x}^*$. Now $\mathbf{x}^*$ can always be expressed as the convex combination of some of the extreme points of the bounding hyperplane or line. Therefore, $\mathbf{x}^*$ can be expressed as

$$\mathbf{x}^* = \sum_{i=1}^{k} \mu_i \mathbf{x}_i \geq \mathbf{0}; \qquad \sum_{i=1}^{k} \mu_i = 1, \;\; k \leq n.$$

Now $\mathbf{x}$ can be expressed as

$$\mathbf{x} = \lambda_1\mathbf{x}_1 + \lambda_2\mathbf{x}^*, \quad \lambda_1 \geq 0, \lambda_2 \geq 0, \lambda_1 + \lambda_2 = 1,$$

$$\text{or,} \qquad \mathbf{x} = \lambda_1\mathbf{x}_1 + \lambda_2 \sum_{i=1}^{k} \mu_i \mathbf{x}_i.$$

Here $\lambda_1 \geq 0$, $\lambda_2\mu_i \geq 0$ for $i = 1, 2, \cdots, k$ and

$$\begin{aligned} \lambda + \lambda_2\mu_1 + \lambda_2\mu_2 + \cdots + \lambda_2\mu_k &= \lambda_1 + \lambda_2(\mu_1 + \mu_2 + \cdots + \mu_k) \\ &= \lambda_1 + \lambda_2 = 1. \end{aligned}$$

Therefore, $\mathbf{x}$ can be expressed as the convex combination of some or all the extreme points of the convex polyhedron.

Theorem 5.2.7 *If the convex set of F.S. of a L.P.P. be a convex polyhedron, then any objective function has both finite maximum and minimum corresponding to some extreme points of the convex polyhedron.*

Proof. Let X, convex set of F.S. of a L.P.P. be a convex polyhedron with finite number of extreme points $\mathbf{x}_1, \mathbf{x}_2, \cdots, \mathbf{x}_k$ and the objective function be $z = \mathbf{cx}$.

Therefore, z has definitely finite optimal value [maximum or minimum].

It is required to prove that z attains its optimal value corresponding to one of the values $\mathbf{x}_1, \mathbf{x}_2, \cdots, \mathbf{x}_k$. Let the objective function z is to be maximised and for a point $\hat{\mathbf{x}}$ of $X, \hat{z} = \mathbf{c}\hat{\mathbf{x}}$ If $\hat{\mathbf{x}}$ is an extreme point then there is nothing to prove. If $\hat{\mathbf{x}}$ be not an extreme point then

$$\hat{\mathbf{x}} = \lambda_1 \mathbf{x}_1 + \cdots + \lambda_k \mathbf{x}_k = \sum_{i=1}^{k} \lambda_i \mathbf{x}_i, \lambda_i \geq 0, \sum_{i=1}^{k} \lambda_i = 1.$$

Then

$$\hat{z} = \mathbf{c}\hat{\mathbf{x}} = \mathbf{c}(\lambda_1 \mathbf{x}_1 + \lambda_2 \mathbf{x}_2 + \cdots + \lambda_k \mathbf{x}_k) = \lambda_1 \mathbf{cx}_1 + \lambda_2 \mathbf{cx}_2 + \cdots + \lambda_k \mathbf{cx}_k.$$

If the maximum of $\mathbf{cx}_i$ be $\mathbf{cx}_1$ (say) $[i = 1, 2, \cdots, k]$.

Then

$$\hat{z} \leq \mathbf{cx}_1(\lambda_1 + \lambda_2 + \cdots + \lambda_k), \quad \text{or,} \quad \hat{z} \leq \mathbf{cx}_1.$$

But the maximum value of z is $\hat{z} = \mathbf{c}\hat{\mathbf{x}}$.

Then $\mathbf{c}\hat{\mathbf{x}} = \mathbf{cx}_1$ and therefore $\hat{\mathbf{x}} = \mathbf{x}_1$ (an extreme point).

Hence the maximum of the objective function z is attained at one of the extreme points which proves the theorem.

Similar is the proof when the objective function z is to be minimized.

Note (1): When the convex set of F.S. of a L.P.P. is strictly bounded, i.e., a convex polyhedron, any objective function has both finite maximum and minimum values which will be obtained from the extreme points of the convex set of F.S. But if the convex set of F.S. is bounded from below only, then an objective function may have either maximum or minimum but not the both, which will also be obtained from the extreme points of the set. Further a particular objective function may not have any optimum value, maximum or minimum, when the convex set is a convex polytope.

Note (2): In the syllabus of C.U. statement of the theorem is wrong. The convex set will be definitely a convex polyhedron. Otherwise, the theorem is meaningless (See Theorem 2, page 51 Linear Programming: Saul I. Gass).

Definition 5.2.1 *Multiple optimal solutions: If a L.P.P. has more than one optimal solution, then the problem is said to have multiple optimal solutions.*

Theorem 5.2.8 *If a L.P.P. has at least two optimal feasible solutions, then there are infinite number of optimal solutions, which are the convex combination of the initial optimal solutions.*

Let $\mathbf{x}_1$ and $\mathbf{x}_2$ be $(\mathbf{x}_1 \neq \mathbf{x}_2)$ be two optimal feasible solutions of a L.P.P. which will maximize the objective function $z = \mathbf{cx}$ subject to $A\mathbf{x} = \mathbf{b}$, $\mathbf{x} \geq \mathbf{0}$.

Then $\hat{z} = \mathbf{cx}_1$ and $\hat{z} = \mathbf{cx}_2$ and $A\mathbf{x}_1 = \mathbf{b}$, $A\mathbf{x}_2 = \mathbf{b}$.

Let

$$\mathbf{x}_3 = \lambda\mathbf{x}_1 + (1-\lambda)\mathbf{x}_2, \quad 0 \le \lambda \le 1.$$

$$A\mathbf{x}_3 = \lambda A\mathbf{x}_1 + (1-\lambda)A\mathbf{x}_2 = \mathbf{b}$$

which indicates that $\mathbf{x}_3$ is also a solution set of $A\mathbf{x} = \mathbf{b}$.

Again

$$\mathbf{c}\mathbf{x}_3 = \lambda\mathbf{c}\mathbf{x}_1 + (1-\lambda)\mathbf{c}\mathbf{x}_2 = \lambda\hat{z} + (1-\lambda)\hat{z} = \hat{z}$$

and $\mathbf{x}_3 \ge \mathbf{0}$ as $\mathbf{x}_1$, $\mathbf{x}_2 \ge \mathbf{0}$ and $0 \le \lambda \le 1$.

Therefore, $\mathbf{x}_3$ is also an optimal feasible solution of the L.P.P. which is the convex combination of two distinct points $\mathbf{x}_1$ and $\mathbf{x}_2$ which indicates that there are infinite optimal solutions.

Note: From this we can say that if a L.P.P. has $k (k \ge 2)$ optimal feasible solutions then the problem has an infinite optimal solutions which are the convex combination of k initial optimal solutions. This statement is valid both for basic and non-basic optimal feasible solutions.

▶ **Example 5.2.2** *Prove that in E^2, the set $S = \{(x_1, x_2)/x_1 - 2x_2 = 2\}$ is a convex set.*

Solution: The set is not a null set. Let (x_{11}, x_{21}) and (x_{12}, x_{22}) be any two points of the set. Then

$$\left.\begin{aligned} x_{11} - 2x_{21} &= 2 \\ \text{and} \quad x_{12} - 2x_{22} &= 2 \end{aligned}\right\}. \qquad \text{(i)}$$

Now the convex combination of the two points is a point which is

$$[\lambda x_{11} + (1-\lambda)x_{12}, \lambda x_{21} + (1-\lambda)x_{22}], \quad 0 \le \lambda \le 1.$$

Now

$$\begin{aligned} &\lambda x_{11} + (1-\lambda)x_{12} - 2\{\lambda x_{21} + (1-\lambda)x_{22}\} \\ = \ &\lambda(x_{11} - 2x_{21}) + (x_{12} - 2x_{22}) - \lambda(x_{12} - 2x_{22}) \\ = \ &2\lambda + 2 - 2\lambda = 2 \quad \text{[From (i)]} \end{aligned}$$

Hence the point, which is the convex combination of any two points is a point of the set. Thus the set is a convex set.

▶ **Example 5.2.3** *Prove that in E^2, the set $X = \{(x, y)|x + 2y \le 5\}$ is a convex set.*

Solution: The set is not a null set. Let (x_1, y_1) and (x_2, y_2) be any two points of the set X.

Then

$$\left.\begin{aligned} x_1 + 2y_1 &\le 5 \\ \text{and} \quad x_2 + 2y_2 &\le 5 \end{aligned}\right\}. \qquad \text{(i)}$$

Now the convex combination of the two points is a point given by

$$[\lambda x_1 + (1-\lambda)x_2, \lambda y_1 + (1-\lambda)y_2], \quad 0 \le \lambda \le 1.$$

Now

$$\begin{aligned}\lambda x_1 + (1-\lambda)x_2 + 2\{\lambda y_1 + (1-\lambda)y_2\} &= \lambda(x_1 + 2y_1) + x_2 + 2y_2 - \lambda(x_2 + 2y_2)\\ &\le 5\lambda + 5 - 5\lambda = 5 \quad [\because 0 \le \lambda \le 1].\end{aligned}$$

that the point is a point of the set. Thus the set is a convex set.

▶ **Example 5.2.4** *Prove that the set defined by* $X = \{\mathbf{x} : |\mathbf{x}| \le 2\}$ *is a convex set.*

Solution: The set X is not a null set. Let x_1 and x_2 be any two points belonging to the set X. Then

$$|x_1| \le 2, \quad |x_2| \le 2. \tag{i}$$

The convex combination of the points x_1 and x_2 is a point

$$\mathbf{x}^* = \lambda x_1 + (1-\lambda)x_2, \quad 0 \le \lambda \le 1.$$

Now

$$\begin{aligned}|\lambda x_1 + (1-\lambda)x_2| &\le |\lambda x_1| + |(1-\lambda)x_2|\\ &= \lambda|x_1| + (1-\lambda)|x_2| \quad [\therefore \lambda, 1-\lambda \ge 0]\\ &\le 2\lambda + 2(1-\lambda) = 2 \quad \text{[From (i)]}\end{aligned}$$

$\Rightarrow |\lambda x_1 + (1-\lambda)x_2| \le 2$ which indicates that $\mathbf{x}^* \in X$ and the set is a convex set.

▶ **Example 5.2.5** *Prove that the point* $(1, 0, 2, -2)$ *is a point of the open half space* $2x_1 + 3x_2 + x_3 - 3x_4 < 13$ *but the point* $(2, 2, -1, -4)$ *is not a point of the open half space* $2x_1 + 3x_2 + x_3 - 3x_4 < 13$.

Solution: For the point $(1, 0, 2, -2)$, the value of the expression

$$\begin{aligned}2x_1 + 3x_2 + x_3 - 3x_4 &= 2 \times 1 + 3 \times 0 + 1 \times 2 - 3(-2)\\ &= 10 < 13.\end{aligned}$$

Hence the point lies on the open half space

$$2x_1 + 3x_2 + x_3 - 3x_4 < 13.$$

For the point $(2, 2, -1 - 4)$, the value of the expression

$$\begin{aligned}2x_1 + 3x_2 + x_3 - 3x_4 &= 2 \times 2 + 3 \times 2 - 1 - 3 \times (-4)\\ &= 21 > 13.\end{aligned}$$

Hence the point $(2, 2, -1, -4)$ does not belong to the open half space

$$2x_1 + 3x_2 + x_3 - 3x_4 < 13.$$

► **Example 5.2.6** *Prove that in E^2, the set $X = \{(x, y)/x^2 + y^2 \leq 4\}$ is a convex set.*

(i) **Geometrical approach:** The set of points represents a circle of radius 2 with the boundary and all its interior points. Hence the set is a convex set.

(ii) **Algebraical approach:** Let (x_1, y_1) and (x_2, y_2) be any two points of the set X.

Then

$$\left.\begin{array}{l} x_1^2 + y_1^2 \leq 4 \\ x_2^2 + y_2^2 \leq 4 \end{array}\right\}. \qquad \text{(i)}$$

Now any convex combination of the points is a point which is

$$\{\lambda x_1 + (1 - \lambda)x_2, \lambda y_1 + (1 - \lambda)y_2\}, 0 \leq \lambda \leq 1.$$

Again,

$$\begin{aligned} & \{\lambda x_1 + (1 - \lambda)x_2\}^2 + \{\lambda y_1 + (1 - \lambda)y_2\}^2 \\ = \; & \lambda^2(x_1^2 + y_1^2) + (1 - \lambda)^2(x_2^2 + y_2^2) + 2\lambda(1 - \lambda)(x_1x_2 + y_1y_2) \\ \leq \; & \lambda^2(x_1^2 + y_1^2) + (1 - \lambda)^2(x_2^2 + y_2^2) + 2\lambda(1 - \lambda)\left[\frac{x_1^2 + x_2^2 + y_1^2 + y_2^2}{2}\right] \\ & \left[\because \quad x_1x_2 \leq \frac{x_1^2 + x_2^2}{2}, \quad y_1y_2 \leq \frac{y_1^2 + y_2^2}{2}\right] \\ \leq \; & 4\lambda^2 + 4(1 - \lambda)^2 + 8\lambda(1 - \lambda) \quad [\because \quad x_1^2 + x_2^2 + y_1^2 + y_2^2 \leq 8] \\ = \; & 4\lambda^2 + 4 + 4\lambda^2 - 8\lambda + 8\lambda - 8\lambda^2 \\ = \; & 4 \Rightarrow \text{the point} \in X. \text{ Thus the set is a convex set.} \end{aligned}$$

► **Example 5.2.7** *Prove that in E^2, set $X = \{(x, y)/|x| \leq 2, |y| \leq 1\}$ is a convex set.* (1) **Geometrical approach:** The set represents a rectangle with all its interior points having the four extreme points $(2, 1)(2, -1)$, $(-2, -1)$ and $(-2, 1)$. Hence the set is a convex set.

(2) **Algebraical approach:** The set is not empty. Let (x_1, y_1) and (x_2, y_2) be any two points belonging to the set X. Then

$$|x_1| \leq 2, \; |x_2| \leq 2 \qquad \text{(i)}$$

$$\text{and} \quad |y_1| \leq 1, \; |y_2| \leq 1. \qquad \text{(ii)}$$

The convex combination of the points which is a point given by

$$\{\lambda x_1 + (1 - \lambda)x_2, \; \lambda y_1 + (1 - \lambda)y_2\}, \quad 0 \leq \lambda \leq 1.$$

Now,

$$\begin{aligned} |\lambda x_1 + (1 - \lambda)x_2| & \leq |\lambda x_1| + |(1 - \lambda)x_2| \qquad \text{(iii)} \\ & = \lambda|x_1| + (1 - \lambda)|x_2| \quad [\because \quad \lambda \text{ and } 1 - \lambda \geq 0] \\ & \leq 2\lambda + 2(1 - \lambda) \\ & = 2 \\ & \Rightarrow |\lambda x_1 + (1 - \lambda)x_2| \leq 2 \end{aligned}$$

In the same way we can prove that

$$|\lambda y_1 + (1-\lambda)y_2| \leq 1 \qquad \text{(iv)}$$

Thus the convex combination of any two points $\in X$. Thus the set X is a convex set.

Short Answer Type Questions with Answers

1. Define: (a) point set, (b) line, (c) line segment, (d) hyper-plane in E^n, (e) Euclidean space, (f) Hyper-sphere, (g) Convex-polyhedron, (h) Convex hull. [*Ans.* See 5.1 and 5.2]

2. What is a convex set? What are its extreme points? [*Ans.* See 5.2]

3. (a) Give two examples of convex sets in R^2. [C.U.'82]

   (b) Give two examples of convex set in R^3.

   [*Ans.* (a) Triangle, Rectangle, square, circle with their interior points are the examples of convex set in E^2.

   (b) Cube and sphere with all its interior points.]

4. Give an example of a convex set which has no extreme point.

   [*Ans.* Hyper-plane, line etc.]

5. (a) Give an example of a convex set whose all boundary points are extreme points.

   [*Ans.* Circle with all its interior points and boundary points. Here all boundary points are extreme points.]

   (b) Explain the terms: (i) convex hull, (ii) convex polyhedron and (iii) Simplex. State giving reasons which of the following are the examples of one or more of these: A line segment, a circle, a triangle, a quadrilateral. [C.U.(H)'83]

   [*Ans.* Line segment, triangle; convex polyhedron [as well as simplex]. circle [convex hull], convex quadrilateral [Convex polyhedron].

6. What is a simplex? Name one simplex each in two and three dimension respectively.

7. Find out the extreme points (if any) of the following convex sets:

   (a) $S = \{(x,y)|x^2 + y^2 \leq 25\}$ [C.U.'82]

   (b) $S = \{(x,y)||x|| \leq a, |y| \leq b\}$

   (c) $S = \{(x,y)||x| \leq 1, y \leq 1\}$

   (d) Set $\mathbf{cx} = k$. [C.U.'82]

   [*Ans.* (a) A point (x_1, y_1) which satisfy the condition $x_1^2 + y_1^2 = 25$ is an extreme point. From that it is evident that this set has infinite number extreme points

   (b) $[x_1 = a, y_1 = b]$, $[x_2 = a,\ y_2 = -b]$, $[x_3 = -a,\ y_3 = b]$, $[x_4 = -a,\ y_4 = -b]$.

   (c) $[x_1 = 1,\ y_1 = 1]$, $[x_2 = -1, y_2 = 1]$.

   (d) This set has no extreme point at all.]

8. (a) Prove that hyperplane is a convex set.

(b) Prove that set of F.S. of $A\mathbf{x} = \mathbf{b}$, $x \geq \mathbf{0}$ is a closed convex set.

[*Ans.* See theorem 5.2.4]

9. What is the relation between B.F.S. and extreme point solution of a convex set of the feasible solutions of a L.P.P.?

[*Ans.* Each B.F.S. is a extreme point solution and vice-versa and there is one to one correspondance between a B.F.S. and extreme point solution in the absence of degeneracy.]

10. Prove that extreme points are finite in number. [*Ans.* See theorem 5.2.8]

11. A hyperplane is given by the equation $3x_1 + 2x_2 + 4x_3 + 6x_4 = 7$. Find in which half space are the points $(-6, 1, 7, 2)$ and $(1, 2, -4, 1)$? [C.U.(P)'83,'99]

[*Ans.* First point is in the space $3x_1 + 2x_2 + 4x_3 + 6x_4 > 7$ and the second point is in the space $3x_1 + 2x_2 + 4x_3 + 6x_4 < 7$.]

12. Are the points $(4, 1, 0, 2)$ and $(-3, 2, 4, 4)$ of the same side of the hyperplane $2x_1 + x_1 + 3x_3 + x_4 = 10$? [*Ans.* Yes]

13. (a) Is the point $(1, 10)$ in the convex set of F.S. determined by the constraints $2x_1 + 5x_2 \leq 40, x_1 + x_2 \leq 11, x_2 \geq 4, x_1 \geq 0, x_2 \geq 0$? Give reasons for your answer.

[*Ans.* No. Because the point satisfies the second and third constraints but does not satisfy the first constraint.]

(b) Is the point $(3, 1)$ in the convex set of F.S. determined by the constraints $x_1 + x_2 \leq 4$, $2x_1 + x_2 \leq 6$, $4x_1 + 3x_2 \leq 28, x_1 \geq 0, x_2 \geq 0$? Give reasons for your answer.

[*Ans.* No. The point does not satisfy the second constraint.]

(c) Is the point $(4, 3)$ in the convex set of feasible solution determined by the set of constraints $2x_1 + 7x_2 \geq 22$, $x_1 + x_2 \geq 6$, $3x_1 + x_2 \geq 10$, $x_1 \geq 0$, $x_2 \geq 0$?

[*Ans.* Yes. Because the point satisfies all the five constraints.]

14. Write a short note about the nature of the convex set of the feasible solutions of a L.P.P.

[*Ans.* The convex set of F.S. of a L.P.P. is either strictly bounded convex polyhedron or a convex set bounded from below only. Of course, the convex set may be a null set.

15. A particular objective function has both finite maximum and finite minimum subject to certain constraints. What is the nature of the convex set of feasible solutions of the L.P.P.?

[*Ans.* Strictly bounded convex polyhedron.]

16. The convex set of F.S. of a particular L.P.P. is a strictly bounded convex polyhedron with the extrene points $(0, 4)$, $(7, 4)$, $(5, 6)$, $(1, 8)$. The objective function is $2x_1 - 3x_2$. Find the maximum and minimum value of the objective function.

[*Ans.* $\max z = 2$ at (7, 4) and $\min z = -22$ at (1, 8).]

17. The convex set of F.S. of a particular L.P.P. is bounded from below only. It has three extreme points (5, 1), (1, 6) and (4, 8). The objective function $x_1 - 3x_2$ is known to have finite maximum. Find the maximum value of the objective function.
[*Ans.* max $z = 2$ at $(5, 1)$.]

18. The convex set of feasible solutions of a L.P.P. is bounded from below only and it has three extreme points $(4, 2)$, $(1, 5)$ and $(4, 8)$. The objective function $x_1 - x_2$ is known to have finite minimum. Find the minimum value of the objective function.

[*Ans.* min $z = -4$ at $(1, 5)$ and $(4, 8)$ which indicates that infinite optimal solutions exist.

19. Find graphically the feasible space, if any for the following case,

$$2x_1 + 3x_2 \leq 6,\ 2x_1 + 3x_2 \geq 6,\ x_1,\ x_2 \geq 0.$$ [C.U.(P)'83]

[*Ans.* Segment of the line $2x_1 + 3x_2 = 6$ with extreme points $(3, 0)$ and $(0, 2)$.]

Exercise 5

1. Define the boundary points and extreme points of a convex set. Prove that all the extreme points are boundary points but the converse is not necessarily true. Can there be any convex set without any extreme point?

2. Give an example of convex set, where all the boundary points are extreme points.

3. Prove that the set $X = \{(x_1, x_2) : x_1x_2 \leq 1, x, \geq 0, x_2 \geq 0\}$ is not a convex set.

4. (a) Prove that the set $z = \{(x_1, x_2) : x_1^2 + x_2^2 \leq 25\}$ is a convex set,.
   (b) Prove that the set $S = \{(x_1, x_2) | x_1^2 + x_2^2 = 4\}$ is not a convex set.
   (c) Prove that $[2, 0, 1, 0]$ is an extreme point of the convex set of F.S. of the set of equations

$$\begin{aligned} 3x_1 - 5x_2 + x_3 - 2x_4 &= 7 \\ 6x_1 - 10x_2 - x_3 + 5x_4 &= 11, \quad x_j \geq 0, \quad j - 1, \cdots, 4. \end{aligned}$$

5. Prove that extreme points of the convex set of F.S. of a L.P.P. are finite in number [number of variables are finite].

6. A hyperplane is given by the equation

$$2x_1 + 3x_2 + 4x_3 - x_4 = 6.$$

   Find in which half space are the points $(4, -3, 2, 1)$, $(1, 2 - 3, 1)$?

7. (a) Prove that hyperplane is a convex set.
   (b) Prove that the open half spaces $A\mathbf{x} < \mathbf{b}$ and $A\mathbf{x} > \mathbf{b}$ are convex sets.
   (c) Prove that the convex polyhedron is a convex set.

8. The vertices of a triangle are $(0,0)$, $(0,4)$, $(3,0)$. Express the points $(1,1)$, $(2,1)$, $(1,1.5)$ as the convex combination of the extreme points.

9. Prove that every extreme point of the convex set of F.S. to $A\mathbf{x} = \mathbf{b}$ is a B.F.S. and conversely. [C.U.M.Sc.(Appl.Math.)'77]

10. Prove that if $\mathbf{x}_1, \mathbf{x}_2, \cdots, \mathbf{x}_k$ be k different optimal feasible solutions to a L.P.P. then any convex combination of $\mathbf{x}_1, \mathbf{x}_2, \cdots, \mathbf{x}_k$ is also an optimal solution. [C.U.M.Sc.(Appl.Math.)'76]

11. Find the extreme points of the convex set of F.S. given by

$$\begin{aligned} 2x_1 + x_2 + 3x_3 + 2x_4 &= 11 \\ 3x_1 - 3x_2 + 5x_3 + x_4 &= 17, \quad x_j \geq 0,\ j = 1, \cdots, 4. \end{aligned}$$

[*Ans.* $[4,0,1,0]$, $[0,2/7,25/7,0]$, $[0,0,23/7,4/7]$]

Chapter 6

Fundamental Properties of Simplex Method

6.1 Fundamental theorem of Linear programming

Theorem 6.1.1 *If a L.P.P. Optimize,* $z = \mathbf{cx}$ *subject to* $A\mathbf{x} = \mathbf{b}$, $\mathbf{x} \geq \mathbf{0}$, *where* A *is the* $(m \times n)$ *coefficient matrix* $(m < n)$ *and* $r(A) = m$, *has an optimal solution then there exists at least one B.F.S., which will be optimal.*

▶ **Example 6.1.1** *Maximize,* $z = x_1 - x_2 + x_3$ *subject to*

$$\begin{array}{rcl} x_1 - 2x_2 + x_3 & = & 3 \\ 2x_1 + x_2 \qquad + x_4 & = & 2, \quad x_j \geq 0, \; j = 1, 2, \cdots, 4 \end{array}$$

with the assumption that the optimal solution exists.

Solution: As the problem has an optimal solution then at least one B.F.S. is optimal. The B.S. are

$$\mathbf{x}_1 = \left[\frac{7}{5}, \; -\frac{4}{5}, \; 0, \; 0\right], \quad \mathbf{x}_2 = [1, \; 0, \; 2, \; 0], \quad \mathbf{x}_3 = [3, \; 0, \; 0, \; -4],$$

$$\mathbf{x}_4 = [0, \; 2, \; 7, \; 0], \quad \mathbf{x}_5 = \left[0, \; -\frac{3}{2}, \; 0, \; \frac{7}{2}\right], \quad \mathbf{x}_6 = [0, \; 0, \; 3, \; 2]$$

(B.S. are to be determined by the method given previously)

Of these, only $\mathbf{x}_2$, $\mathbf{x}_4$, and $\mathbf{x}_6$ are B.F.S.

$$z = 1 - 0 + 2 = 3 \text{ for } \mathbf{x} = \mathbf{x}_2$$
$$z = 0 - 2 + 7 = 5 \text{ for } \mathbf{x} = \mathbf{x}_4$$
$$z = 0 - 0 + 3 = 3 \text{ for } \mathbf{x} = \mathbf{x}_6.$$

Hence the optimal value of $z = 5$ for $\mathbf{x} = \mathbf{x}_4 = [0, \; 2, \; 7, \; 0]$.

▶ **Example 6.1.2** *Find the optimal value and the optimal solution or solutions of the problem,*

$$\textit{Maximize} \quad z = x_1 - x_2 + 2x_3 + 3x_4$$

subject to

$$\begin{array}{rcrcrcrcl} 2x_1 & + & x_2 & + & 3x_3 & + & 2x_4 & = & 11 \\ 3x_1 & - & 3x_2 & + & 5x_3 & + & x_4 & = & 17 \end{array}, \quad x_j \geq 0,\ j = 1, 2, \cdots, 4$$

with the assumption that finite optimal solution exists.

Solution: The set of constraints are

$$\mathbf{a}_1 x_1 + \mathbf{a}_2 x_2 + \mathbf{a}_3 x_3 + \mathbf{a}_4 x_4 = \mathbf{b}, \tag{i}$$

where $\mathbf{a}_1 = \begin{bmatrix} 2 \\ 3 \end{bmatrix}$, $\mathbf{a}_2 = \begin{bmatrix} 1 \\ -3 \end{bmatrix}$, $\mathbf{a}_3 = \begin{bmatrix} 3 \\ 5 \end{bmatrix}$, $\mathbf{a}_4 = \begin{bmatrix} 2 \\ 1 \end{bmatrix}$ and $\mathbf{b} = \begin{bmatrix} 11 \\ 17 \end{bmatrix}$.

Six square matrices taking two at a time from the vectors $\mathbf{a}_1, \cdots, \mathbf{a}_4$, etc, are

$$B_1 = (\mathbf{a}_1, \mathbf{a}_2) = \begin{bmatrix} 2 & 1 \\ 3 & -3 \end{bmatrix}, \quad B_2 = (\mathbf{a}_1, \mathbf{a}_3) = \begin{bmatrix} 2 & 3 \\ 3 & 5 \end{bmatrix},$$

$$B_3 = (\mathbf{a}_1, \mathbf{a}_4) = \begin{bmatrix} 2 & 2 \\ 3 & 1 \end{bmatrix}, \quad B_4 = (\mathbf{a}_2, \mathbf{a}_3) = \begin{bmatrix} 1 & 3 \\ -3 & 5 \end{bmatrix},$$

$$B_5 = (\mathbf{a}_2, \mathbf{a}_4) = \begin{bmatrix} 1 & 2 \\ -3 & 1 \end{bmatrix}, \quad B_6 = (\mathbf{a}_3, \mathbf{a}_4) = \begin{bmatrix} 3 & 2 \\ 5 & 1 \end{bmatrix},$$

$$\det B_1 = -9 \neq 0, \quad \det B_2 = 1 \neq 0, \quad \det B_3 = -4 \neq 0,$$

$$\det B_4 = 14 \neq 0, \quad \det B_5 = 7 \neq 0, \quad \det B_6 = -7 \neq 0.$$

The six square sub-matrices are non-singular and therefore all the matrices are basis matrices and there exist six basic solutions.

For B_1, B.S. is

$$\mathbf{x}_1 = B_1^{-1}\mathbf{b} = -\frac{1}{9}\begin{bmatrix} -3 & -1 \\ -3 & 2 \end{bmatrix}\begin{bmatrix} 11 \\ 17 \end{bmatrix}\begin{bmatrix} 50/9 \\ -1/9 \end{bmatrix}.$$

Thus

$$\mathbf{x}_1 = \left[\frac{50}{9},\ -\frac{1}{9},\ 0,\ 0\right].$$

Other basic solutions which are to be calculated in the same manner given above are

$$\mathbf{x}_2 = [4,\ 0,\ 1,\ 0], \quad \mathbf{x}_3 = \left[\frac{23}{4},\ 0,\ 0,\ -\frac{1}{4}\right], \quad \mathbf{x}_4 = \left[0,\ \frac{2}{7},\ \frac{25}{7},\ 0\right],$$

$$\mathbf{x}_5 = \left[0,\ -\frac{23}{7},\ 0,\ \frac{50}{7}\right] \quad \text{and} \quad \mathbf{x}_6 = \left[0,\ 0,\ \frac{23}{7},\ \frac{4}{7}\right].$$

Of all B.S. only $\mathbf{x}_2$, $\mathbf{x}_4$ and $\mathbf{x}_6$ are feasible.

The value of z for $\mathbf{x}_2$ is $1 \times 4 - 1 \times 0 + 2 \times 1 + 3 \times 0 = 6$;

for $\mathbf{x}_4$ is $1 \times 0 - 1 \times \frac{2}{7} + 2 \times \frac{25}{7} + 3 \times 0 = \frac{48}{7}$;

for $\mathbf{x}_6$ is $1 \times 0 - 1 \times 0 + 2 \times \frac{23}{7} + 3 \times \frac{4}{7} = \frac{58}{7}$.

Thus $\max z = \frac{58}{7}$ for $x_1 = 0$, $x_2 = 0$, $x_3 = \frac{23}{7}$, $x_4 = \frac{4}{7}$, i.e., the optimal value is $\frac{58}{7}$ and the optimal solution is $[0,\ 0,\ \frac{23}{7},\ \frac{4}{7}]$.

N.B. The problem has been solved in Example (8.5.1) in the simplex theory, where it has been shown that the problem has finite optimal value and no assumption regarding the existence of finite optimal solution is needed there.

6.2 Reduction of a F.S. to a B.F.S. (Reduction Theorem)

Important theorem 6.2.1 *If a linear programming problem, $A\mathbf{x} = \mathbf{b}$, $\mathbf{x} \geq \mathbf{0}$, where A is the $m \times n$ coefficient matrix ($n > m$), $r(A) = m$ has one feasible solution then it has at least one basic feasible solution.*

Proof: Let

$$\mathbf{x} = [x_1, x_2, \cdots, x_n]$$

be a feasible solution to the set of equations $A\mathbf{x} = \mathbf{b}$, $\mathbf{x} \geq \mathbf{0}$. Out of n components of the feasible solution, let k components be positive and the remaining $(n-k)$ components be zero $(1 \leq k \leq n)$ and let us also make an assumption that the first k components are positive and the last $(n-k)$ components are zero.

Then

$$\mathbf{x} = [x_1, x_2, \cdots, x_k, \overbrace{0, \cdots, 0}^{n-k}]$$

and if $\mathbf{a}_1$, $\mathbf{a}_2$, $\cdots$, $\mathbf{a}_k$ be the column vectors corresponding to the variables x_1, x_2, $\cdots$, x_k then

$$x_1\mathbf{a}_1 + x_2\mathbf{a}_2 + \cdots + x_k\mathbf{a}_k = \mathbf{b}, \quad \text{or,} \quad \sum_{j=1}^{k} x_j\mathbf{a}_j = \mathbf{b}. \tag{i}$$

Now there are three cases to discuss:

(i) $k \leq m$ and the column vectors $\mathbf{a}_1$, $\mathbf{a}_2$, $\cdots$, $\mathbf{a}_k$ are L.I.

(ii) $k > m$.

(iii) $k \leq m$ and the column vectors $\mathbf{a}_1$, $\mathbf{a}_2$, $\cdots$, $\mathbf{a}_k$ are not L.I.

(i) If $k \leq m$ and the column vectors $\mathbf{a}_1$, $\mathbf{a}_2$, $\cdots$, $\mathbf{a}_k$ are L.I. then by definition, the feasible solution is the basic feasible solution. If $k = m$ then the solution is a non-degenerate B.F.S. and if $k < m$ then the solution is a degenerate B.F.S.

(ii) When $k > m$ and the column vectors $\mathbf{a}_1$, $\mathbf{a}_2$, $\cdots$, $\mathbf{a}_k$ are not L.I. and hence the solution is not basic. But by applying a technique which is given below, the number of positive variables can be reduced and ultimately k will be m, keeping all the m non-zero variables positive and the solution will be non-degenerate B.F.S. provided the column vectors corresponding to the positive variables are L.I.

Procedure: As the column vectors $\mathbf{a}_1, \mathbf{a}_2, \cdots, \mathbf{a}_k$ are not linearly independent then there exist scalar quantities λ_j $[j = 1, 2, \cdots, k]$ not all zero such that

$$\lambda_1\mathbf{a}_1 + \lambda_2\mathbf{a}_2 + \cdots + \lambda_k\mathbf{a}_k = \mathbf{0}, \quad \text{or,} \quad \sum_{j=1}^{k} \lambda_j\mathbf{a}_j = \mathbf{0}. \tag{ii}$$

Now at least one λ_j is positive [if it is negative, multiply both sides by -1 to make it positive].

Let

$$\nu = \max_j \left(\frac{\lambda_j}{x_j}\right) \quad [j = 1, 2, \cdots, k].$$

As all $x_j > 0$ and $\max \lambda_j > 0$ then ν is essentially a positive quantity.

From (i) and (ii) we get

$$\sum_{j=1}^{k} \left(x_j - \frac{\lambda_j}{\nu}\right)\mathbf{a}_j = \mathbf{b}. \tag{iii}$$

which indicates that

$$\mathbf{x}' = \left[x_1 - \frac{\lambda_1}{\nu},\ x_2 - \frac{\lambda_2}{\nu},\ \cdots,\ x_k - \frac{\lambda_k}{\nu}, 0, \cdots, 0\right]$$

is a solution set of the equations $A\mathbf{x} = \mathbf{b}$.

Now

$$\nu \geq \frac{\lambda_j}{x_j} \quad \text{[the sign of equality holds for maximum } \lambda_j/x_j]$$

then

$$x_j \geq \frac{\lambda_j}{\nu}, \quad \text{or,} \quad x_j - \frac{\lambda_j}{\nu} \geq 0 \quad \text{and} \quad x_j - \frac{\lambda_j}{\nu} = 0 \ \text{ for at least one } j.$$

Then

$$x_j' = x_j - \frac{\lambda_j}{\nu} \geq 0 \quad [j = 1, 2, \cdots, k]$$

and at least one of them is equal to zero.

Therefore

$$\mathbf{x}' = [x_1', x_2', \cdots, x_k', 0, \cdots, 0,]$$

is also a feasible solution of $A\mathbf{x} = \mathbf{b}$ with maximum number of positive variables $k-1$. By applying the method repeatedly, ultimately a feasible solution with only m positive values of the variables will be obtained. Now if the column vectors corresponding to the positive variables [we assume it for the time being] are L.I. then the solution will be a non-degenerate B.F.S.

(iii) If $k = m$ and the column vectors $\mathbf{a}_1, \mathbf{a}_2, \cdots, \mathbf{a}_m$ are not L.I. then the feasible solution is not a basic solution. Applying the procedure given in case (ii) the number of non-zero values of variables can be reduced to $p(p < m)$ of which all the non-zero values are positive and there exists a value $p(1 \leq p < m)$, such that

$\mathbf{a}_1, \mathbf{a}_2, \cdots, \mathbf{a}_p$ are linearly independent and the solution is a degenerate B.F.S. This procedure is applicable also to the case (ii) when $\mathbf{a}_1, \mathbf{a}_2, \cdots, \mathbf{a}_k (k = m)$ are not L.I.

Alternative statement: If a set of m independent linear equations with n variables $(n > m)$ has a F.S. then it has at least one B.F.S.

Note (a): $x_1 = 4$, $x_2 = 0$, $x_3 = 1$ is a feasible solution of the set of equations

$$\begin{aligned} 2x_1 - x_2 - x_3 &= 7 \\ \text{and} \quad 4x_1 + 3x_2 - 2x_3 &= 14 \end{aligned}$$

Here the number of positive variables in F.S. is 2 and the number of equations is 2, i.e., $k = 2 = m \cdot r(A) = 2$ but the column vectors of the positive variables x_1 and x_3, i.e., $\mathbf{a}_1$ and $\mathbf{a}_3$ are not linearly independent.

Hence the solution $x_1 = 4$, $x_2 = 0$ and $x_3 = 1$ is not a basic solution.

$$\mathbf{a}_1 = \begin{bmatrix} 2 \\ 4 \end{bmatrix} \quad \text{and} \quad \mathbf{a}_3 = \begin{bmatrix} -1 \\ -2 \end{bmatrix}.$$

Then

$$\mathbf{a}_1 = -2\mathbf{a}_3, \quad \text{or,} \quad \mathbf{a}_1 + 2\mathbf{a}_3 = \mathbf{0}.$$

Then $\lambda_1 = 1$ and $\lambda_3 = 2$,

$$\nu = \max_j \left(\frac{\lambda_j}{x_j}\right) = \max\left[\frac{1}{4}, \frac{2}{1}\right] = 2.$$

Hence

$$x_1' = 4 - \frac{1}{2} = \frac{7}{2}, \quad x_3' = 1 - \frac{2}{2} = 0$$

and therefore $\mathbf{x}' = [\frac{7}{2}, 0, 0]$ is a degenerate basic feasible solution of the set of equations

$$\begin{aligned} 2x_1 - x_2 - x_3 &= 7 \\ \text{and} \quad 4x_1 + 3x_2 - 2x_3 &= 14 \end{aligned}$$

with x_1 and x_2 as basic and x_3 non-basic variables. [$\mathbf{a}_1$ and $\mathbf{a}_2$ are linearly independent.]

Note (b): If all λ_j are non-negative or non-positive only one non-degenerate B.F.S. can be obtained by using the theory.

An important theorem (without proof):

Theorem 6.2.2 *The necessary and the sufficient condition, that all basic solutions will exist and non-degenerate, is that every set of m column vectors of the augmented matrix $[A\mathbf{b}]$ is L.I.*

▶ **Example 6.2.1** $x_1 = 1$, $x_2 = 3$, $x_3 = 2$ *is a feasible solution of the equations*

$$\begin{aligned} 2x_1 + 4x_2 - 2x_3 &= 10 \\ 10x_1 + 3x_2 + 7x_3 &= 33 \end{aligned}$$

Reduce the above F.S. to a B.F.S. by reduction theorem.

Solution: Three vectors

$$\mathbf{a}_1 = \begin{bmatrix} 2 \\ 10 \end{bmatrix}, \quad \mathbf{a}_2 = \begin{bmatrix} 4 \\ 3 \end{bmatrix} \quad \text{and} \quad \mathbf{a}_3 = \begin{bmatrix} -2 \\ 7 \end{bmatrix},$$

$$A = \begin{bmatrix} 2 & 4 & -2 \\ 10 & 3 & 7 \end{bmatrix} \quad \text{and} \quad r(A) = 2.$$

Hence the two equations are linearly independent. But the three vectors $\mathbf{a}_1$, $\mathbf{a}_2$ and $\mathbf{a}_3$ are not linearly independent. Hence there exist three constants λ_1, λ_2, λ_3 (at least one of them non-zero) such that

$$\lambda_1 \mathbf{a}_1 + \lambda_2 \mathbf{a}_2 + \lambda_3 \mathbf{a}_3 = \mathbf{0} \text{ will be satisfied,}$$

$$\text{or,} \qquad \lambda_1 \begin{bmatrix} 2 \\ 10 \end{bmatrix} + \lambda_2 \begin{bmatrix} 4 \\ 3 \end{bmatrix} + \lambda_3 \begin{bmatrix} -2 \\ 7 \end{bmatrix} = \mathbf{0}. \qquad \text{(i)}$$

From (i) we get

$$2\lambda_1 + 4\lambda_2 - 2\lambda_3 = 0$$
$$10\lambda_1 + 3\lambda_2 + 7\lambda_3 = 0.$$

By cross-multiplication we get

$$\frac{\lambda_1}{34} = \frac{\lambda_2}{-34} = \frac{\lambda_3}{-34} = k = -\frac{1}{34} \text{ (say).}$$

Then we get $\lambda_1 = -1$, $\lambda_2 = 1$, $\lambda_3 = 1$.

Hence

$$-\mathbf{a}_1 + \mathbf{a}_2 + \mathbf{a}_3 = \mathbf{0}. \qquad \text{(ii)}$$

Taking $\lambda_1 = -1$, $\lambda_2 = 1$ and $\lambda_3 = 1$ we have

$$\nu = \max_j \left(\frac{\lambda_j}{x_j}\right) = \max\left(-\frac{1}{1}, \frac{1}{3}, \frac{1}{2}\right) = \frac{1}{2}.$$

Hence a feasible solution given by

$$\mathbf{x}_1' = \left[x_1 - \frac{\lambda_1}{\nu},\ x_2 - \frac{\lambda_2}{\nu},\ x_3 - \frac{\lambda_3}{\nu}\right] \quad \left[\textbf{Note: } |B| = \begin{vmatrix} 2 & 4 \\ 10 & 3 \end{vmatrix} = -34 \neq 0\right]$$

$$= [1 + 2,\ 3 - 2,\ 2 - 2]$$

$$= [3,\ 1,\ 0]$$

is a basic feasible solution, because the positive variables are two and the column vectors $\mathbf{a}_1$ and $\mathbf{a}_2$ corresponding to the positive variables x_1 and x_2 are linearly independent.

Again (ii) can be written as

$$\mathbf{a}_1 - \mathbf{a}_2 - \mathbf{a}_3 = \mathbf{0};$$

then $\lambda_1 = 1$, $\lambda_2 = -1$, $\lambda_3 = -1$.

$$\nu = \max_j \left(\frac{\lambda_j}{x_j}\right) = \max\left(\frac{1}{1},\ -\frac{1}{3},\ -\frac{1}{2}\right) = 1.$$

Hence the feasible solution given by

$$\begin{aligned}\mathbf{x}_2' &= \left[x_1 - \frac{\lambda_1}{v},\ \ x_2 - \frac{\lambda_2}{v},\ \ x_3 - \frac{\lambda_3}{v}\right] \\ &= [1-1,\ \ 3+1,\ \ 2+1] \\ &= [0,\ \ 4,\ \ 3]\end{aligned}$$

is a B.F.S. as the column vectors $\mathbf{a}_2$, $\mathbf{a}_3$ corresponding to the positive variables x_2 and x_3 are linearly independent.

Note: There are two basic feasible solutions from the above procedure, both of which are non -degenerate. Compare the answers with the worked out Example 4.3.1 where B.S. are obtained from the stand point of definition.

The next problem (6.4.2) will be solved by using a new technique not shown in the above problem.

▶ **Example 6.2.2** *Given* $(1, 1, 2)$ *is a basic feasible solution to the set of equations*

$$\begin{aligned}x_1 + 2x_2 + 3x_3 &= 9 \\ 2x_1 - x_2 + x_3 &= 3\end{aligned}$$

Reduce the above feasible solution to one or more basic feasible solutions.

[C.U.(P)'86]

Solution: $\mathbf{a}_1 = \begin{bmatrix}1\\2\end{bmatrix}$, $\mathbf{a}_2 = \begin{bmatrix}2\\-1\end{bmatrix}$, $\mathbf{a}_3 = \begin{bmatrix}3\\1\end{bmatrix}$ and $\mathbf{b} = \begin{bmatrix}9\\3\end{bmatrix}$.

The vectors $\mathbf{a}_1$, $\mathbf{a}_2$, $\mathbf{a}_3$ associated with the three non-zero (here positive) variables x_1, x_2, x_3 are L.D. [∵ There are three, two component vectors]. Hence the given feasible solution is not basic.

The set of equations are

$$\mathbf{a}_1 x_1 + \mathbf{a}_2 x_2 + \mathbf{a}_3 x_3 = \mathbf{b}. \tag{i}$$

Now $(1, 1, 2)$ is a solution set. Therefore

$$\mathbf{a}_1 + \mathbf{a}_2 + 2\mathbf{a}_3 = \mathbf{b}. \tag{ii}$$

The three vectors $\mathbf{a}_1$, $\mathbf{a}_2$, $\mathbf{a}_3$ are L.D. Therefore, there exist three scalar quantities λ_1, λ_2 and λ_3 [at least one of them be non-zero] such that

$$\lambda_1\mathbf{a}_1 + \lambda_2\mathbf{a}_2 + \lambda_3\mathbf{a}_3 = \mathbf{0} \quad \text{relation holds}$$

$$\text{or,} \quad \lambda_1\begin{bmatrix}1\\2\end{bmatrix} + \lambda_2\begin{bmatrix}2\\-1\end{bmatrix} + \lambda_3\begin{bmatrix}3\\1\end{bmatrix} = \begin{bmatrix}0\\0\end{bmatrix} \quad \text{holds.} \tag{iii}$$

Equating we get

$$\left.\begin{aligned} \lambda_1 + 2\lambda_2 + 3\lambda_3 &= 0 \\ 2\lambda_1 - \lambda_2 + \lambda_3 &= 0 \end{aligned}\right\}. \tag{iv}$$

By cross-multiplication we get

$$\frac{\lambda_1}{5} = \frac{\lambda_2}{5} = \frac{\lambda_3}{-5} = k \neq 0 = \frac{1}{5} \quad \text{(say)}.$$

Then $\lambda_1 = 1$, $\lambda_2 = 1$, $\lambda_3 = -1$.

Putting the values of λ_1, λ_2, λ_3 in (iii) we get

$$\mathbf{a}_1 + \mathbf{a}_2 - \mathbf{a}_3 = \mathbf{0}. \tag{v}$$

Now

$$\max_j \left(\frac{\lambda_j}{x_j}\right) = \max\left(\frac{\lambda_1}{x_1}, \frac{\lambda_2}{x_2}, \frac{\lambda_3}{x_3}\right) = \max\left(\frac{1}{1}, \frac{1}{1}, \frac{-1}{2}\right) = 1$$

which occurs at $j = 1, 2$.

Thus we shall have to eliminate either $\mathbf{a}_1$ or $\mathbf{a}_2$ from (ii) and (v) to get a feasible solution with number of positive variables less than three.

From (ii) and (v) we get

$$\begin{aligned} \mathbf{a}_1 + \mathbf{a}_2 + 2\mathbf{a}_3 &= \mathbf{b} \\ \mathbf{a}_1 + \mathbf{a}_2 - \mathbf{a}_3 &= \mathbf{0}. \end{aligned}$$

Subtracting

$$0\mathbf{a}_1 + 0\mathbf{a}_2 + 3\mathbf{a}_3 = \mathbf{b}. \tag{vi}$$

From (vi) we get $x_1 = 0$, $x_2 = 0$, $x_3 = 3$ is a feasible solution of the set of equations.

Now the vector $\mathbf{a}_3$ associated with non-zero variable x_3 is L.I.

Hence the solution is a B.F.S. Here [$x_1 = 0$ (basic), $x_2 = 0$ (non-basic), $x_3 = 3$ (basic)] and [$x_1 = 0$ (non-basic), $x_2 = 0$ (basic) $x_3 = 3$ (basic)] are two basic feasible solutions as both $(\mathbf{a}_3, \mathbf{a}_1)$ and $(\mathbf{a}_3, \mathbf{a}_2)$ are L.I.

Again (v) can be written as

$$-\mathbf{a}_1 - \mathbf{a}_2 + \mathbf{a}_3 = \mathbf{0} \tag{vii}$$

and

$$\max_j \left(\frac{\lambda_j}{x_j}\right) = \max\left(\frac{-1}{1}, \frac{-1}{1}, \frac{1}{2}\right) = \frac{1}{2}$$

which occurs at $j = 3$.

Then the third vector is to be eliminated from (ii) and (vii) and to get a F.S.

From (ii) and (vii) we get

$$\begin{aligned} \mathbf{a}_1 + \mathbf{a}_2 + 2\mathbf{a}_3 &= \mathbf{b} \\ -2\mathbf{a}_1 - 2\mathbf{a}_2 + 2\mathbf{a}_3 &= \mathbf{0} \end{aligned}$$

Subtracting,

$$3\mathbf{a}_1 + 3\mathbf{a}_2 + 0\mathbf{a}_3 = \mathbf{b}. \qquad \text{(viii)}$$

From (viii) we get $x_1 = 3$, $x_2 = 3$, $x_3 = 0$ is a feasible solution of the problem and the solution is basic as the vectors $\mathbf{a}_1$ and $\mathbf{a}_2$ associated with the positive variables are L.I.

▶ **Example 6.2.3** *$x_1 = 1$, $x_2 = 2$, $x_3 = 1$ and $x_4 = 0$ is a F.S. to the set of equations*

$$\begin{aligned} 11x_1 + 2x_2 - 9x_3 + 4x_4 &= 6 \\ 15x_1 + 3x_2 - 12x_3 + 6x_4 &= 9 \end{aligned}$$

Reduce the F.S. to more than one B.F.S. and prove that one of them is non-degenerate and the other solutions are degenerate.

$$A = \begin{bmatrix} 11 & 2 & -9 & 4 \\ 15 & 3 & -12 & 6 \end{bmatrix} \quad \text{and} \quad r(A) = 2.$$

Solution: Hence the equations are independent. Again the set of equations can be written as

$$\mathbf{a}_1x_1 + \mathbf{a}_2x_2 + \mathbf{a}_3x_3 + \mathbf{a}_4x_4 = \mathbf{b}, \qquad \text{(i)}$$

where

$$\mathbf{a}_1 = \begin{bmatrix} 11 \\ 15 \end{bmatrix}, \quad \mathbf{a}_2 = \begin{bmatrix} 2 \\ 3 \end{bmatrix}, \quad \mathbf{a}_3 = \begin{bmatrix} -9 \\ -12 \end{bmatrix},$$

$$\mathbf{a}_4 = \begin{bmatrix} 4 \\ 6 \end{bmatrix} \quad \text{and} \quad \mathbf{b} = \begin{bmatrix} 6 \\ 9 \end{bmatrix}$$

$x_1 = 1$, $x_2 = 2$, $x_3 = 1$ and $x_4 = 0$ is a solution set of (i) and hence (i) can be written as

$$\mathbf{a}_1 + 2\mathbf{a}_2 + \mathbf{a}_3 = \mathbf{b}. \qquad \text{(ii)}$$

The solution set given abobe is not basic because the number 3 of positive variables are greater than 2, the number of independent equations and evidently $\mathbf{a}_1$, $\mathbf{a}_2$ and $\mathbf{a}_3$ are linearly dependent. The vectors $\mathbf{a}_1$, $\mathbf{a}_2$ and $\mathbf{a}_3$ (associated with the positive variables) can be expressed as

$$\mathbf{a}_1 - \mathbf{a}_2 + \mathbf{a}_3 = \mathbf{0}. \qquad \text{(iii)}$$

Hence $\lambda = 1$, $\lambda_2 = -1$ and $\lambda_3 = 1$.

$$\max_{j=1,2,3}\left(\frac{\lambda_j}{x_j}\right) = \max\left(\frac{\lambda_1}{x_1}, \frac{\lambda_2}{x_2}, \frac{\lambda_3}{x_3}\right) = \max\left(\frac{1}{1}, \frac{-1}{2}, \frac{1}{1}\right) = 1$$

which happen for $j = 1$ and 3. Hence eliminating either $\mathbf{a}_1$ or $\mathbf{a}_3$ from (ii) and (iii) we get a new F.S. Now eliminating $\mathbf{a}_1$ we get

$$3\mathbf{a}_2 = \mathbf{b}. \qquad \text{(iv)}$$

Hence $x_1 = 0$, $x_2 = 3$, $x_3 = 0$, $x_4 = 0$ is a new F.S. which we obtain from (iv) [The coefficient of $\mathbf{a}_2$ is 3 which is x_2; and all other variables are zero]. The solutions is basic as the vector $\mathbf{a}_2$ associated with the positive variable x_2 is linearly independent.

Vectors $\mathbf{a}_2$, $\mathbf{a}_1$ are L.I.; Vectors $\mathbf{a}_2$, $\mathbf{a}_3$ are L.I.; Vectors $\mathbf{a}_2$, $\mathbf{a}_4$ are not L.I. Hence the basic feasible solutions are [$x_1 = 0$, $x_2 = 3$; $x_3 = x_4 = 0$ (non-basic)]; [$x_1 = 0$ (non-basic)], $x_2 = 3$, $x_3 = 0$, $x_4 = 0$ (non-basic). The variables x_2 and x_4 together cannot be considered as basic variables. All B.F.S. are degenerate.

Again (iii) can be written as

$$-\mathbf{a}_1 + \mathbf{a}_2 - \mathbf{a}_3 = \mathbf{0}. \tag{v}$$

Thus $\lambda_1 = -1$, $\lambda_2 = 1$, $\lambda_3 = -1$.

$$\max_j \left(\frac{\lambda_j}{x_j}\right) = \max\left(-\frac{1}{1}, \frac{1}{2}, \frac{-1}{1}\right) = \frac{1}{2}$$

which happens for $j = 2$.

Thus eliminating vectors $\mathbf{a}_2$ from (ii) and (v) we get

$$3\mathbf{a}_1 + 3\mathbf{a}_3 = \mathbf{b} \tag{vi}$$

from which we get a new F.S.

$x_1 = 3$, $x_2 = 0$, $x_3 = 3$, $x_4 = 0$. The solution is basic also as the vectors $\mathbf{a}_1$ and $\mathbf{a}_3$ associated with positive variables are L.I. and the solution is non-degenerate.

▶ **Example 6.2.4** $x_1 = 1$, $x_2 = 2$, $x_3 = 1$ *is a feasible solution of the following set of linearly independent equations.*

$$\begin{aligned} 2x_1 + 3x_2 + 5x_3 &= 13 \\ 3x_1 - x_2 + 3x_3 &= 4 \end{aligned}$$

Reduce the F.S. to a B.F.S.

Solution: Here

$$\mathbf{a}_1 = \begin{bmatrix} 2 \\ 3 \end{bmatrix}, \quad \mathbf{a}_2 = \begin{bmatrix} 3 \\ -1 \end{bmatrix}, \quad \mathbf{a}_3 = \begin{bmatrix} 5 \\ 3 \end{bmatrix} \quad \text{and} \quad \mathbf{b} = \begin{bmatrix} 13 \\ 4 \end{bmatrix}.$$

The set of equations can be written as

$$\mathbf{a}_1 x_1 + \mathbf{a}_2 x_2 + \mathbf{a}_3 x_3 = \mathbf{b}. \tag{i}$$

Since $x_1 = 1$, $x_2 = 2$, $x_3 = 1$ is a solution of (i) then

$$\mathbf{a}_1 + 2\mathbf{a}_2 + \mathbf{a}_3 = \mathbf{b}. \tag{ii}$$

The vectors $\mathbf{a}_1$, $\mathbf{a}_2$, $\mathbf{a}_3$ are linearly dependent.

[Since there are three vectors each having two components]. Thus there exist three scalars, λ_1, λ_2, λ_3 at least one of which is non-zero such that

$$\lambda_1\begin{bmatrix}2\\3\end{bmatrix}+\lambda_2\begin{bmatrix}3\\-1\end{bmatrix}+\lambda_3\begin{bmatrix}5\\3\end{bmatrix}=\begin{bmatrix}0\\0\end{bmatrix}\text{ be satisfied.}$$

Now, equating we get

$$\begin{aligned}2\lambda_1+3\lambda_2+5\lambda_3&=0\\3\lambda_1-\lambda_2+3\lambda_3&=0\end{aligned}$$

By cross multiplication, we get

$$\frac{\lambda_1}{9+5}=\frac{\lambda_2}{15-6}=\frac{\lambda_3}{-2-9}=k=1\text{ (say)}.$$

Then $\lambda_1=14$, $\lambda_2=9$, $\lambda_3=-11$ and hence we can write

$$14\mathbf{a}_1+9\mathbf{a}_2-11\mathbf{a}_3=\mathbf{0}. \tag{iii}$$

Now

$$\max_{j=1,2,3}\left(\frac{\lambda_j}{x_j}\right)=\left(\frac{14}{1},\ \frac{9}{2},\ -\frac{11}{1}\right)=14$$

which occurs for $j=1$. Then the first vector is to be removed from (i) and (iii). We have from (ii)

$$\left.\begin{aligned}14\mathbf{a}_1+28\mathbf{a}_2+14\mathbf{a}_3&=14\mathbf{b}\\\text{and}\quad 14\mathbf{a}_1+9\mathbf{a}_2-11\mathbf{a}_3&=\mathbf{0}\end{aligned}\right\}. \tag{iv}$$

Subtracting we get

$$\begin{aligned}19\mathbf{a}_2+25\mathbf{a}_3&=14\mathbf{b}\\\text{or,}\quad 0\mathbf{a}_1+19\mathbf{a}_2+25\mathbf{a}_3&=14\mathbf{b}\\\text{or,}\quad \frac{0}{14}\mathbf{a}_1+\frac{19}{14}\mathbf{a}_2+\frac{25}{14}\mathbf{a}_3&=\mathbf{b}\end{aligned} \tag{v}$$

From (v) we can say that $(0, 19/14, 25/14)$ is a feasible solution set of the above equations. Now the equation is basic also, since the vectors, $\mathbf{a}_2$ and $\mathbf{a}_3$ associated with the positive variables are

$$\mathbf{a}_2=\begin{bmatrix}3\\-1\end{bmatrix},\quad \mathbf{a}_3=\begin{bmatrix}5\\3\end{bmatrix}\quad\text{and}\quad\det\begin{vmatrix}3&5\\-1&3\end{vmatrix}=14\neq 0.$$

Thus the solution $(0, 19/14, 25/14)$ is a basic feasible solution.

Again (iii) can be written in the manner

$$-14\mathbf{a}_1-9\mathbf{a}_2+11\mathbf{a}_3=\mathbf{0}. \tag{vi}$$

$$\max_j\left(\frac{\lambda_j}{x_j}\right)=\max\left(\frac{-14}{1},\ \frac{-9}{2},\ \frac{11}{2}\right)=\frac{11}{2}$$

which occurs for $j=3$.

Therefore, the third vector is to be eliminated from (ii) and (iii)

$$\begin{aligned} 11\mathbf{a}_1 + 22\mathbf{a}_2 + 11\mathbf{a}_3 &= 11\mathbf{b} \\ 14\mathbf{a}_1 + 9\mathbf{a}_2 - 11\mathbf{a}_3 &= \mathbf{0} \end{aligned}$$

Adding we get

$$25\mathbf{a}_1 + 31\mathbf{a}_2 + 0\mathbf{a}_3 = 11\mathbf{b}$$

$$\text{or,} \quad \frac{25}{11}\mathbf{a}_1 + \frac{31}{11}\mathbf{a}_2 + \frac{0}{11}\mathbf{a}_3 = \mathbf{b}$$

from which we conclude that $(25/11, 31/11, 0)$ is a feasible solution of the set of equations, and the solution is also basic, since the vectors $\mathbf{a}_1$ and $\mathbf{a}_2$ associated with the positive variables are L.I.

▶ **Example 6.2.5** *Prove that $x_1 = 3$, $x_2 = 2$, $x_3 = 4$, $x_4 = 0$ is a feasible solution of the set of equations but not basic.*

$$\begin{aligned} 2x_1 + 5x_2 - 3x_3 + x_4 &= 4 \\ 6x_1 + 16x_2 - 9x_3 + 5x_4 &= 14 \end{aligned}$$

Reduce the F.S. to a B.F.S. by using reduction theory and prove that by using the theory, it is not possible to find more than one B.F.S. though the problem may have more than one B.F.S.

Solution: The set of equations are

$$\mathbf{a}_1x_1 + \mathbf{a}_2x_2 + \mathbf{a}_3x_3 + \mathbf{a}_4x_4 = \mathbf{b}, \tag{i}$$

where

$$\mathbf{a}_1 = \begin{bmatrix} 2 \\ 6 \end{bmatrix}, \quad \mathbf{a}_2 = \begin{bmatrix} 5 \\ 16 \end{bmatrix}, \quad \mathbf{a}_3 = \begin{bmatrix} -3 \\ -9 \end{bmatrix},$$

$$\mathbf{a}_4 = \begin{bmatrix} 1 \\ 5 \end{bmatrix}, \quad \text{and} \quad \mathbf{b} = \begin{bmatrix} 4 \\ 14 \end{bmatrix}.$$

The three vectors associated with positive variables are linearly dependent. Hence the solution is not a basic solution though the solution is feasible since all components of the solution set are non-negative and satisfy the two equations.

Putting the values of $x_1, x_2, \cdots, x_4$ in (i) we get

$$3\mathbf{a}_1 + 2\mathbf{a}_2 + 4\mathbf{a}_3 = \mathbf{b}. \tag{ii}$$

Again only by mere inspection, we can say that

$$3\mathbf{a}_1 + 0\mathbf{a}_2 + 2\mathbf{a}_3 = \mathbf{0}. \tag{iii}$$

Thus $\lambda_1 = 3$, $\lambda_2 = 0$, $\lambda_3 = 2$.

$$\max_{j=1,2,3,} \left(\frac{\lambda_j}{x_j}\right) = \max\left(\frac{\lambda_1}{x_1}, \frac{\lambda_2}{x_2}, \frac{\lambda_3}{x_3}\right) = \max\left(\frac{3}{3}, \frac{0}{2}, \frac{2}{4}\right) = 1$$

which occurs for $j = 1$.

Then the first vector is to be eliminated from (ii) and (iii).

We have

$$3\mathbf{a}_1 + 2\mathbf{a}_2 + 4\mathbf{a}_3 = \mathbf{b}$$
$$3\mathbf{a}_1 + 0\mathbf{a}_2 + 2\mathbf{a}_3 = \mathbf{0}$$

Subtracting we get

$$0 \cdot \mathbf{a}_1 + 2\mathbf{a}_2 + 2\mathbf{a}_3 = \mathbf{b} \qquad \text{(iv)}$$

which implies that $(0, 2, 2, 0)$ is a feasible solution of the set of equations. The vectors associated with the positive variables are $\mathbf{a}_2$ and $\mathbf{a}_3$ which are linearly independent since

$$\begin{vmatrix} 5 & -3 \\ 16 & -9 \end{vmatrix} = 3 \neq 0.$$

Hence the solution $(0, 2, 2, 0)$ obtained by reduction method is basic feasible.

Note (1) Since all λ_j are non-negative then by reduction method, it is not possible to obtain more than one B.F.S. though $[0,\ 2/3,\ 0,\ 2/3]$ is another B.F.S.

▶ **Example 6.2.6** *Consider the system of equations,*

$$x_1 + 2x_2 + 4x_3 + x_4 = 7$$
$$2x_1 - x_2 + 3x_3 - 2x_4 = 4$$

Here $x_1 = 1$, $x_2 = 1$, $x_3 = 1$ and $x_4 = 0$ is a feasible solution.

Reduce the feasible solution to two different basic feasible solutions.

[D.U.B.Sc.(Math.)'77]

Solution: $A = \begin{bmatrix} 1 & 2 & 4 & 1 \\ 2 & -1 & 3 & -2 \end{bmatrix}$ and $R(A) = 2$.

Hence the equations are linearly independent. Again the set of equations can be written as

$$\mathbf{a}_1 x_1 + \mathbf{a}_2 x_2 + \mathbf{a}_3 x_3 + \mathbf{a}_4 x_4 = \mathbf{b}, \qquad \text{(i)}$$

where

$$\mathbf{a}_1 = \begin{bmatrix} 1 \\ 2 \end{bmatrix}, \quad \mathbf{a}_2 = \begin{bmatrix} 2 \\ -1 \end{bmatrix}, \quad \mathbf{a}_3 = \begin{bmatrix} 4 \\ 3 \end{bmatrix},$$

$$\mathbf{a}_4 = \begin{bmatrix} 1 \\ -2 \end{bmatrix} \quad \text{and} \quad \mathbf{b} = \begin{bmatrix} 7 \\ 4 \end{bmatrix}.$$

As $x_1 = 1$, $x_2 = 1$, $x_3 = 1$ and $x_4 = 0$ is a solution set then (i) can be written as

$$\mathbf{a}_1 + \mathbf{a}_2 + \mathbf{a}_3 = \mathbf{b}. \qquad \text{(ii)}$$

Here the vectors $\mathbf{a}_1$, $\mathbf{a}_2$, $\mathbf{a}_3$ associated with the non-zero variables are L.D. Hence the solution given in the problem is not basic.

Now we can find that

$$2\mathbf{a}_1 + \mathbf{a}_2 - \mathbf{a}_3 = \mathbf{0}. \tag{iii}$$

[Find out the values of λ_1, λ_2 etc. by previous knowledge]

Hence $\lambda_1 = 2$, $\lambda_2 = 1$, $\lambda_3 = -1$.

$$\max_j \left(\frac{\lambda_j}{x_j}\right) = \max\left(\frac{2}{1}, \frac{1}{1}, \frac{-1}{1}\right) = 2$$

which occurs at $j = 1$.

Thus eliminating $\mathbf{a}_1$ from (ii) and (iii) we shall get a new feasible solution. From (ii) and (iii) we get

$$\frac{1}{2}\mathbf{a}_2 + \frac{3}{2}\mathbf{a}_3 = \mathbf{b}$$

from which we conclude that

$x_1 = 0$, $x_2 = 1/2$, $x_3 = 3/2$ and $x_4 = 0$ is a new feasible solution to the problem. The solution is basic as the vectors $\mathbf{a}_2$ and $\mathbf{a}_3$ associated with the positive variables are L.I. Again (iii) can be written as

$$-2\mathbf{a}_1 - \mathbf{a}_2 + \mathbf{a}_3 = \mathbf{0}. \tag{iv}$$

Thus $\lambda_1 = -2$, $\lambda_2 = -1$, $\lambda_3 = 1$.

$$\max_j \left(\frac{\lambda_j}{x_j}\right) = \max\left(\frac{-2}{1}, \frac{-1}{1}, \frac{1}{1}\right) = 1$$

which occurs for $j = 3$.

Thus eliminating $\mathbf{a}_3$ from (ii) and (iv) we get

$$3\mathbf{a}_1 + 2\mathbf{a}_2 = \mathbf{b}$$

from which we conclude that $x_1 = 3$, $x_2 = 2$, $x_3 = 0$, $x_4 = 0$ is a new feasible solution and the solution is basic also as the vectors associated with the positive variables are L.I.

Thus two B.F.S. are

$$\mathbf{x}_1 = \left[0, \ \frac{1}{2}, \ \frac{3}{2}, \ 0\right] \quad \text{and} \quad \mathbf{x}_2 = [3, \ 2, \ 0, \ 0].$$

► **Example 6.2.7** *Prove that $x_1 = 3$, $x_2 = 0$, $x_3 = 0$ is a F.S. to the following set of equations*

$$\begin{aligned} 4x_1 + \phantom{\tfrac{3}{2}}x_2 - 3x_3 &= 12 \\ 6x_1 + \tfrac{3}{2}x_2 + x_3 &= 18 \end{aligned}$$

Is the solution basic? If so, which are the basic variables?

Solution: $\mathbf{a}_1 = \begin{bmatrix} 4 \\ 6 \end{bmatrix}$, $\mathbf{a}_2 = \begin{bmatrix} 1 \\ 3/2 \end{bmatrix}$, $\mathbf{a}_3 = \begin{bmatrix} -3 \\ 1 \end{bmatrix}$, and $R(A) = 2$.

Since the solution set $x_1 = 3$, $x_2 = 0$, $x_3 = 0$ satisfy both the equations then the solution is a F.S. [As none of them negative].

The column vector $\mathbf{a}_1$ associated with non-zero variable x_1 is linearly independent.

Hence the solution is basic. Now let us assume tha x_1 and x_2 are basic variables. The matrix of the corresponding vectors is

$$B_1 = (\mathbf{a}_1, \mathbf{a}_2) = \begin{bmatrix} 4 & 1 \\ 6 & 3/2 \end{bmatrix} \text{ and } |B_1| = 0.$$

Hence the matrix is not a basis matrix and therefore x_1 and x_2 together cannot be considered as basic variables.

Now assume x_1 and x_3 are basic variables. The square matrix of corresponding vectors is

$$B_2 = (\mathbf{a}_1, \mathbf{a}_3) = \begin{bmatrix} 4 & -3 \\ 6 & 1 \end{bmatrix} \quad \text{and} \quad |B_2| = 22 \neq 0.$$

Hence the matrix is a basis matrix. Thus x_1 and x_3 are the basic variables. Therefore, the solution is a basic solution with x_1 and x_3 basic variables and x_2 non-basic variable. Of course, the solution is degenerate.

Exercise 6

1. The L.P.P., Minimize $z = 2x_1 - 3x_2 + x_4$

$$\text{subject to} \quad \begin{aligned} 3x_1 + 2x_2 + x_3 \quad\quad &= 15 \\ 2x_1 + 4x_2 \quad\quad + x_4 &= 8, \quad x_j \geq 0, \ j = 1, 2, 3, 4. \end{aligned}$$

has an optimal solution. Using the property of the fundamental theroem of L.P. find the minimum value of z and the corresponding solution set.

2. (a) The L.P.P. Maximize $z = -x_1 + 4x_2 + x_3$

$$\text{subject to} \quad \begin{aligned} x_1 + 3x_2 + x_3 \quad\quad &= 6 \\ -3x_1 + 4x_2 \quad\quad + x_4 &= 4, \quad x_j \geq 0, \ j = 1, 2, 3, 4 \end{aligned}$$

has an optimal solution. Find the maximim value of the objective function after finding all B.F.S.

(b) With the assumption that the finite value of the objective function exists, find out the optimal value of the problem maximize, $z = x_1 + 5x_2 + x_3$

$$\text{subject to} \quad \begin{aligned} x_1 + x_2 + x_3 &= 9 \\ 2x_1 - 4x_2 + 3x_3 &= 4, \quad x_1, x_2, x_3 \geq 0 \end{aligned}$$

3. (a) The L.P.P. Maximize $z = 2x_1 - x_2 + 5x_3$

$$\text{subject to} \quad \begin{aligned} x_1 + 2x_2 + 2x_3 \quad\quad &= 16 \\ 4x_1 + 7x_2 \quad\quad + x_4 &= 6, \quad x_j \geq 0, \ j = 1, 2, 3, 4 \end{aligned}$$

has an optimal solution. Find the maximum value of the objective function from the Basic feasible solutions.

(b) Maximize $6x_1 - 4x_2 + 4x_3 + x_4$

subject to
$$\begin{aligned} 2x_1 - \ x_2 + 3x_3 + \ x_4 &= 6 \\ 4x_1 - 2x_2 - \ x_3 + 2x_4 &= 10, \quad x_j \geq 0, \ j = 1, 2, 3, 4 \end{aligned}$$

[Assume the existence of an optimal solution.]

4. $x_1 = 2$, $x_2 = 3$, $x_3 = 1$ is a feasible solution of the equations

$$\begin{aligned} 2x_1 + x_2 + 4x_3 &= 11 \\ 3x_1 + x_2 + 5x_3 &= 14 \end{aligned}$$

Reduce the F.S. to a B.F.S. [C.U.(H)'90,'92,'95, '99] [V.U.(P)'91]

5. $x_1 = 1$, $x_2 = 3$, $x_3 = 2$ is a feasible solution of the equations

$$\begin{aligned} 5x_1 + 4x_2 - \ x_3 &= 15 \\ 2x_1 + 7x_2 - 4x_3 &= 15 \end{aligned}$$

Reduce the F.S. to B.F.S. [V.B.U.(H)'85; C.U.(H)'87]

6. (a) Prove that $x_1 = 2$, $x_2 = 3$, $x_3 = 0$ is a feasible solution but not a basic feasible of the set of equations

$$\begin{aligned} 3x_1 + \ 5x_2 - 7x_3 &= 21 \\ 6x_1 + 10x_2 + 3x_3 &= 42 \end{aligned}$$

Reduce the feasible solution to a basic feasible solution and show that the solution is degenerate. [C.U.(P)'89]

(b) $x_1 = 2$, $x_2 = 4$, $x_3 = 1$ is a feasible solution to the set of equations

$$\begin{aligned} 2x_1 - \ x_2 + 2x_3 &= 2 \\ x_1 + 4x_2 \qquad &= 18 \end{aligned}$$

Reduce the F.S. to a B.F.S. one. [C.U.(P)'88; V.U.(P)'91]

7. $x_1 = 1$, $x_2 = 0$, $x_3 = 2$, $x_4 = 1$ is a feasible solution to the set of equations

$$\begin{aligned} 2x_1 + 3x_2 + 3x_3 - x_4 &= 7 \\ x_1 + 5x_2 + 2x_3 + x_4 &= 6 \end{aligned}$$

Reduce the feasible solution to one or more basic feasible solutions.

8. $x_1 = 3$, $x_2 = 1$, $x_3 = 1$ and $x_4 = 2$ is a feasible solution of the set of equations

$$\begin{aligned} x_1 + x_2 + \ x_3 + \ x_4 &= 7 \\ 2x_1 + x_2 + 2x_3 + \ x_4 &= 11 \\ x_2 + \ x_3 + 2x_4 &= 6 \end{aligned}$$

Reduce the feasible solution to one or more basic feasible solutions.

9. Prove that $x_1 = 4$, $x_2 = 1$, $x_3 = 3$ is a feasible solution to the set of equations (linearly independent).

$$\begin{aligned} 2x_1 - 3x_2 + \ x_3 &= 8 \\ x_1 + 2x_2 + 3x_3 &= 15 \end{aligned}$$

Is the solution set, a basic feasible solution? If not, reduce the feasible solution to one or more basic feasible solutions.

10. $x_1 = 3$, $x_2 = 1$, $x_3 = 0$, $x_4 = 3$ is a feasible solution to the set of equation (linearly independent)

$$\begin{aligned} 4x_1 + x_2 - x_3 + x_4 &= 16 \\ x_1 + 2x_2 + 3x_3 + 2x_4 &= 11 \end{aligned}$$

From the stand point of theory of reduction of F.S. to B.F.S. how many B.F.S. can be determined? Find all of them.

11. Prove that $x_1 = 1, x_2 = 1, x_3 = 3$ is a basic feasible solution of the following linearly independent set of equations

$$\begin{aligned} 2x_1 + 3x_2 + x_3 &= 8 \\ 5x_1 - 2x_2 \qquad &= 3 \\ 4x_1 \qquad - x_3 &= 1 \end{aligned}$$

12. $x_1 = 4$, $x_2 = -1$, $x_3 = 2$ is a solution set of following L.I. equations

$$\begin{aligned} 4x_1 + 9x_2 + x_3 &= 9 \\ x_1 + x_2 + x_3 &= 5 \end{aligned}$$

Can we reduce the solution to a basic feasible solution? Give reasons.

13. Prove that $x_1 = 2$, $x_2 = 1$ and $x_3 = 3$ is a feasible solution of the set of equations

$$\begin{aligned} 4x_1 + 2x_2 - 3x_3 &= 1 \\ -6x_1 - 4x_2 + 5x_3 &= -1 \end{aligned}$$

Reduce the F.S. to a B.F.S. by reduction theory.

[V.B.U.(H)'83; C.U.(H)'87; C.U.(P)'95]

14. For the system of equations

$$\begin{aligned} x_1 + x_2 + 2x_3 &= 9 \\ 3x_1 + 2x_2 + 5x_3 &= 22 \end{aligned}$$

show that the feasible solution $(2, 3, 2)$ is not basic.

Reduce the feasible solution to one or more basic feasible solutions by using reduction theory.

15. Solve the L.P.P. by direct method [Optimal solution exists]

Maximize, $z = -x_1 - x_2 - x_3$

subject to
$$\begin{aligned} x_1 - x_2 + 2x_3 &= 2 \\ -x_1 + 2x_2 - x_3 &= 1, \quad x_1 \geq 0, x_2 \geq 0, x_3 \geq 0. \end{aligned}$$

16. Prove that $x_1 = 2$, $x_2 = 1$, $x_3 = 0$ is a F.S. to the following set of equations

$$\begin{aligned} 2x_1 - x_2 + 3x_3 &= 3 \\ -6x_1 + 3x_2 + 7x_3 &= -9 \end{aligned}$$

Is the solution basic? Justify your answer. If the solution be not basic, reduce the F.S. to B.F.S. one.

Answers

1. min $z = -6$ at $[0, 2, 11, 0]$. 2. (a) max $z = 7$ at $[0, 1, 3, 0]$; (b) max $z = 155/7$ at $[0, 23/7, 40/7]$. 3. (a) max $z = 40$ at $[0, 0, 8, 6]$; (b) max $z = 116/7$ at $[18/7, 0, 2/7, 0]$. 4. $[3, 5, 0]$, $[1/2, 0, 5/2]$. 5. $[0, 5, 5]$, $[5/3, 5/3, 0]$. 6. (a) $[x_1 = 7,\ x_3 = 0,\ x_2 = 0,$ (non-basic)$]$; $[x_2 = 4\frac{1}{5},\ x_3 = 0,\ x_1 = 0$ (non-basic)$]$; (b) $[26/9, 34/9, 0]$, $[0, 9/2, 13/4]$. 7. $[0, 0, 13/5, 4/5]$, $[13/3, 0, 0, 5/3]$. 8. $[1, 3, 3, 0]$, $[4, 0, 0, 3]$. 9. $[9/5, 0, 22/5]$, $[61/7, 22/7, 0]$. 10. $[3, 0, 0, 4]$, $[3, 4, 0, 0,]$. 12. No, we can only reduce a feasible solution to a basic feasible solution. 13. $[1, 0, 1]$. 14. $[4, 5, 0]$, $[0, 1, 4]$. 15. max $z = -3$ at $[0, 4/3, 5/3]$. 16. No. $[x_1 = 3/2$ (basic), $x_2 = 0$, $x_3 = 0$ (basic)$]$.

Chapter 7

Graphical or Geometrical Method of Solving a L.P.P.

7.1 Introduction

We know from the fundamental theorem of L.P.P. (algebrical approach) that if a L.P.P. admits an optimal solution then at least one of the B.F.S. will be an optimal solution. Again we know that there is one to one correspondence between the basic feasible solutions and extreme points of the convex set of feasible solutions (In the absence of degenaracy). There are two types of convex set of feasible solutions (i) convex polyhedron, which is strictly bounded and has finite number of extreme points (ii) Convex polytope, has finite number of extreme points but not bounded from above. In the case of convex polyhedron, each and every problem has both finite maximum and finite minimum. But this is not true for a convex polytope. Here all objective functions may have finite maximum or finite minimum but not both; there is some problem which has neither finite maximum nor finite minimum and the problem has said to have unbounded solution. We are mainly concerned with the convex polyhedron where the optimum value will be determined from the extreme points only. We are mainly interested with such problems though there are some problems where the convex set is a convex polytope. I have discussed such problems very carefully and scientifically.

► **Example 7.1.1** *Find the convex set of feasible solutions, if any, of the following problems and comment on the nature of the convex sets:*

The constraints are

(a) $x_1 + x_2 \geq 2$
$2x_1 + 3x_2 \leq 6$
$x_1 - x_2 \leq 2, \quad x_1, x_2 \geq 0.$

(b) $x_1 - x_2 \leq 1$
$x_1 + x_2 \geq 4$
$x_1 - 3x_2 \leq 3, \quad x_1, x_2 \geq 0.$

(c) $2x_1 + 3x_2 \leq 30$
$4x_1 + 3x_2 \geq 60, \quad x_1, x_2 \geq 0.$

(d) $-x_1 + x_2 \geq 3$
$x_1 - x_2 \geq 2, \quad x_1, x_2 \geq 0.$

Solution: (a) AB is the straight line represented by $x_1+x_2 = 2$. To determine the region $x_1 + x_2 \geq 2$, take any point not on the line AB, say $(0,0)$ and for

$(0,0)$, the value of the expression $x_1 + x_2$ is 0 (zero) which is less than or equal to 2, which does not satisfy the constraint. Thus the opposite side of the region, which contains the point $(0,0)$, will be the region, determined by $x_1 + x_2 \geq 2$. In the same manner other two regions defined by the constraints $2x_1 + 3x_2 \leq 6$, $x_1 - x_2 \leq 2$ have been determined. $x_1 \geq 0$, $x_2 \geq 0$, absolutely represents the first quadrant. Now the intersection of the regions is the convex set of feasible solutions which is a strictly bounded convex set ABC with three extreme points $A(2,0)$, $B(0,2)$ and $C(12/5, 2/5)$ which are finite in number [The co-ordinates of the point C is the intersection of the line $2x_1+3x_2 = 6$, $x_1 - x_2 = 2$]. Hence the convex set of F.S. is a convex polyhedron which can be generated by the convex combination of three extreme points $(2,0)$, $(0,2)$ and $(12/5, 2/5)$, i.e., any point on the region ABC can be expressed as the convex combination of the extreme points.

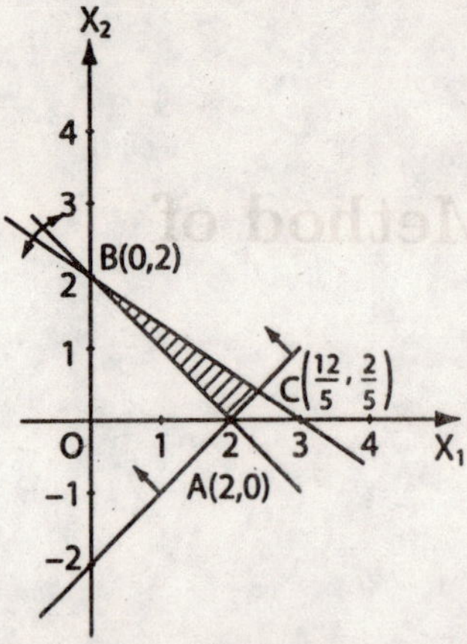

Fig. 7.1

(b) Here the convex set of feasible solutions is the shaded region of the first quadrant which is bounded from below only. It has only two extreme points $C(5/2, 3/2)$ and $D(0,4)$. Here the convex set of F.S. is not a convex polyhedron as it is not possible to express all points of the feasible set as the convex combination of the extreme points $(5/2, 3/2)$ and $(0,4)$. For example, $(4,4)$ is a point belonging to the F.S. set. But the point cannot be expressed as the convex combination of the points $(5/2, 3/2)$ and $(0,4)$. Such type of convex set of F.S. is called a convex polytope.

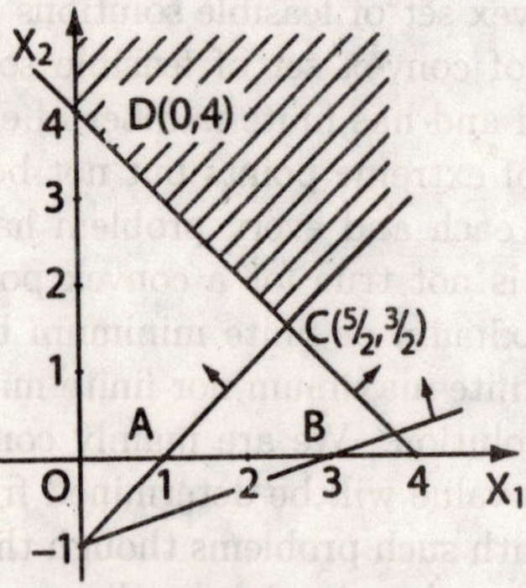

Fig. 7.2

Remark: Here the constraint $x_1 - 3x_2 \leq 3$ is redundant as the presence or absence of this constraint does not change the nature of the convex set of F.S.

(c) Here the convex set of F.S. is stricly a point $A(15,0)$. This is the only extreme point of the set, as this point cannot be expressed as the convex combination of any other two points belonging to the set and this question does not arise at all as there are no other points belonging to the set. This set is a convex polyhedron with only one extreme point.

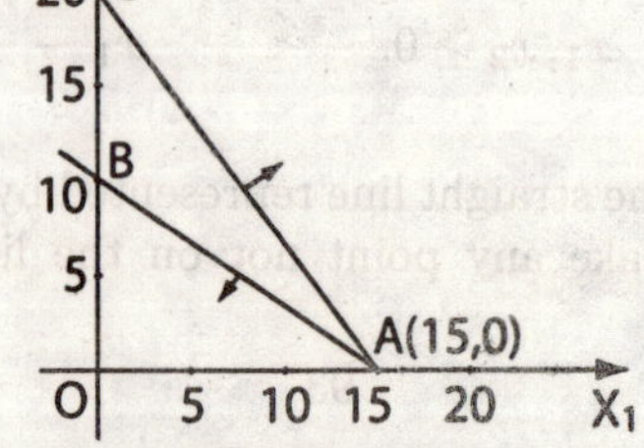

(d) Here the convex set of F.S. is a null set, i.e., there is no feasible solution set defined by the constraints.

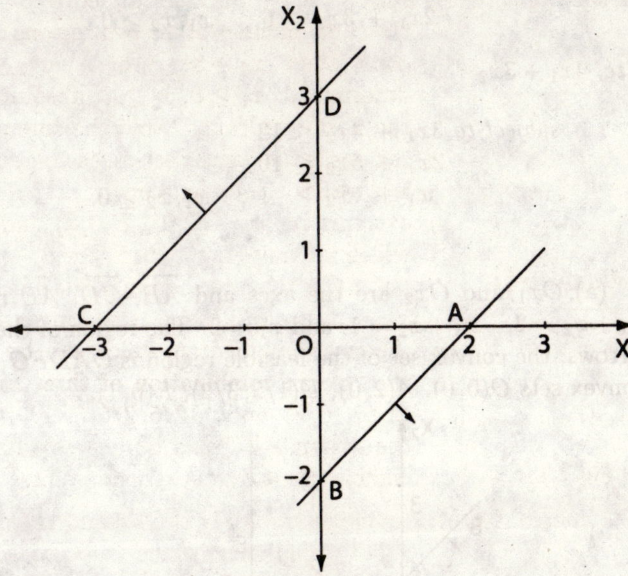

Fig.7.4

▶ **Example 7.1.2** *Draw graphically the feasible space, if any, given by the following L.P.P. and find out the extreme points of the feasible region:*

(a) *Maximize,* $z = 2x_1 + x_2$

$$\text{subject to} \quad x_1 + x_2 \leq 2$$
$$-x_1 + x_2 \leq 1$$
$$x_1 \leq 2\,, \quad x_1, x_2 \geq 0.$$

[C.U.(P)'89]

(b) *Maximize,* $z = 9x_1 + 7x_2$

$$\text{subject to} \quad x_1 + 2x_2 \leq 7$$
$$x_1 - x_2 \leq 4\,, \quad x_1, x_2 \geq 0.$$

[C.U.(P)'87,'92]

(c) *Maximize,* $z = 2x_1 + 5x_2$

$$\text{subject to} \quad 5x_1 + 6x_2 \geq 30$$
$$3x_1 + 2x_2 \leq 21$$
$$x_1 + x_2 \leq 12\,, \quad x_1, x_2 \geq 0.$$

(d) *Minimize,* $z = 4x_1 + 5x_2$

$$\text{subject to} \quad x_1 + 2x_2 \geq 2$$
$$5x_1 + 3x_2 \leq 15\,, \quad x_1, x_2 \geq 0.$$

(e) *Maximize,* $3x_1 + 5x_2$

$$\text{subject to} \quad x_1 + x_2 \leq 5$$
$$x_1 - x_2 \leq 4\,, \quad x_1, x_2 \geq 0.$$

(f) *Minimize,* $z = 3x_1 - x_2$

$$\text{subject to } x_1 + x_2 \le 2$$
$$2x_1 + 3x_2 \ge 6, \quad x_1, x_2 \ge 0.$$

[C.U.(P)'90]

(g) *Maximize,* $4x_1 + 3x_2$

$$\text{subject to } 3x_1 + 4x_2 \le 12$$
$$2x_1 + 5x_2 \le 10$$
$$x_1 + x_2 \ge 1, \quad x_1, x_2 \ge 0.$$

Solution: (a) Ox_1 and Ox_2 are the axes and $\overleftrightarrow{AB}$, $\overleftrightarrow{CD}$, $\overleftrightarrow{AE}$ represent the straight lines $x_1 + x_2 = 2$, $-x_1 + x_2 = 1$, and $x_1 = 2$. The region of the inequations are shown by arrows, the convex set of the feasible region is $OADFO$, the extreme points of the convex sets $O(0,0)$, $A(2,0)$, $D(1/2, 3/2)$, $F(0,1)$.

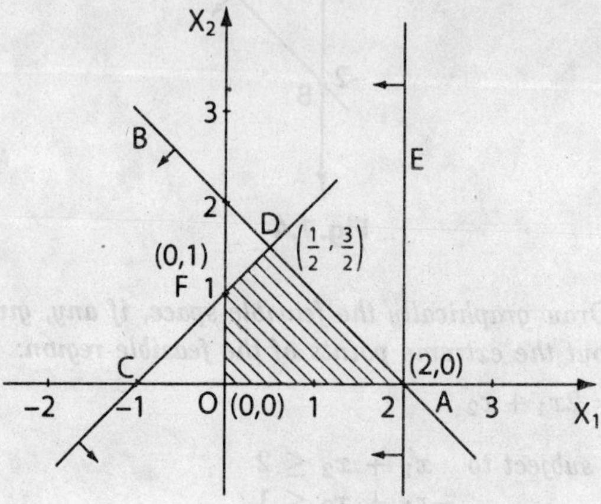

Fig. 7.5

(b) Ox_1 and Ox_2 are the axes and $\overleftrightarrow{AB}$ and $\overleftrightarrow{CD}$ are the straight lines $x_1 + 2x_2 = 7$, $x_1 - x_2 = 4$, since x_1, $x_2 \ge 0$, the convex set of the feasible region $OCDBO$; the extreme points are $O(0,0)$, $C(4,0)$, $D(5,1)$, $B(0,7/2)$.

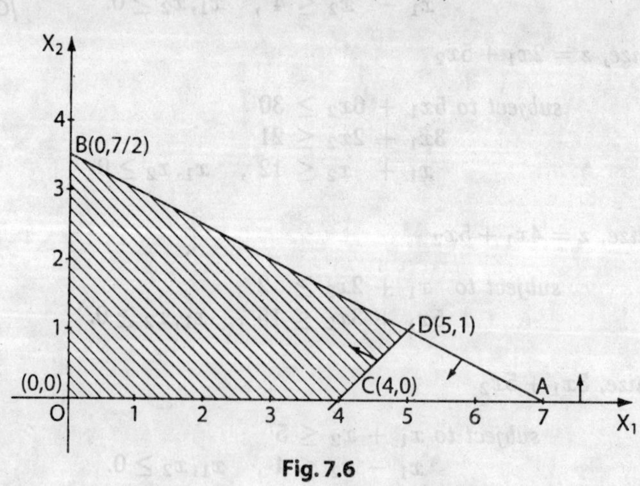

Fig. 7.6

(c) Ox_1 and Ox_2 are the axes and $\overleftrightarrow{AB}$, $\overleftrightarrow{CD}$ and $\overleftrightarrow{EF}$ are the straight lines $5x_1 + 6x_2 = 30$, $3x_1 + 2x_2 = 21$ and $x_1 + x_2 = 12$; since $x_1 \geq 0$, $x_2 \geq 0$ the convex set of the feasible region is $ACDBA$ and the extreme points are, $A(6,0), C(7,0), D(0,21/2), B(0,5)$. Here the constraint $x_1 + x_2 \leq 12$ is redundant.

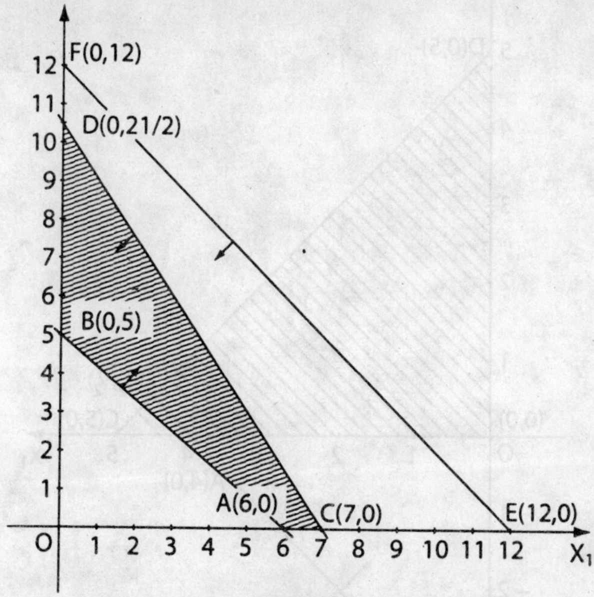

Fig. 7.7

(d) Ox_1 and Ox_2 be the axes and $\overleftrightarrow{AB}$, $\overleftrightarrow{CD}$ are the equations of the straight lines $x_1 + 2x_2 = 2$ and $5x_1 + 3x_2 = 15$ respectively. Since x_1 and $x_2 \geq 0$, the convex set of feasible region is $ACDB$ and the extreme points one $A(2,0), C(3,0), D(0,5), B(0,1)$.

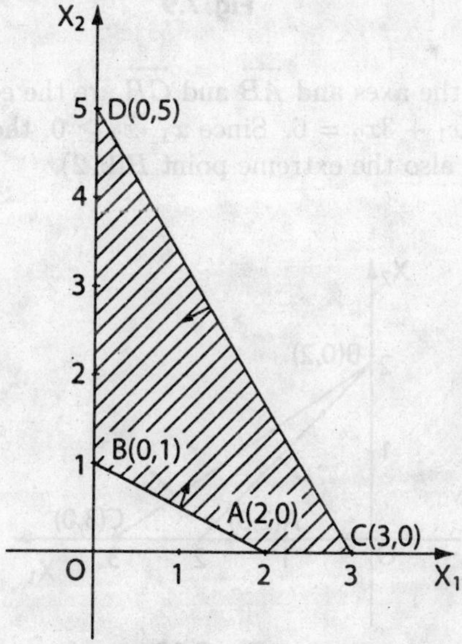

Fig. 7.8

(e) Ox_1 and Ox_2 be the axes and $\overleftrightarrow{AB}$ and $\overleftrightarrow{CD}$ are the straight lines $x_1 - x_2 = 4$, and $x_1 + x_2 = 5$ respectively. Since x_1 and $x_2 \geq 0$ the convex set of the feasible region is $OAEDO$, and the extreme points are $A(4,0)$, $E(9/2, 1/2)$, $D(0,5)$, $O(0,0)$.

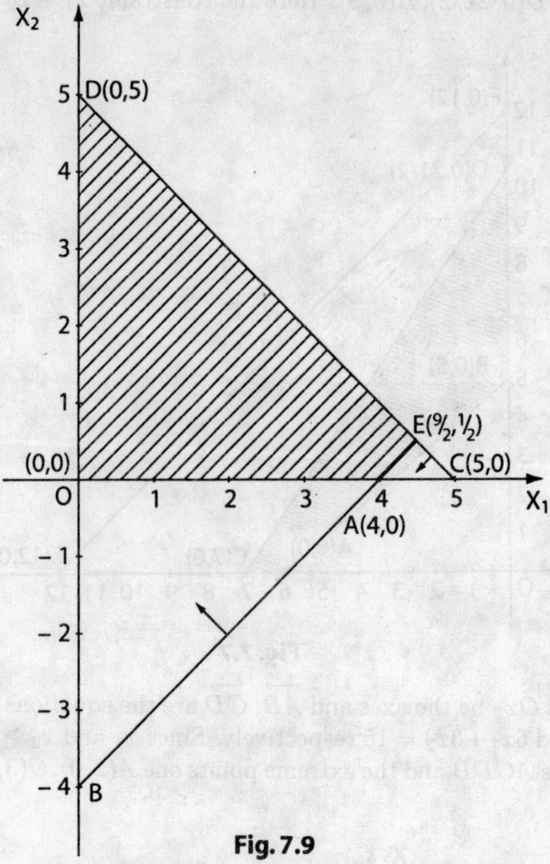

Fig. 7.9

(f) Ox_1 and Ox_2 be the axes and $\overleftrightarrow{AB}$ and $\overleftrightarrow{CB}$ are the equations of the straight lines $x_1 + x_2 = 2$ and $2x_1 + 3x_2 = 6$. Since x_1, $x_2 \geq 0$, the only feasible region is the point $(0,2)$ which is also the extreme point $B(0,2)$.

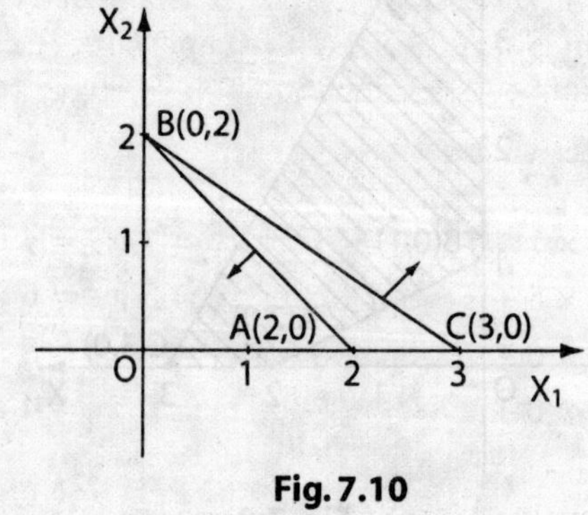

Fig. 7.10

(g) Ox_1 and Ox_2 be the axes and $\overleftrightarrow{AB}$, $\overleftrightarrow{CD}$ and $\overleftrightarrow{EF}$ are the equations of the straight lines $3x_1 + 4x_2 = 12$, $2x_1 + 5x_2 = 10$ and $x_1 + x_2 = 1$, since $x_1, x_2 \geq 0$, the convex set of the feasible region is $EAKDFE$ and the extreme points are $E(1,0), A(4,0), K(20/7, 6/7), D(0,2), F(0,1)$.

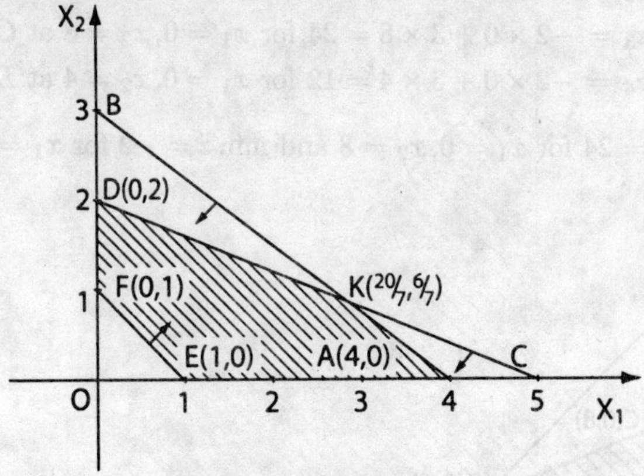

Fig. 7.11

► **Example 7.1.3** *Solve geometrically the L.P.P.*

$$\text{Maximize,} \quad z = 4x_1 + 7x_2$$

subject to

$$\begin{aligned} 2x_1 + 5x_2 &\leq 40 \\ x_1 + \ \ x_2 &\leq 11 \\ x_2 &\geq 4\,, \quad x_1 \geq 0, x_2 \geq 0. \end{aligned}$$

[C.U.(P)'97]

Solution: As $x_2 \geq 4$, then the condition $x_2 \geq 0$ is redundant.

Let $\overleftrightarrow{AB}$, $\overleftrightarrow{BC}$, $\overleftrightarrow{DA}$ and $\overleftrightarrow{DC}$ denote the straight lines given by the equations $x_1 + x_2 = 11$, $2x_1 + 5x_2 = 40$, $x_2 = 4$ and $x_1 = 0$ respectively.

The convex set of feasible solutions of L.P.P. is the convex region $ABCD$. This region is also known as an admissible region. It is a strictly bounded region. The four extreme points are $A(7,4)$, $B(5,6)$, $C(0,8)$ and $D(0,4)$. The region is a convex polyhedron with finite number of extreme points.

The value of the objective function z is

$$\begin{aligned} z_1 &= 4 \times 7 + 7 \times 4 = 56 \text{ for } x_1 = 7, x_2 = 4 \text{ at } A \\ z_2 &= 4 \times 5 + 7 \times 6 = 62 \text{ for } x_1 = 5, x_2 = 6 \text{ at } B \\ z_3 &= 4 \times 0 + 7 \times 8 = 56 \text{ for } x_1 = 0, x_2 = 8 \text{ at } C \\ z_4 &= 4 \times 0 + 7 \times 4 = 28 \text{ for } x_1 = 0, x_2 = 4 \text{ at } D \end{aligned}$$

Hence the maximum value of $z = 62$ at $x_1 = 5, x_2 = 6$.

Note: (1) With the same constraints, the optimum value of the different objective functions can be determined. For example, if the objective function is $z = -2x_1 + 3x_2$ then

$$\begin{aligned} z_1 &= -2 \times 7 + 3 \times 4 = -2 \text{ for } x_1 = 7, x_2 = 4 \text{ at } A \\ z_2 &= -2 \times 5 + 3 \times 6 = 8 \text{ for } x_1 = 5, x_2 = 6 \text{ at } B \\ z_3 &= -2 \times 0 + 3 \times 8 = 24 \text{ for } x_1 = 0, x_2 = 8 \text{ at } C \\ z_4 &= -2 \times 0 + 3 \times 4 = 12 \text{ for } x_1 = 0, x_2 = 4 \text{ at } D \end{aligned}$$

Hence $\max z = 24$ for $x_1 = 0, x_2 = 8$ and $\min z = -2$ for $x_1 = 7, x_2 = 4$

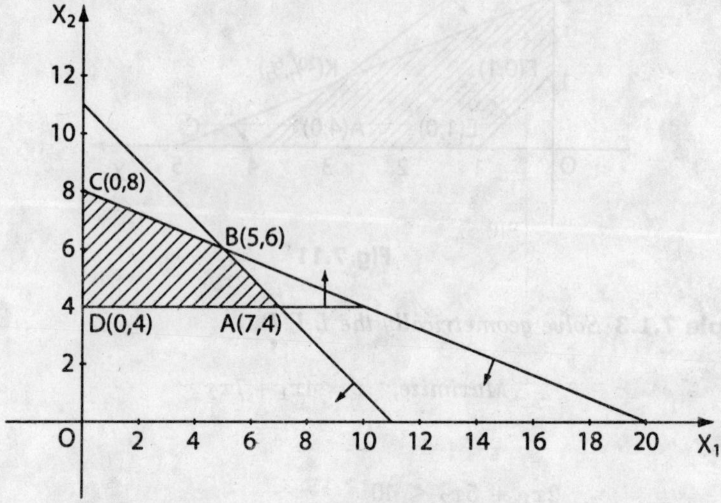

Fig. 7.12

(ii) Objective function $z = -x_1 - 4x_2$. Then

$$\begin{aligned} z_1 &= -1 \times 7 - 4 \times 4 = -23 \text{ for } x_1 = 7, x_2 = 4 \text{ at } A \\ z_2 &= -1 \times 5 - 4 \times 6 = -29 \text{ for } x_1 = 5, x_2 = 6 \text{ at } B \\ z_3 &= -1 \times 0 - 4 \times 8 = -32 \text{ for } x_1 = 0, x_2 = 8 \text{ at } C \\ z_4 &= -1 \times 0 - 4 \times 4 = -16 \text{ for } x_1 = 0, x_2 = 4 \text{ at } D \end{aligned}$$

Hence $\max z = -16$ for $x_1 = 0, x_2 = 4$ and $\min z = -32$ for $x_1 = 0, x_2 = 8$.

(iii) Objective function $z = 2x_1 + 5x_2$. Then

$$\begin{aligned} z_1 &= 34 \text{ at } x_1 = 7, x_2 = 4; \\ z_2 &= 40 \text{ at } x_1 = 5, x_2 = 6; \\ z_3 &= 40 \text{ at } x_1 = 0, x_2 = 8; \\ z_4 &= 20 \text{ at } x_1 = 4, x_2 = 0. \end{aligned}$$

Thus $\max z = 40$ at $x_1 = 5$, $x_2 = 6$ and $x_1 = 0$, $x_2 = 8$. Hence alternative optimal solutions exist.

2. Here the convex set of feasible solutions is strictly bounded, i.e., convex polyhedron. That is why any objective function has both finite maximum and minimum.

▶ **Example 7.1.4** *Solve graphically the L.P.P.*

$$\text{Maximize,} \quad z = 5x_1 - 2x_2$$

subject to

$$5x_1 + 6x_2 \geq 30$$
$$9x_1 - 2x_2 = 72$$
$$x_2 \leq 9, \quad x_1, x_2 \geq 0.$$

Solution: Let $\overleftrightarrow{AD}$, $\overleftrightarrow{CD}$ and $\overleftrightarrow{DE}$ denote the straight lines given by the equations, $5x_1 + 6x_2 = 30$, $x_2 = 9$ and $9x_1 - 2x_2 = 72$ respectively.

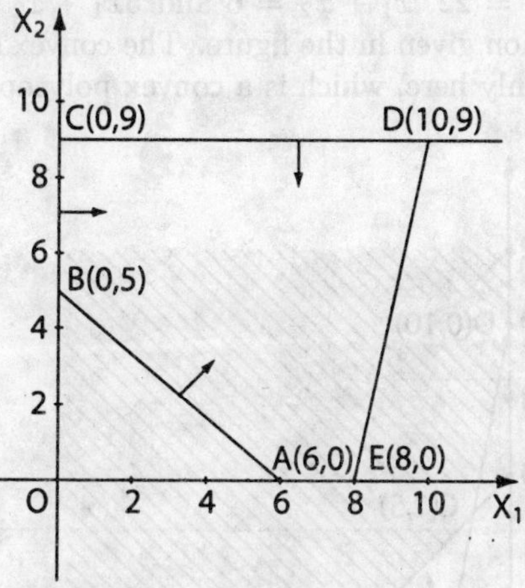

Fig. 7.13

The line AB, and the right hand portion of it, is the region represented by $5x_1 + 6x_2 \geq 30$; $x_1 \geq 0$, is the region RHS of the line BC (including the line).

x_1 axis and upward portion of it, is the region, $x_2 \geq 0$.

CD line and the downward portion of it is the region $x_2 \leq 9$. **And DE line is the line represented by the equation** $9x_1 - 2x_2 = 72$. Hence the **convex region** satisfied by all the constraints, is the segment DE of the line DE. Therefore, the extreme points of the convex region are only $D(10, 9)$ and $E(8, 0)$. Then

$$z_1 = 5 \times 10 - 2 \times 9 = 32 \text{ for } x_1 = 10, x_2 = 9 \text{ at } D$$

$$z_2 = 5 \times 8 - 2 \times 0 = 40 \text{ for } x_1 = 8, x_2 = 0 \text{ at } E$$

Hence $\max z = z_2 = 40$ for $x_1 = 8$, $x_2 = 0$ at E.

Note: If the objective function $z = -x_1 + x_2$ then

$$z_1 = -10 + 9 = -1 \text{ for } x_1 = 10, x_2 = 9 \text{ at } D$$

$$z_2 = -8 + 0 = -8 \text{ for } x_1 = 8, x_2 = 0 \text{ at } E$$

Hence $\max z = z_1 = -1$ for $x_1 = 10, x_2 = 9$ at D.

Note. The constraint $5x_1 + 6x_2 \geq 30$ has no contribution in determining the convex set of feasible solutions. This constraint is known as a **redundant constraint**.

▶ **Example 7.1.5** *Solve graphically the L.P.P.*

$$\text{Maximize,} \quad z = 2x_1 + 3x_2$$

subject to

$$\begin{aligned} 2x_1 + 7x_2 &\geq 22 \\ x_1 + \ x_2 &\geq \ 6 \\ 5x_1 + \ x_2 &\geq 10\ , \quad x_1, x_2 \geq 0. \end{aligned}$$

[C.U.(P)'94;'99]

Solution: Let $\overleftrightarrow{AB}$, $\overleftrightarrow{BC}$, $\overleftrightarrow{CD}$ denote respectively the straight lines given by the equations $2x_1 + 7x_2 = 22$, $x_1 + x_2 = 6$ and $5x_1 + x_2 = 10$. The admissible region is the dotted portion given in the figure. The convex set of feasible solutions is bounded from below only here, which is a convex polytope.

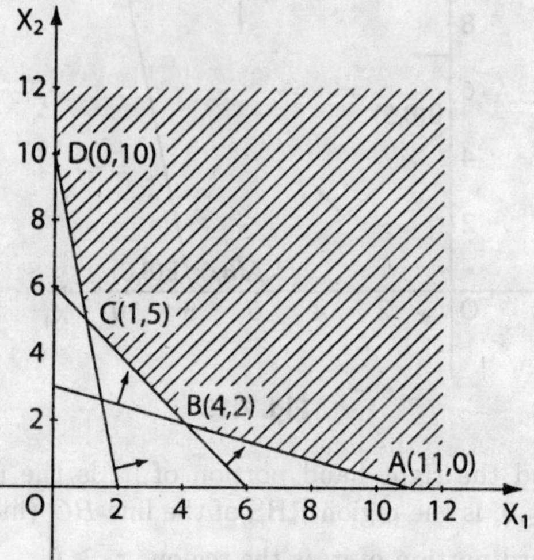

Fig. 7.14

The four extreme points are $A(11, 0), B(4, 2), C(1, 5)$ and $D(0, 10)$. Then

$$z_1 = 2 \times 11 + 3 \times 0 = 22 \text{ for } x_1 = 11, x_2 = 0 \text{ at } A$$

$$z_2 = 2 \times 4 + 3 \times 2 = 14 \text{ for } x_1 = 4, x_2 = 2 \text{ at } B$$

$$z_3 = 2 \times 1 + 3 \times 5 = 17 \text{ for } x_1 = 1, x_2 = 5 \text{ at } C$$

$$z_4 = 2 \times 0 + 3 \times 10 = 30 \text{ for } x_1 = 0, x_2 = 10 \text{ at } D$$

Hence $\min(z_1, z_2, z_3, z_4) = \min(22, 14, 17, 30) = 14$ for $x_1 = 4, x_2 = 2$ at B.

To verify that the minimum is finite take any two points $(4.7, 1.8)$ and $(3.9, 2.1)$ very close to the extreme point B [extreme point at which the objective function attains its lowest] on the line segments BA and BC respectively and the value of the objective function corresponding to the points $(4.7, 1.8)$ and $(3.9, 2.1)$ are $2 \times 4.7 + 3 \times 1.8 = 14.8$ and $2 \times 3.9 + 3 \times 2.1 = 14.1$ both of which are greater than the lowest value 14 at B. Thus the objective function attains its minimum value 14 at $B(x_1 = 4, x_2 = 2)$.

Note. (a) If $z = 2x_1 + 3x_2$ and the problem is to be maximized then no finite value of z will be obtained. Here $\max(z_1, z_2, z_3, z_4) = 30$ at $D(0, 10)$. But for points on Dx_2, eg, at $(0, 11)$ the value of the objective function is 33 which is greater than 30 and in the same way it can be established that the objective function has no finite maximum. In that case, the problem is said to have an **unbounded solution.**

(b) Any objective function $lx_1 - mx_2$ (both l, m +ve or both −ve) has no finite maximum or minimum subject to the constraints given in the problem (7.1.5).

▶ **Example 7.1.6** *Solve graphically the L.P.P.*

$$\textit{Minimize,} \quad z = -2x_1 + x_2$$

subject to

$$\begin{aligned} x_1 + \ x_2 &\geq \ 6 \\ 3x_1 + 2x_2 &\geq 16 \\ x_2 &\leq \ 9\ , \quad x_1 \geq 0, x_2 \geq 0. \end{aligned}$$

Solution: The feasible region is given in the figure (shaded region). The region is bounded from below only. The four extreme points are $A(6,0)$, $P(4,2)$, $B(0,8)$ and $C(0,9)$.

$$z_1 = -12 \text{ at } A,$$
$$z_2 = -6 \text{ at } P,$$
$$z_3 = 8 \text{ at } B \text{ and}$$
$$z_4 = 9 \text{ at } C.$$

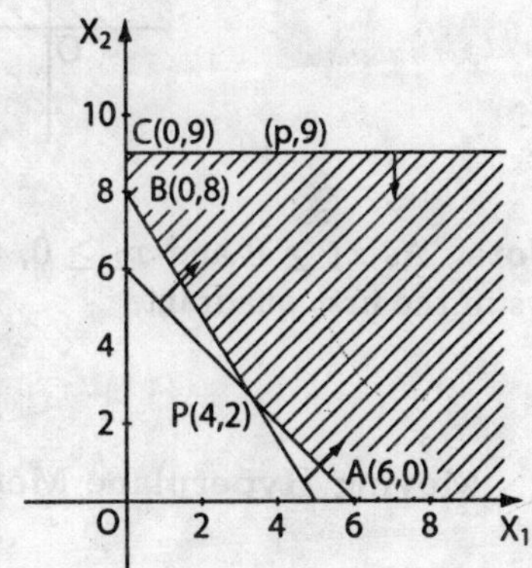

Fig. 7.15

Min. $(z_1, z_2, z_3, z_4) = -12$ at $A(6, 0)$. But for points very close to A on the line Ax_1, eg, at $(7, 0)$ the value of the objective function is -14 which is less than -12 and in the same way it can be established that the objective function has no finite minimum. In that case the problem is said to have an **unbounded solution.**

Note. (a) If $z = -2x_1 + x_2$ and the problem is to maximize it, then finite value of z will be obtained. Here $\max(z_1, z_2, z_3, z_4) = 9$ at $C(0, 9)$. To verify that maximum is finite, take any two points $(0, 8)$ and $(p, 9)$ in the feasible region very close to C on CB segment and on line $x_2 = 9$. The values of z are 8 and $9 - 2p (p > 0)$ both of which are less than 9. Hence $\max z = 9$ at $C(0, 9)$.

(b) This objective function has finite maximum but no finite minimum (in this convex set of F.S. which is bounded from below only). Again if the objective function be $2x_1 - x_2$ then it has finite minimum at $C(0, 9)$ but no finite maximum.

▶ **Example 7.1.7** *Solve graphically the L.P.P.*

$$\textit{Minimize,} \quad z = -2x_1 + 5x_2$$

subject to

$$\begin{aligned} x_1 + x_2 &\geq 7 \\ 3x_1 + x_2 &\leq 6\ , \quad x_1 \geq 0, x_2 \geq 0. \end{aligned}$$

[**Hints.** The region restricted by the constraints $x_1 + x_2 \geq 7$ and $3x_1 + x_2 \leq 6$ is the shaded portion given in the figure 7.16 and the region restricted by the constraints $x_1 \geq 0$, $x_2 \geq 0$ is the first quadrant. Therefore in two dimensional plane, there exists no region satisfying all the four restrictions or constraints. Hence there is no **admissible** or **feasible** region. Therefore the problem has no feasible solution and no question of finding out minimum value of the objective function arises then.]

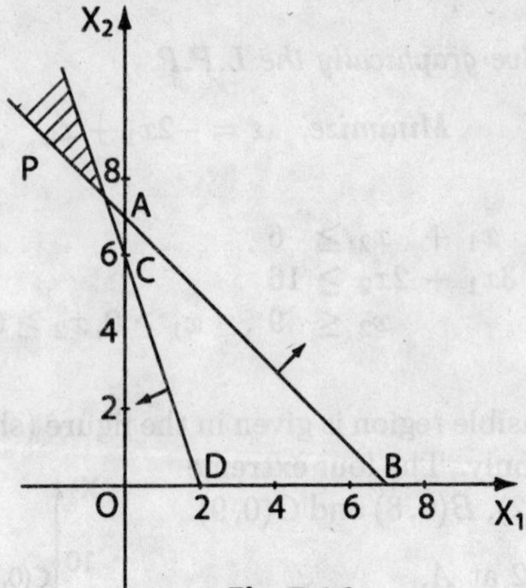

Fig. 7.16

Note. As $x_1 \geq 0$ and $x_2 \geq 0$, then the admissible or feasible region must be always in the first quadrant.

7.1.1 Moving Hyperplane Method

There is another geometrical method of solving a L.P.P. which is called the moving hyperplane method. But judging its difficulty this method may be considered as suitable only when the convex set is a polytope. Below given is an example of that type.

▶ **Example 7.1.8** *Minimize,* $z = 3x_1 - 2x_2$ *subject to*

$$\begin{aligned} 3x_1 + 4x_2 &\geq 12 \\ x_1 - 3x_2 &\leq 6 \\ x_1 - 2x_2 &\leq -4\,, \quad x_1 \geq 0, x_2 \geq 0. \end{aligned}$$

Solution: The convex set of the feasible region is given by the shaded region which is a convex polytope. The three extreme points of the feasible region are $A(6,0), B(4,0)$ and $C(\frac{4}{5}, \frac{12}{5})$.

For $z = k = 3x_1 - 2x_2$, the objective function represents a straight line and for different values of k, we get a system of parallel straight lines.

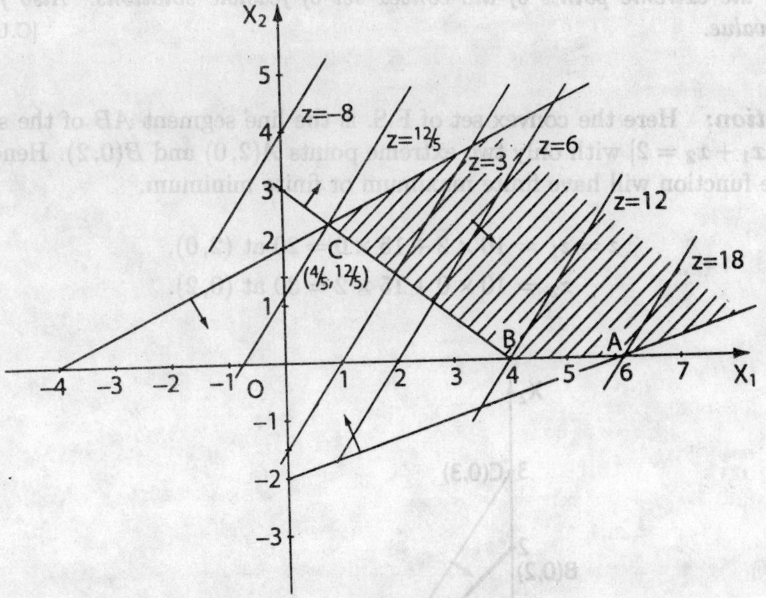

Fig. 7.17

Now on the feasible region, the value of the objective function increases as it moves parallel to itself in the direction given in the figure (by arrow). On the other hand if it moves in the opposite direction parallel to itself, the value of the objective function gradually decreases and the minimum value of the objective function is $\min z = -\frac{12}{5}$ when it passes through the extreme point $C(\frac{4}{5}, \frac{12}{5})$ of the convex set of F.S. The value of the objective function further decreases as given in the fig. 7.17. But now no part of the straight line is in the feasible region. Thus the minimum value of the objective function is $-\frac{12}{5}$ at $x_1 = \frac{4}{5}, x_2 = \frac{12}{5}$.

Note. (i) But if the problem be maximize, $z = 3x_1 - 2x_2$ with the same set of constraints, then within the feasible region, the value of the objective function gradually increases as the line of the objective function moves parallel to itself (as given by arrow) and the objective function has no finite maximum and the problem is said to have an unbounded solution.

(ii) If the problem has a bounded optimal solution, then one of the extreme points gives the optimal solution which we have seen in the above problem, but if the problem has no finite value of the objective function, then mere presence of one or more extreme points do not give any assurance of having a finite optimal solution which we have seen in the above case.

▶ **Example 7.1.9** *Represent geometrically the set of constraints of following L.P.P.*

$$\textit{Maximize,} \quad z = 10x_1 + 15x_2$$

subject to

$$x_1 + x_2 = 2$$
$$3x_1 + x_2 \leq 6 , \quad x_1, x_2 \geq 0,$$

and find the extreme points of the convex set of feasible solutions. Also find the optimal value. [C.U.(P)'83]

Solution: Here the convex set of F.S. is the line segment AB of the straight line $AB[x_1 + x_2 = 2]$ with only two extreme points $A(2,0)$ and $B(0,2)$. Hence each objective function will have finite maximum or finite minimum.

$$z = z_1 = 10 \times 2 + 15 \times 0 = 20 \text{ at } (2,0),$$
$$z_2 = 10 \times 0 + 15 \times 2 = 30 \text{ at } (0,2).$$

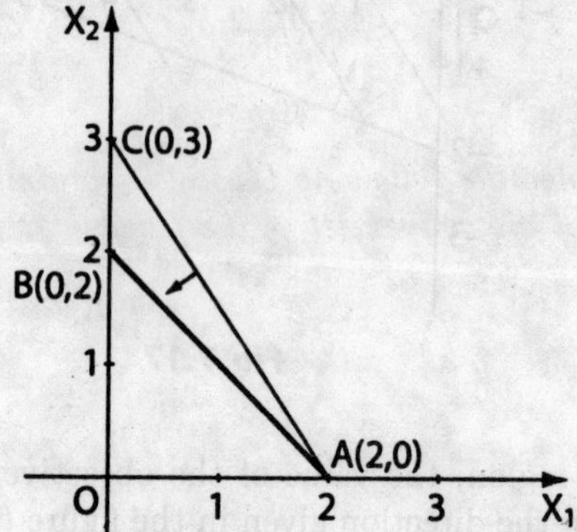

Fig. 7.18

Hence max $z = 30$ at $x_1 = 0, x_2 = 2$.

▶ **Example 7.1.10** *Solve the following L.P.P. by graphical method.*

$$\textit{Minimize,} \quad z = 4x_1 + x_2$$

subject to

$$x_1 + 2x_2 \le 3$$
$$4x_1 + 3x_2 = 6$$
$$3x_1 + x_2 \ge 3, \quad x_1, x_2 \ge 0.$$

[C.U.(P)'97]

Solution: Here the convex set of F.S. is the line segment CD, the extreme points are $C(\frac{3}{2},0), D(\frac{3}{5},\frac{6}{5})$. Thus

$$z_1 = 4 \times \frac{3}{2} + 1 \times 0 = 6 \text{ at } C\left(\frac{3}{2},0\right)$$

$$z_2 = 4 \times \frac{3}{5} + 1 \times \frac{6}{5} = \frac{18}{5} \text{ at } D\left(\frac{3}{5},\frac{6}{5}\right)$$

Hence $\min z = \frac{18}{5}$ at $x_1 = \frac{3}{5}, x_2 = \frac{6}{5}$.

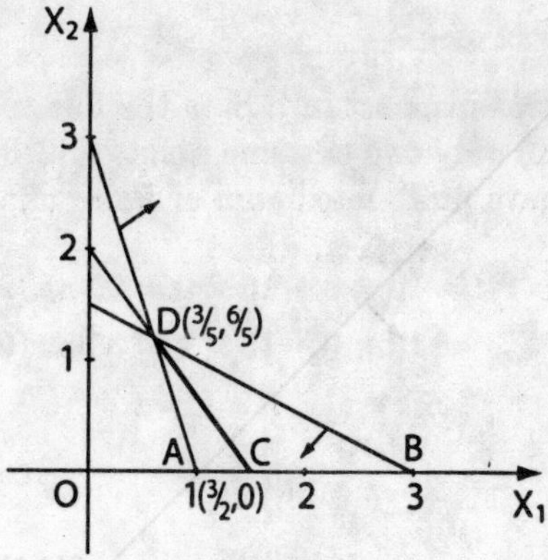

Fig. 7.19

▶ **Example 7.1.11** *Solve the following Linear Programming Problems after finding the extreme points of the convex set of the feasible solutions.*

(a) *Maximize,* $z = 3x_1 + 5x_2$

$$\text{subject to} \quad \begin{aligned} x_1 - 3x_2 &\leq 4 \\ -x_1 + x_2 &\geq -4 \\ x_1 + x_2 &\leq 8, \quad x_1, x_2 \geq 0. \end{aligned}$$

(b) *Maximize,* $z = 4x_1 + 7x_2$

$$\text{subject to} \quad \begin{aligned} 12x_1 + 7x_2 &\leq 42 \\ 5x_1 + 4x_2 &\leq 20 \\ 2x_1 + 3x_2 &\geq 6, \quad x_1, x_2 \geq 0. \end{aligned}$$

(c) *Maximize,* $z = x_1 + 3x_2$

$$\text{subject to} \quad \begin{aligned} 3x_1 + x_2 &\leq 11 \\ x_1 - x_2 &\leq 1 \\ 2x_1 - x_2 &\geq -1, \quad x_1, x_2 \geq 0. \end{aligned}$$

(d) *Maximize,* $z = 2x_1 - x_2$

$$\text{subject to} \quad \begin{aligned} x_1 - x_2 &\leq 1 \\ x_1 &\leq 3, \quad x_1, x_2 \geq 0. \end{aligned}$$ [C.U.(P)'95]

Solution: (a) Ox_1 and Ox_2 be the axes. $\overleftrightarrow{AB}$, $\overleftrightarrow{AC}$ and $\overleftrightarrow{LN}$ represent the straight lines $x_1 - 3x_2 = 4$, $-x_1 + x_2 = -4$ and $x_1 + x_2 = 8$ respectively. The convex set of the feasible region is $ABCA$ where the extreme points are $A(4,0)$, $B(7,1)$, $C(6,2)$ and the feasible region is a convex polyhedron. Therefore any objective function has finite maximum and minimum.

$$z_1 = 3 \times 4 + 5 \times 0 = 12 \text{ at } (4,0)$$
$$z_2 = 3 \times 7 + 5 \times 1 = 26 \text{ at } (7,1)$$
$$z_3 = 3 \times 6 + 5 \times 2 = 28 \text{ at } (6,2)$$

Then $\max z = 28$ at $C(6,2)$.

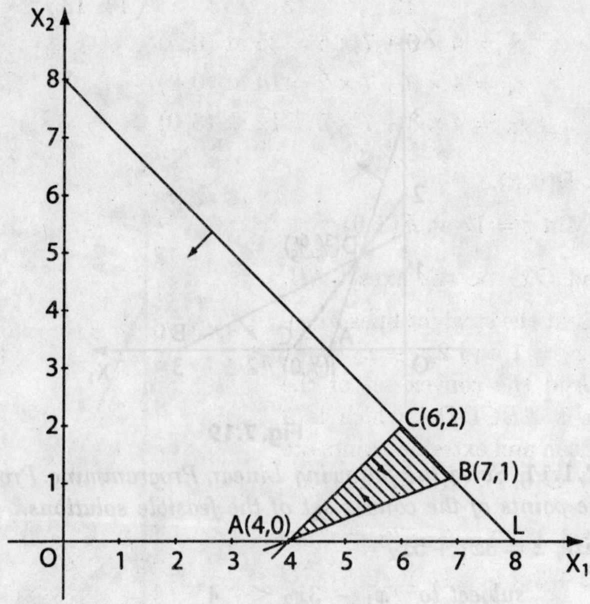

Fig. 7.20

Remark: Similarly min $z = 12$ at $A(4,0)$.

(b) Ox_1 and Ox_2 be the axes. $\overleftrightarrow{AB}$, $\overleftrightarrow{CD}$, $\overleftrightarrow{EF}$ represent the straight lines $12x_1 + 7x_2 = 42$, $5x_1 + 4x_2 = 20$ and $2x_1 + 3x_2 = 6$ respectively. And the convex set of the feasible region is $AKDFEA$ and extreme points are $A(\frac{7}{2}, 0)$, $K(\frac{28}{13}, \frac{30}{13})$, $D(0,5)$, $F(0,2)$, $E(3,0)$ which is a convex polyhedron.

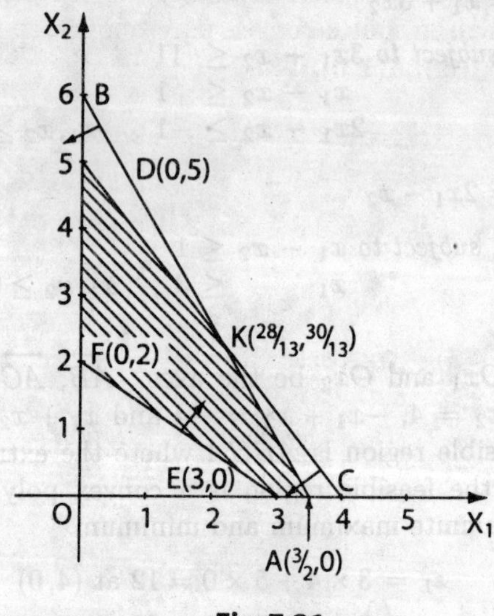

Fig. 7.21

$$z_1 = 4 \times \frac{7}{2} + 7 \times 0 = 14 \text{ at } \left(\frac{7}{2}, 0\right)$$

$$z_2 = 4 \times \frac{28}{13} + 7 \times \frac{30}{13} = \frac{322}{13} \text{ at } \left(\frac{28}{13}, \frac{30}{13}\right)$$
$$z_3 = 4 \times 0 + 7 \times 5 = 35 \text{ at } (0,5)$$
$$z_4 = 4 \times 0 + 7 \times 2 = 14 \text{ at } (0,2)$$
$$z_5 = 4 \times 3 + 7 \times 0 = 12 \text{ at } (3,0)$$

Max $z = 35$ at $D(0,5)$.

Remark: Min $z = 12$ at $E(3,0)$.

(c) Ox_1 and Ox_2 be the axes. $\overleftrightarrow{AB}$, $\overleftrightarrow{BC}$, $\overleftrightarrow{DA}$ represent the straight lines $3x_1 + x_2 = 11$, $x_1 - x_2 = 1$ and $2x_1 - x_2 = -1$ respectively. And the convex set of the feasible region is $ABCODA$ which is a convex polyhedron and extreme points are $A(3,2)$, $B(2,5)$, $C(0,1)$, $O(0,0)$, $D(1,0)$.

$$z_1 = 1 \times 3 + 3 \times 2 = 9 \text{ at } (3,2)$$
$$z_2 = 1 \times 2 + 3 \times 5 = 17 \text{ at } (2,5)$$
$$z_3 = 1 \times 0 + 3 \times 1 = 3 \text{ at } (0,1)$$
$$z_4 = 1 \times 0 + 3 \times 0 = 0 \text{ at } (0,0)$$
$$z_5 = 1 \times 1 + 3 \times 0 = 1 \text{ at } (1,0)$$

$\max z = 17$ at $B(2,5)$.

Remark: Min $z = 0$ at $O(0,0)$.

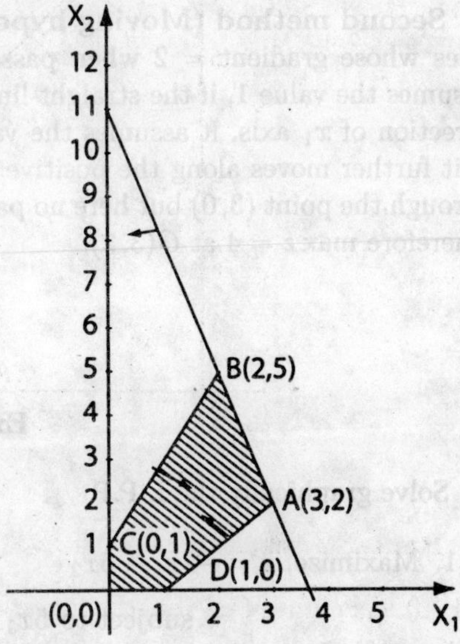

Fig. 7.22

(d) **First method:** Ox_1 and Ox_2 be the set of axes. AC and BC are the straight lines $x_1 - x_2 = 1$ and $x_1 = 3$ respectively. Here the convex set of the feasible region is not a convex polyhedron, but a convex polytope which is unbounded above; the extreme points are $O(0,0), A(1,0), C(3,2)$.

$$z_1 = 2 \times 0 - 1 \times 0 = 0 \text{ at } (0,0)$$
$$z_2 = 2 \times 1 - 1 \times 0 = 2 \text{ at } (1,0)$$
$$z_3 = 2 \times 3 - 1 \times 2 = 4 \text{ at } (3,2)$$

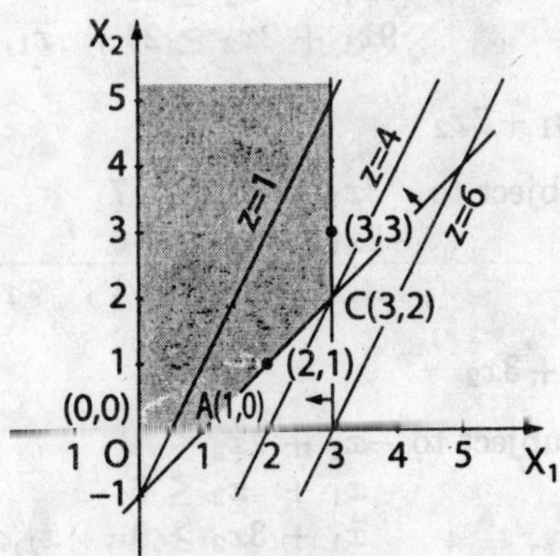

Fig. 7.23

Maximum value of z attains at $C(3,2)$. Now to test that maximum value is finite at $C(3,2)$, we take two neighbouring points $(2,1)$ and $(3,3)$ on the straight lines AC and BC on the feasible region.

$$z_4 = 2 \times 2 - 1 \times 1 = 3 \text{ at } (2,1)$$
$$z_5 = 2 \times 3 - 1 \times 3 = 3 \text{ at } (3,3)$$

both of which are less then 4, then $\max z = 4$ at $C(3,2)$.

Second method (Moving hyperplane method): We draw a set of straight lines whose gradient $= 2$ when passes through the point $(0,-1)$, $z = 2x_1 - x_2$ assumes the value 1, if the straight line moves parallel to itself towards the positive direction of x_1 axis, it assumes the value $z = 4$ when passes through the point C. If it further moves along the positive direction it assumes the value 6 when passes through the point $(3,0)$ but here no part of the straight line is on the feasible region. Therefore $\max z = 4$ at $C(3,2)$.

Exercise 7

Solve graphically the L.P.P.

1. Maximize, $z = -2x_1 + 5x_2$

$$\begin{aligned} \text{subject to } 5x_1 + 2x_2 &\le 45 \\ 4x_1 + 5x_2 &\le 53 \\ x_1 \qquad &\ge 2, \quad x_1, x_2 \ge 0. \end{aligned}$$

Find also the minimum value of z.

2. Minimize, $z = 4x_1 - 3x_2$

$$\begin{aligned} \text{subject to } 2x_1 - \quad x_2 &\ge 4 \\ 4x_1 + 3x_2 &\le 28, \quad x_1, x_2 \ge 0. \end{aligned}$$

3. Maximize, $z = x_1 + 3x_2$

$$\begin{aligned} \text{subject to } x_1 + \quad x_2 &\le 5 \\ 6x_1 - \quad x_2 &\le 30 \\ 9x_1 + 2x_2 &\ge 24, \quad x_1, x_2 \ge 0. \end{aligned}$$

4. Minimize, $z = -2x_1 + 7x_2$

$$\begin{aligned} \text{subject to } \quad 3x_1 + 2x_2 &\le 17 \\ -2x_1 + 3x_2 &\le 6 \\ x_2 &\ge 1, \quad x_1, x_2 \ge 0. \end{aligned}$$

5. Minimize, $z = 2x_1 + 3x_2$

$$\begin{aligned} \text{subject to } -x_1 + 2x_2 &\le 4 \\ x_1 + \quad x_2 &\le 6 \\ x_1 + 3x_2 &\ge 9, \quad x_1, x_2 \ge 0. \end{aligned}$$

6. Maximize, $z = -7x_1 + 4x_2$

$$\text{subject to } 5x_1 - 2x_2 \geq 0$$
$$x_1 - 6x_2 \leq 0$$
$$x_1 + x_2 \leq 7, \quad x_1, x_2 \geq 0.$$

7. Minimize, $z = x_1 + 5x_2$

$$\text{subject to } x_1 + x_2 \leq 4$$
$$2x_1 + x_2 \leq 6$$
$$4x_1 + 3x_2 \geq 28, \quad x_1, x_2 \geq 0.$$

8. Maximize, $z = 3x_1 + 4x_2$

$$\text{subject to } x_1 - x_2 \leq -1$$
$$-x_1 + x_2 \leq 0, \quad x_1, x_2 \geq 0.$$

9. Maximize, $z = 7x_1 + 3x_2$

$$\text{subject to } 4x_1 + 5x_2 \leq 40$$
$$x_1 \geq 3$$
$$x_2 \geq 4, \quad x_1 \geq 0, x_2 \geq 0.$$

10. Maximize, $z = 2x_1 - 6x_2$

$$\text{subject to } 3x_1 + 2x_2 \leq 6$$
$$x_1 - x_2 \geq -1$$
$$-x_1 - 2x_2 \geq 1, \quad x_1, x_2 \geq 0.$$

11. Maximize, $z = x_1 + 3x_2$

$$\text{subject to } x_1 - x_2 \geq -1$$
$$x_1 + x_2 \geq 3, \quad x_1, x_2 \geq 0.$$

12. Maximize, $z = 3x_1 + 2x_2$

$$\text{subject to } 3x_1 + 2x_2 \leq 10$$
$$3x_1 + 2x_2 \geq 20, \quad x_1 \geq 0, x_2 \geq 0.$$

13. Minimize, $z = 4x_1 - x_2$

$$\text{subject to } x_1 + 2x_2 \geq 10$$
$$x_1 \leq 12, \quad x_1 \geq 0, x_2 \geq 0.$$

14. (a) Maximize, $z = \frac{1}{7}x_1 + 2x_2$

$$\text{subject to } 3x_1 + 4x_2 \geq 12$$
$$x_2 \leq 10, \quad x_1 \geq 0, x_2 \geq 0.$$

(b) Maximize, $z = x_1 + 5x_2$

$$\text{subject to } 3x_1 + 4x_2 \leq 6$$
$$x_1 + 3x_2 \geq 3, \quad x_1 \geq 0, x_2 \geq 0.$$ [C.U.(H)'80]

(c) Maximize, $z = 3x_1 + 2x_2$

$$\text{subject to } \begin{aligned} -2x_1 + x_2 &\leq 1 \\ x_1 \quad &\leq 2 \\ x_1 + x_2 &\leq 3, \quad x_1, x_2 \geq 0. \end{aligned}$$ [C.U.(P)'89]

15. Maximize, $z = 2x_1 + x_2$

$$\text{subject to } \begin{aligned} 4x_1 + 3x_2 &\leq 12 \\ 4x_1 + x_2 &\leq 8 \\ 4x_1 - x_2 &\leq 8, \quad x_1 \geq 0, x_2 \geq 0. \end{aligned}$$

Find out the extreme points of the convex set of feasible solutions and hence find out the maximum value of the objective function. [C.U.(P)'81]

16. Represent geometrically the set of constraints of the following L.P.P.

$$\text{Maximize,} \quad z = 10x_1 + 15x_2$$

$$\text{subject to } \begin{aligned} x_1 + x_2 &= 2 \\ 3x_1 + 2x_2 &\leq 6, \quad x_1, x_2 \geq 0. \end{aligned}$$

Find the extreme points of the convex set of feasible solutions. Also find the optimal value. [C.U.(P)'83]

17. Minimize, $z = x_1 + 7x_2$

$$\text{subject to } \begin{aligned} -x_1 + 2x_2 &\leq 8 \\ x_1 - x_2 &\leq 4, \quad x_1, x_2 \geq 0. \end{aligned}$$ [C.U.(P)'92]

18. Maximize, $z = 2x_1 + 4x_2$

$$\text{subject to } \begin{aligned} 2x_1 + 3x_2 &\leq 48 \\ x_1 + 3x_2 &\leq 42 \\ x_1 + x_2 &\leq 21, \quad x_1, x_2 \geq 0. \end{aligned}$$ [C.U.(P)'93]

19. Maximize, $z = 500x_1 + 400x_2$

$$\text{subject to } \begin{aligned} 10x_1 + 8x_2 &\leq 800 \\ x_1 \quad &\leq 60 \\ x_2 &\leq 75, \quad x_1, x_2 \geq 0. \end{aligned}$$

Prove that alternative optimal solutions exist. Find at least two optimal solutions where the value of the variables are all integers (Solved example 2.3.4). [C.U.(H)'91,'94,'96]

Solution: Here the straight line AB, CA and DB represent the straight lines $10x_1 + 8x_2 = 800$, $x_1 = 60$, $x_2 = 75$, and the convex set of the feasible region is $OCABDO$ and the extreme points are $O(0,0)$, $C(60,0)$, $A(60,25)$, $B(20,75)$, $D(0,75)$, which is a convex polyhedron.

$$\begin{aligned} z_1 &= 500 \times 0 + 400 \times 0 = 0 \text{ at } (0,0) \\ z_2 &= 500 \times 60 + 400 \times 0 = 30{,}000 \text{ at } (60,0) \\ z_3 &= 500 \times 60 + 400 \times 25 = 40{,}000 \text{ at } (60,25) \\ z_4 &= 500 \times 20 + 400 \times 75 = 40{,}000 \text{ at } (20,75) \text{ and} \\ z_5 &= 500 \times 0 + 400 \times 75 = 30{,}000 \text{ at } (0,75) \end{aligned}$$

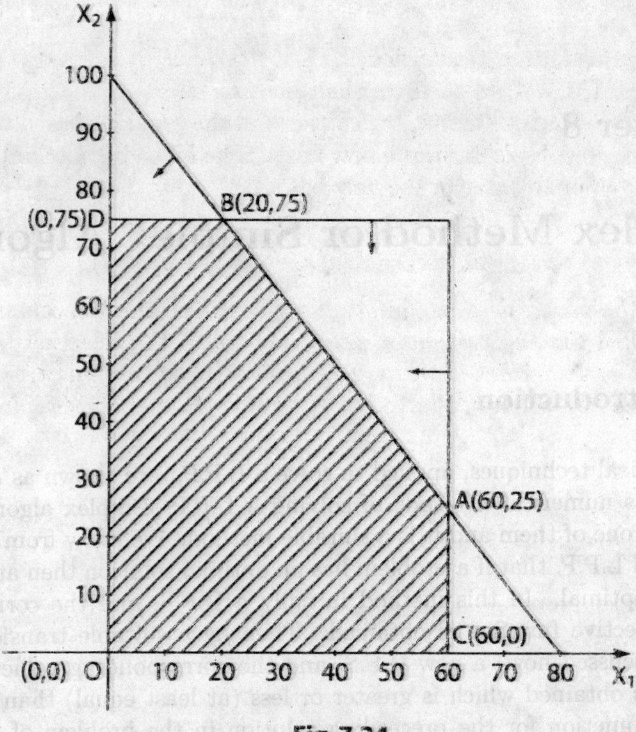

Fig. 7.24

Thus $\max z =$ Rs. 40,000 at $(x_1 = 60, x_2 = 25)$ and $(x_1 = 20, x_2 = 75)$. Therefore multiple optimal solutions exist and the two values are $(x_1 = 60, x_2 = 25)$ and $(x_1 = 20, x_2 = 75)$, where all values of the solution are integers.

Remark. Since there are more than one optimal solutions then any convex combination of them are also solutions but they may or may not be integers. For example, one solution $x_1 = \frac{1}{2}60 + \frac{1}{2} \times 20 = 40$ and $x_2 = \frac{1}{2} \times 25 + \frac{1}{2} \times 75 = 50$, which is also an integer solution.

Answers

1. $\max z = 41$ at $(x_1 = 2, x_2 = 9)$, $\min z = -18$ at $(x_1 = 9, x_2 = 0)$ 2. $\min z = 4$ at $(x_1 = 4, x_2 = 4)$ 3. $\max z = 11$ at $(x_1 = 2, x_2 = 3)$ 4. $\min z = -3$ at $(x_1 = 5, x_2 = 1)$ 5. $\min z = 10\frac{1}{5}$ at $(x_1 = \frac{8}{5}, x_2 = \frac{13}{5})$ 6. $\max z = 6$ at $(x_1 = 2, x_2 = 5)$ 7. No. F.S. 8. No F.S. 9. $\max z = 47$ at $(x_1 = 5, x_2 = 4)$ 10. No F.S. 11. Unbounded solution 12. No. F.S. 13. Unbounded solution. 14. (a) Unbounded solution. (b) $\max z = 7\frac{1}{2}$ at $(x_1 = 0, x_2 = \frac{3}{2})$. (c) $\max z = 8$ at $(x_1 = 2, x_2 = 1)$ 15. Extreme points are $(0, 0)$, $(2, 0)$, $(\frac{3}{2}, 2)$ and $(0, 4)$; $\max z = 5$ at $(x_1 = \frac{3}{2}, x_2 = 2)$. The constraint $4x_1 - x_2 \leq 8$ is redundant. 16. Extreme points are $(2, 0)$, $(0, 2)$ and $\max z = 30$ at $(0, 2)$. 17. $\min z = 0$ at $(x_1 = 0, x_2 = 0)$ 18. $\max z = 60$ at $(x_1 = 6, x_2 = 12)$.

Chapter 8

Simplex Method or Simplex Algorithm (I)

8.1 Introduction

All numerical techniques, applied to solve a L.P.P., are known as *algorithm*. There are various numerical methods of solving a L.P.P. Simplex algorithm or simplex method is one of them and it is a suitable method. We know from the fundamental theorem of L.P.P. that if a problem has an optimal solution then at least one B.F.S. must be optimal. In this method initially a B.F.S. and the corresponding value of the objective function is obtained. Then by a suitable transformation (which will be discussed now) a new B.F.S. and the corresponding value of the objective function is obtained which is greater or less (at least equal) than the value of the objective function for the preceeding solution in the problem of maximization or minimization respectively. The process is continued until an optimal value of the objective function is attained (if it exists at all). The extra advantage of this method is that, it gives correct information about the existence of the optimal solution.

8.2 Procedure

After the introduction of the slack and surplus variables and by proper adjustment of z, let a L.P.P. be, optimize $z = \mathbf{cx}$

$$\text{subject to} \quad A\mathbf{x} = \mathbf{b}, \quad \mathbf{x} \geq 0, \cdots, [A]_{m\times n}, \tag{8.2.1}$$

where $\mathbf{c} = (c_1, c_2, \cdots, c_r, \overbrace{0, 0, \cdots, 0}^{n-r})$ an n component row vector.

$\mathbf{x} = [x_1, x_2, \cdots, x_r, x_{r+1}, \cdots, x_n]$ an n component column vector.

The components $x_{r+1}, x_{r+2}, \cdots, x_n$ are either slack or surplus variables.

We make an assumption $(m < n)$ [m, the number of constraints] and further make an assumption that all components of $\mathbf{b}$ are non-negative by proper adjustment. Of course, the *second assumption is non-restrictive.*

The coefficient matrix A will be an $m \times n$ matrix and $A = (\mathbf{a}_1, \mathbf{a}_2, \cdots, \mathbf{a}_n)$, where $\mathbf{a}_j$ is the jth column vector of the coefficient matrix A, which are called the *activity vectors* associated with the variable x_j $[j = 1, 2, \cdots, n]$. As none of the m

converted equations is redundant then there exists at least one set of m-column vectors of the coefficient matrix A, which are linearly independent. Let β_1, β_2, $\cdots$, β_m be a set of m linearly independent column vectors taken from $\mathbf{a}_1$, $\mathbf{a}_2, \cdots$, $\mathbf{a}_n \in A$. Then one basis matrix B is given by

$$B = B(\beta_1, \beta_2, \cdots, \beta_m).$$

Let $x_{B1}, x_{B2}, \cdots, x_{Bm}$ be the basic variables associated with the column vectors β_1, β_2, $\cdots$, β_m respectively. Then the basic variable vector is

$$\mathbf{x}_B = [x_{B1}, x_{B2}, \cdots, x_{Bm}]. \tag{8.2.2}$$

The solution set corresponding to the basic variables is

$$\hat{\mathbf{x}}_B \text{ or simply } \mathbf{x}_B = B^{-1}\mathbf{b} \tag{8.2.3}$$

We assume that $\mathbf{x}_B \geq \mathbf{0}$, i.e., the solution is a B.F.S. and in that case, the basis B is called an *admissible basis* in the simplex theory.

Let $c_{B1}, c_{B2}, \cdots, c_{Bm}$ be the coefficients of $x_{B1}, x_{B2}, \cdots, x_{Bm}$, respectively in the objective function $z = \mathbf{cx}$ then

$$\mathbf{c}_B = (c_{B1}, c_{B2}, \cdots, c_{Bm}) \tag{8.2.4}$$

is an m-component row vector which is known as the *associated cost vector*.

Now a value z_B is defined as

$$z_B = c_{B1}x_{B1} + c_{B2}x_{B2} + \cdots + c_{Bm}x_{Bm} = \mathbf{c}_B\mathbf{x}_B \tag{8.2.5}$$

z_B, the inner product of $\mathbf{c}_B$ (row vector) and $\mathbf{x}_B$ (column vector), is the value of the objective function, corresponding to the B.F.S., where the basis matrix is B. z_B does not depend on non-basic variables (as each of them of zero).

Now as $(\beta_1, \beta_2, \cdots, \beta_m)$ are linearly independent, then all the column vectors $\mathbf{a}_j$ can be expressed as the linear combination of $\beta_1, \beta_2, \cdots, \beta_m$.

Let

$$\mathbf{a}_j = \beta_1 y_{1j} + \beta_2 y_{2j} + \cdots + \beta_m y_{mj} = B\mathbf{y}_j \tag{8.2.6}$$

the inner product of B and $\mathbf{y}_j$, where $\mathbf{y}_j$ is an m-component column vector given by

$$\mathbf{y}_j = [y_{1j}, y_{2j}, \cdots, y_{mj}]$$

From (8.2.6) we get

$$\mathbf{y}_j = B^{-1}\mathbf{a}_j. \tag{8.2.7}$$

Net evaluation: Evaluation is the inner product of the row vector $\mathbf{c}_B$ and the column vector $\mathbf{y}_j$ which is usually denoted by z_j and z_j is given by

$$z_j = \mathbf{c}_B\mathbf{y}_j = \mathbf{c}_B B^{-1}\mathbf{a}_j = c_{B1}y_{1j} + c_{B2}y_{2j} + \cdots + c_{Bm}y_{mj} \tag{8.2.8}$$

and $z_j - c_j$ is called as *net evaluation*.

The function $z_j - c_j$ $[j = 1, 2, \cdots, n]$ (net evaluation) plays a very important role in determining the optimal stage, in the case of solving the L.P.P. by the simplex method.

If the coefficient matrix A contains m unit column vectors which are linearly independent, then this set of vectors constitute a basis matrix. Let $\mathbf{e}_1, \mathbf{e}_2, \cdots, \mathbf{e}_i, \cdots, \mathbf{e}_m$ be m independent unit vectors of the coefficient matrix, all of which may not be placed consecutively in the ascending order of $i (i = 1, 2, \cdots, m)$. For example, $\mathbf{e}_1, \mathbf{e}_2$ and $\mathbf{e}_3$ may occur at 5th, 7th and 3rd column of the matrix A, respectively. But still if we consider a basis matrix B in the form given by $B = B(\mathbf{e}_1, \mathbf{e}_2, \cdots, \mathbf{e}_m)$ then B is an *identity matrix.* Hence the components of the solution set corresponding to the basic variables are $x_{Bi} = b_i$ $(i = 1, 2, \cdots, m)$. But if the unit vector $\mathbf{e}_i$ of the basis matrix B (identity matrix) occur at the jth column of the coefficient matrix A, then x_j, the jth component of the B.F.S. is equal to x_{Bi} and c_j, the coefficient of x_j in the obejctive function is equal to c_{Bi} which is nothing but the coefficient of x_{Bi}. For example, if $\mathbf{e}_1$ occurs at 5th column, then $x_{B1} = x_5 = b_1$ and $c_{B1} = c_5$. Similarly, if $\mathbf{e}_2$ occurs at 3rd column then $x_{B2} = x_3 = b_2$ and $c_{B2} = c_3$ and so on.

If B is an identity matrix then from (8.2.7)

$$\mathbf{y}_j = B^{-1}\mathbf{a}_j = \mathbf{a}_j \cdots, \tag{8.2.9}$$

i.e., the vector $\mathbf{y}_j$ is nothing but the column vector $\mathbf{a}_j$ due to this transformation.

N.B. (1) In the simplex method, generally all equations are to be adjusted in such a way that there exist m independent unit vectors $\mathbf{e}_1, \mathbf{e}_2, \cdots, \mathbf{e}_m$ in the coefficient matrix which together constitute the initial basis matrix and $\mathbf{b} \geq \mathbf{0}$ so that $B^{-1}\mathbf{b} = I_m^{-1}\mathbf{b} = \mathbf{b} \geq \mathbf{0}$. So the identity matrix is taken to be initial basis to start the simplex procedure. But this is not strictly necessary from the theoretical point of view which has been discussed in the solved Example 8.2.1.

(2) For a basis matrix B, $\mathbf{x}_B = B^{-1}\mathbf{b}$ and the value of the objective function corresponding to $\mathbf{x}_B$ is $z_B = \mathbf{c}_B\mathbf{x}_B$. The value of the objective function does not depend on the values of the non-basic variables (each of them is zero). Hence for calculation of the value of the objective function z, $\mathbf{x}_B = B^{-1}\mathbf{b}$ may be considered as a B.F.S.

▶ **Example 8.2.1** *Find a basic feasible solution, if there be any, of the following set of linearly independent equations and if such solution exists, taking that basis as an admissible basis, calculate all* $\mathbf{y}_j$, $z_j - c_j$ $[j = 1, 2, \cdots, 4]$ *and the value of the objective function corresponding to that B.F.S.*

Maximize, $z = 2x_1 - 4x_2 - x_3 + 4x_4$ *subject to*

$$\begin{aligned} 3x_1 - 5x_2 + x_3 - 2x_4 &= 7 \\ 6x_1 - 10x_2 - x_3 + 5x_4 &= 11, \quad x_j \geq 0,\ j = 1, 2, \cdots, 4. \end{aligned}$$

Solution: $\mathbf{a}_1 = \begin{bmatrix} 3 \\ 6 \end{bmatrix}$, $\mathbf{a}_2 = \begin{bmatrix} -5 \\ -10 \end{bmatrix}$, $\mathbf{a}_3 = \begin{bmatrix} 1 \\ -1 \end{bmatrix}$, $\mathbf{a}_4 = \begin{bmatrix} -2 \\ 5 \end{bmatrix}$ and $\mathbf{b} = \begin{bmatrix} 7 \\ 11 \end{bmatrix}$. One square matrix $(\mathbf{a}_1, \mathbf{a}_2) = \begin{bmatrix} 3 & -5 \\ 6 & -10 \end{bmatrix}$, taking two at a time from the four vectors $\mathbf{a}_1, \cdots, \mathbf{a}_4$ and det of the matrix is zero. Hence the above square matrix cannot be considered as a basis matrix. Now another square matrix $B = (\mathbf{a}_1, \mathbf{a}_3) = \begin{bmatrix} 3 & 1 \\ 6 & -1 \end{bmatrix}$ and det $B = -9 \neq 0$. Hence B can be considered as a basis matrix and basic solution corresponding to the basis B, is $\mathbf{x}_B = B^{-1}\mathbf{b} = \frac{1}{-9}\begin{bmatrix} -1 & -1 \\ -6 & 3 \end{bmatrix}\begin{bmatrix} 7 \\ 11 \end{bmatrix} = \begin{bmatrix} 2 \\ 1 \end{bmatrix} \geq \mathbf{0}$. The solution is feasible and hence B can be taken as an *admissible basis*.

Thus $\mathbf{x}_B = [x_{B1}, x_{B2}] = [x_1, x_3] = [2, 1]$.

$c = (c_1, c_2, c_3, c_4) = (2, -4, -1, 4), \mathbf{c}_B = (c_{B1}, c_{B2}) = (c_1, c_3) = (2, -1)$, the cost components corresponding to the basic variables x_1 and x_3.

Now

$$\begin{aligned}[\mathbf{y}_1, \mathbf{y}_2, \mathbf{y}_3, \mathbf{y}_4] &= B^{-1}[\mathbf{a}_1, \mathbf{a}_2, \mathbf{a}_3, \mathbf{a}_4] \\ &= \frac{1}{-9}\begin{bmatrix} -1 & -1 \\ -6 & 3 \end{bmatrix}\begin{bmatrix} 3 & -5 & 1 & -2 \\ 6 & -10 & -1 & 5 \end{bmatrix} \\ &= \begin{bmatrix} 1 & -5/3 & 0 & 1/3 \\ 0 & 0 & 1 & -3 \end{bmatrix}.\end{aligned}$$

Hence

$$\mathbf{y}_1 = \begin{bmatrix} 1 \\ 0 \end{bmatrix}, \mathbf{y}_2 = \begin{bmatrix} -5/3 \\ 0 \end{bmatrix}, \mathbf{y}_3 = \begin{bmatrix} 0 \\ 1 \end{bmatrix} \text{ and } \mathbf{y}_4 = \begin{bmatrix} 1/3 \\ -3 \end{bmatrix},$$

Then

$$\mathbf{y}_1 = \begin{bmatrix} y_{11} \\ y_{21} \end{bmatrix} = \begin{bmatrix} 1 \\ 0 \end{bmatrix}, \quad \mathbf{y}_2 = \begin{bmatrix} y_{12} \\ y_{22} \end{bmatrix} = \begin{bmatrix} -5/3 \\ 0 \end{bmatrix},$$

$$\mathbf{y}_3 = \begin{bmatrix} y_{13} \\ y_{23} \end{bmatrix} = \begin{bmatrix} 0 \\ 1 \end{bmatrix} \text{ and } \mathbf{y}_4 = \begin{bmatrix} y_{14} \\ y_{24} \end{bmatrix} = \begin{bmatrix} 1/3 \\ -3 \end{bmatrix}.$$

$$\begin{aligned}
z_1 - c_1 &= \mathbf{c}_B\mathbf{y}_1 - c_1 = c_{B1}y_{11} + c_{B2}y_{21} - c_1 = 2 \times 1 + (-1) \times 0 - 2 = 0 \\
z_2 - c_2 &= \mathbf{c}_B\mathbf{y}_2 - c_2 = c_{B1}y_{12} + c_{B2}y_{22} - c_2 = 2 \times (-5/3) + (-1) \times 0 + 4 = 2/3 \\
z_3 - c_3 &= \mathbf{c}_B\mathbf{y}_3 - c_3 = c_{B1}y_{13} + c_{B2}y_{23} - c_3 = 2 \times 0 + (-1) \times 1 - (-1) = 0 \\
z_4 - c_4 &= \mathbf{c}_B\mathbf{y}_4 - c_4 = c_{B1}y_{14} + c_{B2}y_{24} - c_4 = 2 \times 1/3 + (-1) \times (-3) - 4 = -1/3 \\
z_0 &= \mathbf{c}_B\mathbf{x}_B = c_{B1}x_{B1} + c_{B2}x_{B2} = 2 \times 2 + (-1) \times 1 = 3
\end{aligned}$$

Note: It is important to note that $\mathbf{y}_1$ and $\mathbf{y}_3$ corresponding to the basis vectors $\mathbf{a}_1$ and $\mathbf{a}_3$ are unit vectors and they can be arranged to form a unit matrix and $z_1 - c_1 = z_3 - c_3 = 0$ corresponding to the basis vectors $\mathbf{a}_1$ and $\mathbf{a}_3$. [See theorem 8.5.1 and the Appendix]

8.3 The Theoretical Development of the Simplex Method or Simplex Algorithm

The theoretical development of the method is very difficult to understand for the students at this stage. We shall just discuss the working principle of the method. Detailed discussion is given in the Appendix. Interested students may go through it.

Let a given maximization problem be

$$\text{Maximize,} \quad z = c_1x_1 + c_2x_2 + \cdots + c_jx_j + \cdots + c_nx_n$$

subject to

$$\begin{array}{ccccccccccc} a_{11}x_1 &+& a_{12}x_2 &+& \cdots &+& a_{1j}x_j &+ \cdots +& a_{1n}x_n &=& b_1 \\ a_{21}x_1 &+& a_{22}x_2 &+& \cdots &+& a_{2j}x_j &+ \cdots +& a_{2n}x_n &=& b_2 \\ \cdots & & & & & & & & & & \\ a_{i1}x_1 &+& a_{i2}x_2 &+& \cdots &+& a_{ij}x_j &+ \cdots +& a_{in}x_n &=& b_i \\ \cdots & & & & & & & & & & \\ a_{m1}x_1 &+& a_{m2}x_2 &+& \cdots &+& a_{mj}x_j &+ \cdots +& a_{mn}x_n &=& b_m \end{array}$$

where $x_j \geq 0$, $j = 1, 2, \cdots, n$ $(n > m)$.

Some of the variables may be slack or surplus variables.

In matrix notation, the problem is

$$\text{Maximize,} \quad z = \mathbf{cx}$$

subject to $\mathbf{a}_1x_1 + \mathbf{a}_2x_2 + \cdots + \mathbf{a}_jx_j + \cdots + \mathbf{a}_nx_n = \mathbf{b}$.

[The vector $\mathbf{a}_j$; $j = 1, 2, \cdots, n$ are called *activity vectors*], where

$$\begin{aligned} \mathbf{c} &= (c_1, c_2, \cdots, c_j, \cdots, c_n) \to \text{ a row vector} \\ \mathbf{x} &= [x_1, x_2, \cdots, x_j, \cdots, x_m] \to \text{ a column vector} \\ \mathbf{a}_j &= [a_{1j}, a_{2j}, \cdots, a_{ij}, \cdots, a_{mj}] \to \text{ a column vector} \\ \mathbf{b} &= [b_1, b_2, \cdots, b_j, \cdots, b_m] \to \text{ a column vector.} \end{aligned} \tag{8.3.1}$$

8.3.1 Search for a basis which will give a B.F.S.

Out of the vectors $\mathbf{a}_1$, $\mathbf{a}_2$, $\cdots$, $\mathbf{a}_j$, $\cdots$, $\mathbf{a}_n$, we shall have to select arbitrarily m vectors which are linearly independent (there exists always at least one set of such vectors, since the equations are linearly independent) which form a basis matrix B. With that basis, find out the basic solution and let us assume that the B.S., $\mathbf{x}_B = B^{-1}\mathbf{b} \geq \mathbf{0}$, i.e., the solution is a B.F.S. Such basis is called an *admissible basis* to start the simplex method.

In all practical problems, there always exists an identity matrix I_m and $\mathbf{b} \geq \mathbf{0}$ and

$$\mathbf{x}_B = I_m^{-1}\mathbf{b} = I_m\mathbf{b} = \mathbf{b} \geq \mathbf{0}. \tag{8.3.2}$$

So it will not be difficult to find an initial B.F.S.

Let the B.F.S.

$$\mathbf{x}_B = [x_{B1}, x_{B2}, \cdots, x_{Bm}]$$

which are some m variables corresponding to the m activity vectors belongs to the basis B. Let the cost components of the variables $[x_{B1}, x_{B2}, \cdots, x_{Bm}]$ be $(c_{B1}, c_{B2}, \cdots, c_{Bm})$ which are some m components belongs to $\mathbf{c}$. Thus corresponding to the basic variables.

$$\mathbf{c}_B = (c_{B1}, c_{B2}, \cdots, c_{Bm}),$$

an m component row vector.

Now we calculate the column vectors $\mathbf{y}_1, \mathbf{y}_2, \cdots, \mathbf{y}_j, \cdots, \mathbf{y}_n$, where

$$\mathbf{y}_j = [y_{1j}, y_{2j}, \cdots, y_{ij}, \cdots, y_{mj}]$$

by using the formula

$$\mathbf{y}_j = B^{-1}\mathbf{a}_j, \quad [j = 1, 2, \cdots, n] \tag{8.3.3}$$

Since in all practical purposes, the initial basis is I_m, thus

$$\mathbf{y}_j = I_m^{-1}\mathbf{a}_j = I_m\mathbf{a}_j = \mathbf{a}_j.$$

Thus initially,

$$y_{ij} = a_{ij} \tag{8.3.4}$$

But this is not essential, which has been shown in a problem (8.5.1) to understand the real theory lies behind this method.

Now we find out

$$z_j = \mathbf{c}_B\mathbf{y}_j = \mathbf{c}_B B^{-1}\mathbf{a}_j.$$

Thus,

$$z_j = c_{B1}y_{1j} + c_{B2}y_{2j} + \cdots + c_{Bi}y_{ij} + \cdots + c_{Bm}y_{mj} \tag{8.3.5}$$

and the value of the objective function corresponding to the basis B which is denoted by $z_B = z_o$ given by

$$z_o = c_{B1}x_{B1} + c_{B2}x_{B2} + \cdots + c_{Bi}x_{Bi} + \cdots + c_{Bm}x_{Bm} \tag{8.3.6}$$

Now we calculate all

$$z_j - c_j, \quad [j = 1, 2, \cdots, n] \quad \text{by using (8.3.5).}$$

8.3.2 Optimality test

For a maximization problem, if at any stage, all $z_j - c_j \geq 0$ $[j = 1, 2, \cdots, n]$ the problem is at the optimal stage. If at least one $z_j - c_j < 0$, then the problem is not at the optimal stage and we shall have to proceed further. If at least one $z_j - c_j < 0$ and at least one $y_{ij} > 0$, then the value of the objective function can be improved further or at least remains same. If any $z_j - c_j < 0$ and all $y_{ij} \leq 0$ $[i = 1, 2, \cdots, m]$ the problem has no finite optimal value and the problem is said to have an unbounded solution. [Actually all the data are to be placed in a table and the format of the simplex table will be placed in some later stage.]

8.3.3 Selection of vector to enter in the next basis

If the above condition is satisfied, i.e., at least one $z_j - c_j < 0$ and at least one $y_{ij} > 0$ $[i = 1, 2, \cdots, m]$ then we shall have to select a new basis. Thus one new vector is to be selected from $\mathbf{a}_j$ which is not in the previous basis to replace a vector from the previous basis to form a new basis. The following method is to be followed to select a vector to enter in the new basis.

If

$$z_k - c_k = \min(z_j - c_j, z_j - c_j < 0), \tag{8.3.7}$$

then $\mathbf{a}_k$ is the vector is to enter in the new basis and the kth column of the simplex table is called the *key column* or *pivot column*. If the selection is not unique, we can select any vector out of them arbitrarily and the column corresponding to that vector is the key column, the vector $\mathbf{a}_k$ will replace one vector from the previous basis.

8.3.4 Selection of vector which will leave the previous basis and be replaced by $\mathbf{a}_k$

If $\mathbf{a}_k$ is the vector to enter in the new basis, then the rule of selection of the vector to leave the current basis is given below.

If

$$\min_i \left(\frac{x_{Bi}}{y_{ik}}, y_{ik} > 0 \right) = y_{rk}, \tag{8.3.8}$$

then the vector in the rth position of the basis will be replaced by $\mathbf{a}_k$. rth row of the table is called the *key row* and y_{rk} which is called the *key element,* always positive, lies on the intersection of the kth column and the rth row. If the value of r is not unique, i.e.,

$$\min_i \left(\frac{x_{Bi}}{y_{ik}}, y_{ik} > 0 \right) \tag{8.3.9}$$

occur for more than one value of i, then select any value of such i as r and in that case next solution will be definitely degenerate and the vector in the rth position of the previous basis B will be replaced by $\mathbf{a}_k$ to form a new basis B_1 which will give a new basic feasible solution and here we get more than one basis all of which give basic feasible solutions. All these can be done from the table very easily.

8.3.5 Future Procedure

For the basis B_1, the basic solution $\mathbf{x}'_B = B_1^{-1}\mathbf{b}$ will be also feasible, i.e., $\mathbf{x}'_B \geq \mathbf{0}$.

Now, for B_1, we require to calculate $\mathbf{y}'_j = B_1^{-1}\mathbf{a}_j$.

$z'_j = \mathbf{c}'_B\mathbf{x}'_B$ $[\mathbf{c}'_B$ will be formed by replacing c_r by $c_k]$ and finally $z'_j - c_j$ corresponding to the basis B_1 and all these can be done by using the formula given below.

From the simplex table, the quantities of interest can be calculated easily.

If we take $\mathbf{x}_B = \mathbf{y}_0$, i.e., $x_{Bi} = y_{i0}$ $[i = 1, 2, \cdots, m]$ and $z_0 = \mathbf{c}_B\mathbf{x}_B = y_{m+1}, 0$ and $z_j - c_j = y_{m+1}, j$ $[j = 1, 2, \cdots, n]$.

then the rule of transformation from the present table to the next is given below in a compact form.

$z_j - c_j$ can also be calculated by using the formula

$$z_j - c_j = \mathbf{c}_B \mathbf{y}_j - c_j = c_{B1}y_{1j} + c_{B2}y_{2j} + \cdots + c_{Bm}y_{mj} - c_j. \qquad (8.3.10)$$

But in the initial table the value of $z_j - c_j$ can only be calculated by using the formula given in (8.3.10).

8.4 Rule of Construction of a new table from the previous table

Rule of transformation: Divide the elements y_{rj} $[j = 0, 1, \cdots, n]$ of the rth row of the previous table by y_{rk} $[y^*_{rk}$ = key element of the previous table] to get the rth row of next or current table. Thus

$$y'_{rj} = \frac{y_{rj}}{y^*_{rk}}, \quad i = r, \quad j = [0, 1, 2, \cdots, n]$$

and to get the elements of the other rows of the next table, we have

$$y'_{rj} = y_{ij} - y_{ik}\frac{y_{rj}}{y^*_{rk}} = \frac{y_{ij}y^*_{rk} - y_{rj}y_{ik}}{y^*_{rk}}$$

$$= \left| \begin{matrix} y_{ij} & y_{rj} \\ (1) & (2) \\ y_{ik} & y^*_{rk} \end{matrix} \right| \Big/ y^*_{rk},$$

$$i = 1, 2, \cdots, m, m+1, i \neq r. \qquad (8.4.1)$$

8.5 Nature of the Problems

In solving a L.P.P. by any method (here we consider the simplex method) generally following cases arise.

1. *The problem will be solved and we get the finite value of the objective function with finite solution set.*
2. *Unbounded value of the objective function (No finite value of the objective function exists).*

   To solve a maximization problem, if at any stage, at least one $z_j - c_j < 0$ and all $y_{ij} \leq 0$, $[i = 1, 2, \cdots, m]$, the finite value of the objective function does not exist and the problem is said to have an unbounded solution.
3. *Multiple optimal solutions:* If for a problem, optimal value remains same but there exist more than one solution sets, then we say that the multiple optimal solutions or alternative optimal solutions exist. The condition for the existence of multiple optimal solutions is that, at the optimal stage, at least one $z_j - c_j = 0$ corresponding to a non-basis vector. We know that all $z_j - c_j = 0$ corresponding to the basis vectors at each stage. The proof is beyond the scope of the book (Given in the Appendix).

4. The problem will have no feasible solutions. This case will be discussed in the next chapter (Big M method).

5. *Solution of a minimization problem* will be discussed later on.

Table 8.5.1 : Simplex Table (at any stage)

		$\mathbf{c}$	c_1	c_2	$\cdots$	c_k		c_j	c_n	
Basis Vectors	$\mathbf{c}_B$	$\mathbf{b}$	$\mathbf{a}_1$	$\mathbf{a}_2$	$\cdots$	$\mathbf{a}_k$	$\cdots$	$\mathbf{a}_j$	$\mathbf{a}_n$	$\dfrac{x_{Bi}}{y_{ik}}, y_{ik} > 0$
β_1	c_{B1}	$x_{B1} = y_{10}$	y_{11}	y_{12}		y_{1k}		y_{1j}	y_{1n}	
β_2	c_{B2}	$x_{B2} = y_{20}$	y_{21}	y_{22}		y_{2k}		y_{2j}	y_{2n}	
$\vdots$	$\vdots$	$\vdots$	$\vdots$	$\vdots$		$\vdots$		$\vdots$	$\vdots$	
β_r^*	c_{Br}	$x_{Br} = y_{r0}$	y_{r1}	y_{r2}		y_{rk}	$\cdots$	y_{rj}	y_{rn}	
$\vdots$	$\vdots$	$\vdots$	$\vdots$	$\vdots$		$\vdots$		$\vdots$		
β_m	c_{Bm}	$x_{Bm} = y_{m0}$	y_{m1}	y_{m2}		y_{mk}		y_{mj}	y_{mn}	
$z_j - c_j$		$z = y_{m+1 \cdot 0}$	$z_1 - c_1$ $= y_{m+1 \cdot 1}$	$z_2 - c_2$ $= y_{m+1 \cdot 2}$		$z_k - c_k$ $= y_{m+1 \cdot k}$		$z_j - c_j$ $= y_{m+1 \cdot j}$	$z_n - c_n$ $= y_{m+1 \cdot n}$	

Explanation: In the first column, lists the vectors of the current basis, in the 2nd column is the current associated cost vector. In the third column under $\mathbf{b}$, are the values of the current basic variables. Column vectors under $\mathbf{a}_j$ list the current values of $\mathbf{y}_j$. Last row $\{(m+1)$th row$\}$ gives the current value of z and $z_j - c_j$ $[j = 1, 2, \cdots, n]$. If y_{rk} be the key element then in the next table $\mathbf{a}_k$ will replace β_r, the vector in the rth position of the basis, to form a new basis and c_k will replace c_{Br} to form new associated cost vector. The vector which will be replaced by $\mathbf{a}_k$ is marked with asterisk. rth row is the key row and kth column under $\mathbf{a}_k$ is the key column.

From the format of the simplex table, one pertinent question must arises in mind. If $\mathbf{x}_B$ and $\mathbf{y}_j$ $[j = 1, \cdots, n]$ are placed under $\mathbf{b}$ and $\mathbf{a}_j$, respectively, then why shall we write $\mathbf{b}$ and $\mathbf{a}_j$ on the heading line of the format table instead of $\mathbf{x}_B$ and $\mathbf{y}_j$? The answer is very simple. After the proper choice of the *admissible basis* matrix in each stage, $\mathbf{x}_B$ and $\mathbf{y}_j$ at that stage can be obtained pre-multiplying the basis inverse with $\mathbf{b}$ and $\mathbf{a}_j$ $[j = 1, \cdots, n]$, where $\mathbf{b}$ and $\mathbf{a}_j$ remain unchanged in each and every iteration. That is why, on the heading line, notations $\mathbf{b}$ and $\mathbf{a}_j$ vectors have been kept intact instead of $\mathbf{x}_B$ and $\mathbf{y}_j$ $[j = 1, \cdots, n]$ which will generally change in each iteration. From the above table we can easily understand.

(i) Which vectors are in the current basis? (ii) Which vector will leave the basis? (iii) Which vector will enter in the next basis? The unit $\mathbf{y}_j$ vectors corresponding to the basis vectors are not the vectors of the basis at all.

Rule of construction of the next table from the initial table: If the initial B.F.S. is not an optimal solution, we shall have to proceed further to get a new B.F.S. which may optimize the objective function. To get the new B.F.S. we shall have to construct a new table from the initial simplex table. From (8.4.1) we get the new B.F.S. and the changed values y'_{ij}. If we transform the initial simplex table following the transformation (8.4.1), we at once get $\mathbf{x}'_B, \mathbf{y}'_j$, value of the objective function and $z'_j - c_j$ from the next table. Once again we re-state the formula of transformation. The rule of transformation is given below.

Rule of transformation: Divide the elements y_{rj} $[j = 0, 1, \cdots, n]$ of rth row (key row) of the initial table by $\mathbf{y}_{rk}$ = key element] to get the rth row of the next table. Thus

$$y'_{rj} = \frac{y_{rj}}{y^*_{rk}}, \quad i = r, j = (0, 1, 2, \cdots, n)$$

and to get the elements of the other rows of the next table, we have

$$y'_{ij} = y_{ij} - y_{ik}\frac{y_{rj}}{y^*_{rk}} = \frac{y_{ij}y^*_{rk} - y_{rj}y_{ik}}{y^*_{rk}}$$

$$= \left. \begin{matrix} y_{ij} & y_{rj} \\ 1 & 2 \\ & \\ y_{ik} & y^*_{rk} \end{matrix} \right| \Big/ y^*_{rk} \quad \text{obtained from (8.4.1)}$$

$$i = 1, 2, \cdots, m+1, i \neq r.$$

The elements of $(m+1)$th row represents the value of z, which are z_0 or z_B and $z_j - c_j$.

Since computers use a very large + integer instead of M, but here M, in Big–M method thus here it is better to use the formula

$z_j - c_j = \mathbf{c}_B\mathbf{y}_j - c_j$ to calculate $z_j - c_j$ in each table in Big–M method.

General theoretical procedure in solving a L.P.P. using simplex theory. (Note that this method will not be used in practice in future but has been solved to give an exact idea behind the simplex theory).

► **Example 8.5.1** *Solve the L.P.P. (by using general theory).*

$$\textit{Maximize, } z = x_1 - x_2 + 2x_3 + 3x_4$$

subject to

$$\begin{aligned} 2x_1 + x_2 + 3x_3 + 2x_4 &= 11 \\ 3x_1 - 3x_2 + 5x_3 + x_4 &= 17, \quad x_j \geq 0,\ j = 1, 2, 3, 4. \end{aligned}$$

Solution: **Step 1:** Search for a basis which will produce a feasible solution.

$$A = \begin{bmatrix} 2 & 1 & 3 & 2 \\ 3 & -3 & 5 & 1 \end{bmatrix}, \quad R(A) = 2.$$

Thus the two equations are linearly independent and consistent.

$$\mathbf{a}_1 = \begin{bmatrix} 2 \\ 3 \end{bmatrix}, \ \mathbf{a}_2 = \begin{bmatrix} 1 \\ -3 \end{bmatrix}, \ \mathbf{a}_3 = \begin{bmatrix} 3 \\ 5 \end{bmatrix}, \ \mathbf{a}_4 = \begin{bmatrix} 2 \\ 1 \end{bmatrix} \text{ and } \mathbf{b} = \begin{bmatrix} 11 \\ 17 \end{bmatrix}$$

$$\mathbf{c} = (c_1, c_2, c_3, c_4) = (1, -1, 2, 3).$$

Two vectors are required to form a basis.

First consider a square matrix

$$B = (\mathbf{a}_1, \mathbf{a}_2) = \begin{bmatrix} 2 & 1 \\ 3 & -3 \end{bmatrix}, \quad \det B = -9 \neq 0.$$

Hence B can be considered as a basis.

$$B^{-1} = -\frac{1}{9}\begin{bmatrix} -3 & -1 \\ -3 & 2 \end{bmatrix}.$$

The basic solution corresponding to the basis B is

$$\mathbf{x}_B = B^{-1}\mathbf{b} = -\frac{1}{9}\begin{bmatrix} -3 & -1 \\ -3 & 2 \end{bmatrix}\begin{bmatrix} 11 \\ 17 \end{bmatrix} = \begin{bmatrix} 50/9 \\ -1/9 \end{bmatrix}.$$

Here the solution is not feasible. Thus B cannot be considered as an *admissible basis* to start with the simplex procedure.

Now consider the square matrix

$$B = (\mathbf{a}_1, \mathbf{a}_3) = \begin{bmatrix} 2 & 3 \\ 3 & 5 \end{bmatrix}, \quad \det B = 1 \neq 0.$$

Hence $B = \begin{bmatrix} 2 & 3 \\ 3 & 5 \end{bmatrix}$ is also a basis. The B.S. corresponding to the new basis B is

$$\mathbf{x}_B = B^{-1}\mathbf{b} = \frac{1}{1}\begin{bmatrix} 5 & -3 \\ -3 & 2 \end{bmatrix}\begin{bmatrix} 11 \\ 17 \end{bmatrix} = \begin{bmatrix} 4 \\ 1 \end{bmatrix}$$

which is feasible.

Then $\mathbf{x}_B = [x_{B1}, x_{B2}] = [x_1, x_3] = [4, 1]$ and $B = \begin{bmatrix} 2 & 3 \\ 3 & 5 \end{bmatrix}$ can be considered as an *admissible basis* to start with the simplex procedure.

Step 2: Calculation of $\mathbf{y}_j$ and $z_j - c_j$ $[j = 1, 2, 3, 4]$ corresponding to the basis $B = \begin{bmatrix} 2 & 3 \\ 3 & 5 \end{bmatrix}$. We have

$$\begin{aligned}[\mathbf{y}_1, \mathbf{y}_2, \mathbf{y}_3, \mathbf{y}_4] &= B^{-1}(\mathbf{a}_1, \mathbf{a}_2, \mathbf{a}_3, \mathbf{a}_4) \\ &= \begin{bmatrix} 5 & -3 \\ -3 & 2 \end{bmatrix}\begin{bmatrix} 2 & 1 & 3 & 2 \\ 3 & -3 & 5 & 1 \end{bmatrix} = \begin{bmatrix} 1 & 14 & 0 & 7 \\ 0 & -9 & 1 & -4 \end{bmatrix}.\end{aligned}$$

Thus,

$$\mathbf{y}_1 = \begin{bmatrix} y_{11} \\ y_{21} \end{bmatrix} = \begin{bmatrix} 1 \\ 0 \end{bmatrix}, \quad \mathbf{y}_2 = \begin{bmatrix} y_{12} \\ y_{22} \end{bmatrix} = \begin{bmatrix} 14 \\ -9 \end{bmatrix},$$

$$\mathbf{y}_3 = \begin{bmatrix} y_{13} \\ y_{23} \end{bmatrix} = \begin{bmatrix} 0 \\ 1 \end{bmatrix} \text{ and } \mathbf{y}_4 = \begin{bmatrix} y_{14} \\ y_{24} \end{bmatrix} = \begin{bmatrix} 7 \\ -4 \end{bmatrix},$$

Here $B = (\mathbf{a}_1, \mathbf{a}_3) = (\beta_1, \beta_2)$ say. Thus $\mathbf{c}_B = (c_{B1}, c_{B2}) = (c_1, c_3) = (1, 2)$ the cost components corresponding to the basic variables x_1 and x_3.

$$z_1 - c_1 = \mathbf{c}_B\mathbf{y}_1 - c_1 = c_{B1}y_{11} + c_{B2}y_{21} = 1 \times 1 + 2 \times 0 - 1 = 0$$
$$z_2 - c_2 = \mathbf{c}_B\mathbf{y}_2 - c_2 = c_{B1}y_{12} + c_{B2}y_{22} = 1 \times 14 + 2 \times (-9) + 1 = -3$$
$$z_3 - c_3 = \mathbf{c}_B\mathbf{y}_3 - c_3 = c_{B1}y_{13} + c_{B2}y_{23} = 1 \times 0 + 2 \times 1 - 2 = 0$$
$$z_4 - c_4 = \mathbf{c}_B\mathbf{y}_4 - c_4 = c_{B1}y_{14} + c_{B2}y_{24} = 1 \times 7 + 2 \times (-4) - 3 = -4$$
$$z_0 = \mathbf{c}_B\mathbf{x}_B = c_{B1}x_{B1} + c_{B2}x_{B2} = 1 \times 4 + 2 \times 1 = 6.$$

Conclusion: Here $z_2 - c_2$ and $z_4 - c_4$ are both negative with at least one of y_{i2} and $y_{i4} > 0$ $(i = 1, 2)$. Thus the B.F.S. $\mathbf{x}_B = [x_{B1}, x_{B2}] = [x_1, x_3] = [4, 1]$ corresponding to the basis

$$B = (\mathbf{a}_1, \mathbf{a}_3) = \begin{bmatrix} 2 & 3 \\ 3 & 5 \end{bmatrix}$$

is not an optimal solution of the problem and we shall have to proceed further to obtain an optimal solution [if it exists at all].

Step 3: Search for a new vector to enter in the next basis and the vector which will leave the current basis $(\mathbf{a}_1, \mathbf{a}_3)$.

$$\min_j(z_j - c_j, z_j - c_j < 0) = \min(z_2 - c_2, z_4 - c_4)$$
$$= \min(-3, -4) = -4$$

which occurs for $j = 4$.

Then the fourth vector $\mathbf{a}_4$ will enter in the next basis. Thus, the fourth column is the key column. Again,

$$\min_i\left(\frac{x_{Bi}}{y_{i4}},\ y_{i4} > 0\right) = \min\left(\frac{x_{B1}}{y_{14}}, \frac{x_{B2}}{y_{24}},\ y_{i4} > 0\right)$$
$$= \min\left(\frac{4}{7}, \cdots\right),$$
$$\left[\frac{x_{B2}}{y_{24}} = \frac{1}{-4} \text{ is not considered as } y_{24} = -4 < 0\right]$$

which occurs for $i = 1$. Then first row is the key row and $y_{14} = 7$ is the key element. Then the first vector $\beta_1 = \mathbf{a}_1$ of the previous basis will leave the basis and replaced by $\mathbf{a}_4 = \beta_1^*$ (say) to form a new basis and the new basis will be

$$B = (\mathbf{a}_4, \mathbf{a}_3) = (\beta_1^*, \beta_2) \text{ say } = \begin{bmatrix} 2 & 3 \\ 1 & 5 \end{bmatrix}$$

and $y_{14} = 7$ which is the element at the point of intersection of key row [first row] and the key column [fourth column] is the key element.

Now we solve the problem by using simplex tables.

Simplex tables

		c	$c_1=1$	$c_2=-1$	$c_3=2$	$c_4=3$	
Basis	c_B	b	$\mathbf{a}_1$	$\mathbf{a}_2$	$\mathbf{a}_3$	$\mathbf{a}_4$	$x_{Bi}/y_{i4}, y_{i4}>0$
$\mathbf{a}_1$*	$c_{B1}=1$	$x_{B1}=4$	$y_{11}=1$	$y_{12}=14$	$y_{13}=0$	$y^*_{14}=7^*$	$x_{B1}/y_{14}=4^*/7$
$\mathbf{a}_3$	$c_{B2}=2$	$x_{B2}=1$	$y_{21}=0$	$y_{22}=-9$	$y_{23}=1$	$y_{24}=-4$	...
	z_j-c_j	$z_0=6$	$z_1-c_1=0$	$z_2-c_2=-3$	$z_3-c_3=0$	$z_4-c_4=-4^*$	
$\mathbf{a}_4$	$c_{B1}=3$	$x_{B1}=4/7$	$y_{11}=1/7$	$y_{12}=2$	$y_{13}=0$	$y_{14}=1$	
$\mathbf{a}_3$	$c_{B2}=2$	$x_{B2}=23/7$	$y_{21}=4/7$	$y_{22}=-1$	$y_{23}=1$	$y_{24}=0$	
	z_j-c_j	$z_0=58/7$	$z_1-c_1=4/7$	$z_2-c_2=5$	$z_3-c_3=0$	$z_4-c_4=0$	

All the quantities of interest are marked with asterisk.

Construction of the second table from the first table. First row is the key row and $y_{14} = 7$ is the key element and $\mathbf{a}_4$ will remove $\mathbf{a}_1$ to form a new basis $B = (\mathbf{a}_4, \mathbf{a}_3)$.

8.5.1 Calculation of the first row

Devide all the elements of the key row including x_{B1} by 7 (key elements) to get the first row of the second table.

8.5.2 Calculation of third row, $(z_j - c_j)$ row

$$
\begin{aligned}
z_0 &= y_{30} = \frac{7 \times 6 - (-4) \times 4}{7} = \frac{58}{7}, \\
z_1 - c_1 &= y_{31} = \frac{7 \times 0 - (-4) \times 1}{7} = \frac{4}{7}, \\
z_2 - c_2 &= y_{32} = \frac{7 \times (-3) - (-4) \times 14}{7} = 5, \\
z_3 - c_3 &= y_{33} = 0, \\
z_4 - c_4 &= y_{34} = 0,
\end{aligned}
$$

all $z_j - c_j \geq 0,\ j = 1, 2, 3, 4$ thus we reach at the optimal stage. We need not calculate the second row. But we calculate the second row to verify the problem in another method.

8.5.3 Calculation of the second row

$$
\begin{aligned}
y_{20} &= x_{B2} = \frac{7 \times 1 - (-4) \times 4}{7} = \frac{23}{7}, \\
y_{21} &= \frac{7 \times 0 - (-4) \times 1}{7} = \frac{4}{7},
\end{aligned}
$$

$$y_{22} = \frac{7 \times (-9) - (-4) \times 14}{7} = -1,$$
$$y_{23} = 1,$$
$$y_{24} = 0.$$

Of course, we can calculate the $z_j - c_j$ row after completing the body of the second table by using the formula $z_j - c_j = \mathbf{c}_B \mathbf{y}_j - c_j$ and the value of the objective function need not calculate at each stage.

$$z_0 = 3 \times \frac{4}{7} + 2 \times \frac{23}{7} = \frac{58}{7} = \max z,$$
$$z_1 - c_1 = 3 \times \frac{1}{7} + 2 \times \frac{4}{7} - 1 = \frac{4}{7},$$
$$z_2 - c_2 = 3 \times 2 + 2 \times (-1) + 1 = 5,$$
$$z_3 - c_3 = 0,$$
$$z_4 - c_4 = 0.$$

Thus max $z = 58/7$ at $x_1 = 0$, $x_2 = 0$, $x_3 = 23/7$, $x_4 = 4/7$.

Remark: (1) For B.F.S. $\mathbf{x}_B = [x_1, x_3] = [4, 1]$, the value of the objective function is

$$z_B = \mathbf{c}_B \mathbf{x}_B = (1, 2)[4, 1] = 6 \leq \frac{58}{7} = \max z.$$

(2) $B = (\mathbf{a}_1, \mathbf{a}_3)$ is accepted as an *initial admissible basis* to start the simplex procedure. Of course, other basis which can produce a feasible solution can also be taken as an initial admissible basis. For example, $B = (\mathbf{a}_4, \mathbf{a}_3)$ which produces a feasible solution can be taken as an initial admissible basis and in that case *step 2* will give the optimal value of the problem.

(3) In each stage, the *vectors* $\mathbf{a}_1, \mathbf{a}_2, \cdots, \mathbf{a}_4, \mathbf{b}$ *do not change.* The changes those take place are; replacement of one vector of the basis by another vector which is not in the basis to form a new basis, the corresponding changes of $\mathbf{y}_j$ and B.F.S. $\mathbf{x}_B = B^{-1}\mathbf{b}$ etc. To get $\mathbf{y}_j$ and $\mathbf{x}_B$, we only require to pre-multiply B^{-1} with $\mathbf{a}_j$ and $\mathbf{b}$, where B is the basis at that stage. In this problem, the first admissible basis is $B_1 = (\mathbf{a}_1, \mathbf{a}_2)$. The second basis is $B_2 = (\mathbf{a}_4, \mathbf{a}_3)$ which is obtained by replacing $\mathbf{a}_1$ of the previous basis by a vector $\mathbf{a}_4 \in A$, which is not in the previous basis. Thus to form a new basis from the old one, replacement takes place but no change of the vectors in the basis takes place.

(4) It is important to note that $\mathbf{y}_j$ corresponding to the basis vectors in each stage are unit vectors and they are such that, they can be arranged to form a unit matrix and $z_j - c_j = 0$ corresponding to all basis vectors. In **step 2.** $\mathbf{y}_1$ and $\mathbf{y}_3$ are unit vectors and $[\mathbf{y}_1, \mathbf{y}_3] = \begin{bmatrix} 1 & 0 \\ 0 & 1 \end{bmatrix} = I_2$ and $z_1 - c_1 = z_3 - c_3 = 0$.

Similarly in **step 3,** $\mathbf{y}_3$ and $\mathbf{y}_4$ are unit vectors and $[\mathbf{y}_4, \mathbf{y}_3] = \begin{bmatrix} 1 & 0 \\ 0 & 1 \end{bmatrix} = I_2$ and $z_4 - c_4 = z_3 - c_3 = 0$ etc.

(5) This problem has been solved in Example (6.1.2) by using the *fundamental theorem of L.P.P.* making an assumption that an optimal solution exists. But here we need not make any such assumption.

Relation between of the maximum value and minimum value of the objective function.

Theorem 8.5.1 *Minimum value of z is the negative of the maximum $(-z)$ with the same solution set, i.e.,* $\min z = -\max(-z)$ *with the same solution set.*

Proof: $z = \mathbf{cx}$. Let z attain its minimum at $\mathbf{x} = \mathbf{x}_0$ then $\min z = \mathbf{cx}_0$.

Hence

$$\mathbf{cx} \geq \mathbf{cx}_0, \quad \text{or,} \quad -\mathbf{cx} \leq -\mathbf{cx}_0.$$

Therefore,

$$\max(-\mathbf{cx}) = -\mathbf{cx}_0, \quad \text{or,} \quad \mathbf{cx}_0 = -\max(-\mathbf{cx}),$$

$$\text{or,} \quad \min(z) = -\max(-z) \tag{8.5.1}$$

with the same solution set. Similarly,

$$\max(z) = -\min(-z). \tag{8.5.2}$$

It can be easily verified that in each iteration the column vectors $\mathbf{y}_j$ corresponding to the basis vectors $\mathbf{a}_j$ (which are the basis vectors at that stage) are unit vectors.

Proof: Let at any iteration the basis vectors be $\mathbf{a}_1, \mathbf{a}_2, \cdots, \mathbf{a}_m$. Then the basis at that stage is $B = (\mathbf{a}_1, \mathbf{a}_2, \cdots, \mathbf{a}_m)$. We have $\mathbf{y}_j = B^{-1}\mathbf{a}_j$. Therefore, $\mathbf{y}_j$ corresponding to the vectors $\mathbf{a}_1, \mathbf{a}_2, \cdots, \mathbf{a}_m$ are given by

$$(\mathbf{y}_1, \mathbf{y}_2, \cdots, \mathbf{y}_m) = B^{-1}(\mathbf{a}_1, \mathbf{a}_2, \cdots, \mathbf{a}_m),$$

$$\text{or,} \quad (\mathbf{y}_1, \mathbf{y}_2, \cdots, \mathbf{y}_m) = B^{-1}B = I_m. \tag{8.5.3}$$

Hence the column vectors $\mathbf{y}_j$ corresponding to the basis vectors $\mathbf{a}_j$ are always unit vectors and they can be arranged so as to form a unit matrix. And finally, if the initial basis be unit basis I_m then $\mathbf{y}_j$ vectors under the initial unit basis vectors gives the basis inverse in each iteration because

$$\mathbf{y}_j(j = 1, 2, \cdots, m) = B^{-1}I_m = B^{-1}. \tag{8.5.4}$$

8.6 Determination of the initial B.F.S. to a L.P.P.

Case 1: When all constraints are connected with sign "$\leq$" and all $b_i \geq 0$. First of all convert all constraint inequations into equations by adding slack variables one to each of the inequations. If there are m constraint inequations with initial n variables $x_1, x_2, \cdots, x_n$ the converted equations contain $(n+m)$ variables of which m variables are slack variables.

The converted equations are

$$\begin{array}{l}
a_{11}x_1 + a_{12}x_2 + \cdots + a_{1n}x_n + x_{n+1} = b_1 \\
a_{21}x_1 + a_{22}x_2 + \cdots + a_{2n}x_n + \cdots + x_{n+2} = b_2 \\
\cdots\cdots\cdots\cdots\cdots\cdots\cdots\cdots\cdots\cdots\cdots\cdots \\
a_{m1}x_1 + a_{m2}x_2 + \cdots + a_{mn}x_n + \cdots + \cdots + x_{n+m} = b_m
\end{array}$$

where $x_{n+1}, x_{n+2}, \cdots, x_{n+m}$ are m-slack variables.

The coefficient matrix A of the above set of equations is

$$\begin{bmatrix} a_{11} & a_{12} & \cdots & a_{1n} & \overbrace{1 \quad 0\cdots 0}^{m \text{ slack vectors}} \\ a_{21} & a_{22} & \cdots & a_{2n} & 0 \quad 1\cdots 0 \\ \cdots & \cdots & \cdots & \cdots & \cdots \\ a_{m1} & a_{m2} & \cdots & a_{mn} & 0 \quad 0\cdots 1 \end{bmatrix}$$

It is a $m \times (n+m)$ matrix.

The column vectors associated with slack variables are known as *slack vectors*. All slack vectors are unit vectors which are linearly independent. Hence the m slack vectors constitute a basis matrix B which is always true for this type of problem. $B = (\mathbf{e}_1, \mathbf{e}_2, \cdots, \mathbf{e}_m)$. The B.F.S. $\mathbf{x}_B = B^{-1}\mathbf{b} = \mathbf{b} \geq \mathbf{0}$. But as the unit vector $\mathbf{e}_i$ occurs at the $(n+i)$th position, then basic variable $x_{Bi} = x_{n+i} = b_i$ and hence the B.F.S. is given by

$$\mathbf{x}_B = [x_{n+1}, x_{n+2}, \cdots, x_{n+m}] = [b_1, b_2, \cdots, b_m] \geq \mathbf{0}. \tag{8.6.1}$$

The B.F.S. $\mathbf{x}_B$ is taken as initial B.F.S. in solving the problem by simplex method.

Case 2: When the constraints are of mixed type connected with signs "$\leq$" "$\geq$" and "=".

Initially make all b_i's $(i = 1, 2, \cdots, m)$ positive by suitable adjustment. Then convert all the inequations into equations by introducing slack or surplus variables one in each constraint (whenever necessary). The column vectors associated with the surplus variables are known as *surplus vectors.* It is important to note that in this type of problem initially coefficient matrix does not necessarily contain a basis matrix which is a unit matrix. Let us consider a numerical example. Let the constraints be

$$\begin{aligned} 2x_1 + 3x_2 - x_3 &\leq 4 \\ x_1 + x_2 + 4x_3 &\geq 7 \\ x_1 - 3x_2 + 2x_3 &= 8 \end{aligned}$$

Here all b_i's are already positive. After the addition of slack variable x_4 to first constraint and surplus variable x_5 to the second constraint, the converted equations are

$$\begin{aligned} 2x_1 + 3x_2 - x_3 + x_4 \phantom{{}- x_5} &= 4 \\ x_1 + x_2 + 4x_3 \phantom{{}+ x_4} - x_5 &= 7 \\ x_1 - 3x_2 + 2x_3 + &= 8 \end{aligned}$$

The coefficient matrix A is given by

$$A = \begin{bmatrix} 2 & 3 & -1 & 1 & 0 \\ 1 & 1 & 4 & 0 & -1 \\ 1 & -3 & 2 & 0 & 0 \end{bmatrix} \quad \text{and} \quad \mathbf{b} = \begin{bmatrix} 4 \\ 7 \\ 8 \end{bmatrix} \geq \mathbf{0}.$$

There is no unit basis matrix in A. A new technique has been adopted which is known as *artificial variable technique* to get an unit basis matrix from the coefficient matrix. A set of variables are to be added again to the left hand side of the equations (in each equation whenever necessary) to get a unit matrix out of the coefficient matrix. These variables are known as the *artificial variables* and the column vectors associated with the artificial variables are known as *artificial vectors.* But care should be taken to ensure that a minimum number of artificial variables be inserted to get a unit basis matrix. In the above example the constraint equations are

$$\begin{aligned} 2x_1 + 3x_2 - x_3 + x_4 &= 4 \\ x_1 + x_2 + 4x_3 - x_5 + x_6 &= 7 \\ x_1 - 3x_2 + 2x_3 + x_7 &= 8 \end{aligned}$$

after the introduction of the artificial variables x_6 and x_7. The coefficient matrix is then

$$\begin{bmatrix} 2 & 3 & -1 & 1 & 0 & 0 & 0 \\ 1 & 1 & 4 & 0 & -1 & 1 & 0 \\ 1 & -3 & 2 & 0 & 0 & 0 & 1 \end{bmatrix}$$

which contains a unit basis matrix $B = (\mathbf{e}_1, \mathbf{e}_2, \mathbf{e}_3)$. The fourth column vector is a slack vector and sixth and seventh column vectors are artificial vectors. Now the basic feasible solution (artificial) is given by

$$\mathbf{x}_B = B^{-1}\mathbf{b} = \mathbf{b} = [x_{B1}, x_{B2}, x_{B3}] = [x_4, x_6, x_7] = [4, 7, 8] \tag{8.6.2}$$

as $\mathbf{e}_1, \mathbf{e}_2$ and $\mathbf{e}_3$ occur at fourth, sixth and seventh column respectively. Applying this method to numerical examples, the initial B.F.S. can be determined. The artificial variables are essentially non-negative quantities. The method of solving this type of problems will be discussed later [by Big. M-method].

Note: No artificial variable is added to the first equation.

8.7 Advantages and disadvantages of using unit matrix as the initial basis matrix

We can find the initial B.F.S. at once due to the presence of unit basis provided $\mathbf{b} \geq \mathbf{0}$. Moreover, we can determine the value of y_{ij} for all i and j easily for the initial simplex table because $\mathbf{y}_j = B^{-1}\mathbf{a}_j = \mathbf{a}_j$, i.e., $y_{ij} = a_{ij}$ in the initial simplex table. But the artificial vectors included in the unit basis create some difficulties in solving the problem which we shall discuss later.

Note: In simplex method, the initial B.F.S. can be determined by other method when initial basis is not a unit basis which we have discussed previously in Example 4.3.1 and Example 4.3.2. But this method is too difficult and after determining the initial B.F.S. and the vectors $\mathbf{y}_j$, using the relation $\mathbf{y}_j = B^{-1}\mathbf{a}_j$, the starting simplex table can be constructed and the problem can be solved. But this method is rarely used in practice [See Example 8.5.1].

8.8 Construction of the initial simplex table

We have stated earlier, that to get an initial B.F.S. easily, the constraints are to be converted in such a way that a unit basis matrix B is present in the coefficient matrix with all $b_i \geq 0$. Then, initial B.F.S.

$$\mathbf{x}_B = B^{-1}\mathbf{b} = \mathbf{b} \geq \mathbf{0} \quad \text{and} \quad z = \mathbf{c}_B\mathbf{x}_B = \mathbf{c}_B\mathbf{b}$$

(initial value of the objective function) and

$$\mathbf{y}_j = B^{-1}\mathbf{a}_j = \mathbf{a}_j.$$

Then $y_{ij} = a_{ij}$ for all values of i and j. Hence the elements y_{ij} can be determined easily from the coefficient matrix. Now

$$z_j - c_j = \mathbf{c}_B B^{-1}\mathbf{a}_j - c_j = \mathbf{c}_B\mathbf{y}_j - c_j = \mathbf{c}_B\mathbf{a}_j - c_j.$$

With these data an initial simplex table (chart) can be constructed according to table given in table (8.5.1).

8.9 Computational procedure (when m slack vectors constitute the initial basis $B = I_m$ (problem of maximization) and $\mathbf{b} \geq 0$

1. Convert all constraints into equations by adding m slack vectors one to each constraint.
2. Readjust the objective function accordingly.
3. Calculate the I.B.F.S. $\mathbf{x}_B = B^{-1}\mathbf{b} = \mathbf{b}$, associated cost vector $\mathbf{c}_B$, value of the objective function $z_B = \mathbf{c}_B\mathbf{x}_B = \mathbf{c}_B\mathbf{b}$, $\mathbf{y}_j = B^{-1}\mathbf{a}_j = \mathbf{a}_j$ and $z_j - c_j = \mathbf{c}_B\mathbf{y}_j - c_j = \mathbf{c}_B\mathbf{a}_j - c_j$ for all j.
4. With these data construct the initial simplex table as given in table (8.5.1).
5. If all $z_j - c_j \geq 0$, the initial B.F.S. is optimal and thus the value of the objective function will be the last element of the column headed by **b**. If at least one $z_j - c_j < 0$, the solution is not optimal, then proceed further to construct the next table to get the next B.F.S. which may optimize the objective function. For this, the selection of key element y_{rk} is essential. [If $z_k - c_k$ is minimum (negative), the vector $\mathbf{a}_k$ will enter in the next basis, replacing the vector in the rth position under the column *vector in the basis* and c_k will replace c_{Br}. If minimum is not unique select any one of them. The value of r is to be determined by min ratio rule which has been discussed previously in (8.3.8) and (8.3.9).]
6. Construct the next table following the rule of transformation given in (8.4.1). The value of z and $z_j - c_j$ are given in the last row of the table. If $z_j - c_j \geq 0$ for all j the B.F.S. is optimal. If at least one $z_j - c_j < 0$, then proceed further to construct a new table to get an optimal solution and so on until the conditions $z_j - c_j \geq 0$ are satisfied for all columns. At the optimal stage, the optimal value of the objective function is the last element under the column headed by **b**. Of course, we may calculate the optimal value of the objective function by using the relation $z = \mathbf{c}_B\mathbf{x}_B$. [All values are to be calculated from the final table]. The value of the objective function obtained from these two methods will be same. But this does not ensure the correctness of the solution of the problem. But by using duality theory we may verify the correctness of the solution with a fair degree of accuracy.

7. At any stage if $z_j - c_j < 0$ for at least one column with $y_{ij} \leq 0$ for all i, no finite value of the objective function exists.
   The problem is said to have an unbounded solution. We need not calculate the value of the objective function.
8. If none of $z_j - c_j < 0$ but some $z_j - c_j = 0$ corresponding to non-basis vectors, multiple solutions exist.
9. All discussions are related to the problem maximization of the objective function z. For a minimization problem find out the maximum value of $(-z = z')$ and from mini-max theorem min $z = -\max(-z) = -\max z'$ with the same solution set. Hence the problem can be solved by the method above. Of course, minimization problem can be solved as a minimization problem and in that case the vector to be entered in the basis corresponding to positive maximum of $z_j - c_j$ and at the optimal stage all $z_j - c_j \leq 0$ which has been shown in the worked out Example (8.11.1).

8.10 Solution to a L.P.P. when m slack vectors constitute a unit basis and $b_i \geq 0$ for all i.

▶ **Example 8.10.1** *Solve the L.P.P.*

$$\textit{maximize,} \quad z = 5x_1 + 2x_2 + 2x_3$$

subject to

$$\begin{aligned} x_1 + 2x_2 - 2x_3 &\leq 30 \\ x_1 + 3x_2 + x_3 &\leq 36, \quad x_1, x_2, x_3 \geq 0. \end{aligned}$$

Solution: This is a maximization problem. $b_i \geq 0$ for $i = 1, 2$ and the constraints involved with sign '$\leq$'.

Introducing two slack variables x_4 and x_5 one to each constraint, we get the following converted equations

$$\begin{aligned} x_1 + 2x_2 - 3x_3 + x_4 \quad &= 30 \\ x_1 + 3x_2 + x_3 \quad + x_5 &= 36\,, \end{aligned}$$

$x_1, x_2, x_3 \geq 0, x_4, x_5 \geq 0$ (slack variables).

The adjusted objective function $z = 5x_1 + 2x_2 + 2x_3 + 0x_4 + 0x_5$

$$\mathbf{c} = (5, 2, 2, 0, 0).$$

The coefficient of variables constitute the column vectors,

$$\mathbf{a}_1 = \begin{bmatrix} 1 \\ 1 \end{bmatrix}, \quad \mathbf{a}_2 = \begin{bmatrix} 2 \\ 3 \end{bmatrix}, \quad \mathbf{a}_3 = \begin{bmatrix} -2 \\ 1 \end{bmatrix},$$

$$\mathbf{a}_4 = \begin{bmatrix} 1 \\ 0 \end{bmatrix}, \quad \mathbf{a}_5 = \begin{bmatrix} 0 \\ 1 \end{bmatrix} \quad \text{and} \quad \mathbf{b} = \begin{bmatrix} 30 \\ 36 \end{bmatrix} \geq \mathbf{0}.$$

Here $\mathbf{a}_4(\mathbf{e}_1)$ and $\mathbf{a}_5(\mathbf{e}_2)$, the slack vectors are unit vectors which together constitute a unit basis matrix $B = I_2$ and taking this as initial basis, initial B.S. $\mathbf{x}_B = B^{-1}\mathbf{b} = I_2^{-1}\mathbf{b} = I_2\mathbf{b} = \mathbf{b} \geq \mathbf{0}$ which is feasible. Thus $B = I_2$ is the initial **admissible basis** to start the problem.

$$\mathbf{x}_B = [x_{B1}, x_{B2}] = [x_4, x_5] = [b_1, b_2] = [30, 36]$$
$$\mathbf{c}_B = (c_{B1}, c_{B2}) = (c_4, c_5) = (0, 0) = \mathbf{0}$$
$$z_B = \mathbf{c}_B\mathbf{x}_B = c_{B1}x_{B1} + c_{B2}x_{B2} = 0$$
$$\mathbf{y}_j = \mathbf{B}^{-1}\mathbf{a}_j = I_2\mathbf{a}_j = \mathbf{a}_j, \text{ i.e.,}$$
$$y_{ij} = a_{ij}$$

in the initial table $[j = 1, 2, 3, 4, 5]$ but they are no longer the coefficient of variable x_j; they are now quite different quantities.

$$z_j - c_j = \mathbf{c}_B\mathbf{y}_j - c_j = \mathbf{c}_B\mathbf{a}_j - c_j = \mathbf{0}\mathbf{a}_j - c_j = -c_j \quad [j = 1, 2, \cdots, 5].$$

[**Note:** It is obvious that $z_4 - c_4$ and $z_5 - c_5$ corresponding to the basis vectors $\mathbf{a}_4(\mathbf{e}_1)$ and $\mathbf{a}_5(\mathbf{e}_2)$ are zero].

With the help of the data given above, we can construct the initial simplex table as given in the table (8.5.1) putting column vector $\mathbf{x}_B = \mathbf{y}_0$ under $\mathbf{b}$ and $\mathbf{y}_j$ under $\mathbf{a}_j$ $[j = 1, 2, \cdots, 5]$.

Initial simplex table

		c	5	2	2	0	0	
Basis	c_B	b	a_1	a_2	a_3	$a_4(e_1)$	$a_5(e_2)$	$x_{Bi}/y_{i1}, y_{i1}>0$
a_4*	0	30	1*	2	−2	1	0	30/1=30*
a_5	0	36	1	3	1	0	1	36/1=36
z_j-c_j		z_0=0	−5*	−2	−2	0	0	

As $z_1 - c_1$, $z_2 - c_2$ and $z_3 - c_3$ are all negative and at least one component of $\mathbf{y}_1, \mathbf{y}_2$ and $\mathbf{y}_3$ are positive, therefore the present solution is not optimal. Min $(z_j - c_j, z_j - c_j < 0) = \min(z_1 - c_1 = -5,\ z_2 - c_2 = -2,\ z_3 - c_3 = -2) = -5$ which is $z_1 - c_1$ which occurs for $j = 1$. Then the first column is the key column and the first vector $\mathbf{a}_1$ is the vector to enter in the next basis by replacing one vector from $\mathbf{a}_4$ and $\mathbf{a}_5$ which form the initial basis. $z_1 - c_1 = -5$ is marked with asterisk.

$$\min_{i=1,2}\left(\frac{x_{Bi}}{y_{i1}}, y_{i1} > 0\right) = \min\left(\frac{x_{B1}}{y_{11}}, \frac{x_{B2}}{y_{21}}\right) = \min\left(\frac{30}{1}, \frac{36}{1}\right)$$
$$= 30 \text{ which occurs at } i = 1.$$

Therefore, $r = 1$ and the first row is the key row and $y_{11} = 1$ is the key element and since $r = 1$, therefore the first vector $\mathbf{a}_4$ under the column (basis) will leave the current basis and will be replaced by $\mathbf{a}_1$ and $\mathbf{a}_1, \mathbf{a}_5$ will constitute the basis in the next iteration. Vector $\mathbf{a}_4$, the key element $y_{11} = 1$ and min $\left(\frac{x_{B1}}{y_{11}}\right) = 30$ are marked with asterisk to see the quantities of interest at a glance.

Rule of construction of the second table from the first table: Initially we can calculate $z_0 = y_{30}$ and $z_j - c_j = y_{3j} (j = 1, 2, \cdots, 5)$ elements of $z_j - c_j$ row without computing all other rows. If all $z_j - c_j \geq 0$, we need not calculate all other rows. We only require to calculate all the components of the column under $\mathbf{b}$ which are the components of B.F.S. But if at least one $z_j - c_j < 0$, we shall have to complete the whole table.

Calculation

Third row:

$$\begin{aligned} z_0 &= y_{30} = \frac{1 \times 0 - (30 \times -5)}{1} = 150, \\ z_1 - c_1 &= y_{31} = 0, \\ z_2 - c_2 &= y_{32} = \frac{1 \times (-2) - 2 \times (-5)}{1} = 8, \\ z_3 - c_3 &= y_{33} = \frac{1 \times (-2) - (-5 \times -2)}{1} = -12, \\ z_4 - c_4 &= y_{34} = \frac{1 \times 0 - (-5 \times 1)}{1} = 5, \\ z_5 - c_5 &= y_{35} = 0. \end{aligned}$$

Here at least one $z_j - c_j < 0$. Thus we shall have to complete the table. Of course, after completing the first and second row of the table, we can complete the third row ($z_j - c_j$ row) by using the formula $z_j - c_j = \mathbf{c}_B \mathbf{y}_j - c_j = c_{B1} y_{1j} + c_{B2} y_{2j} - c_j$ $j = 1, 2, \cdots, 5$ and for $j = 0$, $z_0 = c_{B1} x_{B1} + c_{B2} x_{B2}$ etc. But the initial rule is followed by all modern computers.

First row (the row corresponding to the key row of the first table): The elements of the first row of the second table will be obtained by dividing all elements of the key row of the first table by $y_{11} = 1$ (key element), starting from the element $x_{B1} = 30$ under the column $\mathbf{b}$.

Second row:

$$\begin{aligned} y_{20} &= x_{B2} = \frac{1 \times 36 - 1 \times 30}{1} = 6, \\ y_{21} &= 0, \\ y_{22} &= \frac{1 \times 3 - 1 \times 2}{1} = 1, \\ y_{23} &= \frac{1 \times 1 - 1 \times (-2)}{1} = 3, \\ y_{24} &= \frac{1 \times 0 - 1 \times 1}{1} = -1, \\ y_{25} &= \frac{1 \times 1 - 0 \times 1}{1} = 1. \end{aligned}$$

The same notation x_{Bi} and y_{ij} have been used in each table. But they differ from table to table. With these data, the second table is to be made complete.

2nd Simplex table: (1st iteration)

		c	5	2	2	0	0	
Basis	c_B	b	a_1	a_2	a_3	$a_4(e_1)$	$a_5(e_2)$	Min ratio $x_{Bi}/y_{i3}, y_{i3}>0$
a_1	5	30	1	2	−2	1	0	...
a_5*	0	6	0	1	3*	−1	1	6/3=2*
z_j-c_j		150	0	8	−12*	5	0	

As $\mathbf{a}_1$ is the first basis vector, the $c_1 = 5$ will replace $c_{B1} = c_4 = 0$ to form new $\mathbf{c}_B$ in the second table.

Note: Of course we can first complete the table and finally the $z_j - c_j$ row or the third row can be calculated by using the formula.

$z_0 = \mathbf{c}_B\mathbf{x}_B$ and $z_j - c_j = \mathbf{c}_B\mathbf{y}_j - c_j$ (verify it). But using the first rule of calculation of z_0 and $z_j - c_j$ we get some advantage and there lies the sweetness of the formula and this formula will be extremely helpful if there are three or more than three constraints and this transform formula is being used in all modern computers.

In the second table, $z_3 - c_3 = -12$ with at least one $y_{i3} > 0$. Hence the solution is not optimal. $\mathbf{a}_3$ is the vector to enter in the next basis. The third column under $\mathbf{a}_3$ is the key column.

$$\min_{i=1,2}\left(\frac{x_{Bi}}{y_{i3}}, y_{i3} > 0\right) = \min\left(\cdots, \frac{x_{B2}}{y_{23}}\right) = \left(\cdots, \frac{6}{3}\right)$$

$$= 2 \text{ which occurs at } i = 2.$$

Therefore, $r = 2$ and the second row in the key row and $y_{23} = 3$ which is marked with asterisk is the key element. Key row and key column are blocked by dotted lines and all quantities of interest are marked with asterisks. Since $r = 2$, the second vector $\mathbf{a}_5$ under the column 'basis' will leave the basis will be replaced by $\mathbf{a}_3$ and the next basis will be $(\mathbf{a}_1, \mathbf{a}_3)$. As the solution is not optimal, the next table is to be made to find the optimal solution.

Rule of construction of the third table from the second table

Third row ($z_j - c_j$ row):

$$y_{30} = z_0 = \frac{3 \times 150 - 6 \times (-12)}{3} = 174,$$

$$z_1 - c_1 = y_{31} = 0,$$

$$z_2 - c_2 = y_{32} = \frac{3 \times 8 - 1 \times (-12)}{3} = 12,$$

$$z_3 - c_3 = y_{33} = 0,$$

$$z_4 - c_4 = y_{34} = \frac{3 \times 5 - (-12 \times -1)}{3} = 1,$$

$$z_5 - c_5 = y_{35} = \frac{3 \times 0 - 1 \times (-12)}{3} = 4.$$

Here all $z_j - c_j \geq 0$, $(j = 1, 2, \cdots, 5)$. Hence the third table is the optimal table and the optimal value of the objective function, $\max z = 174$ which can also be found by making the table complete.

Second row (row corresponding to the key row of the second table): Divide all elements of the second row of the second table by the key element to get the second row of the third table.

First row:

$$y_{10} = x_{B1} = \frac{3 \times 30 - 6 \times (-2)}{3} = 34,$$

$$y_{11} = 1,$$

$$y_{12} = \frac{3 \times 2 - 1 \times (-2)}{3} = \frac{8}{3},$$

$$y_{13} = 0,$$

$$y_{14} = \frac{3 \times 1 - (-2 \times -1)}{3} = \frac{1}{3},$$

$$y_{15} = \frac{3 \times 0 - (1 \times -2)}{3} = \frac{2}{3}$$

and the table is complete:

Third simplex table (2nd iteration)

		c	5	2	2	0	0
Basis	c_B	**b**	$\mathbf{a}_1$	$\mathbf{a}_2$	$\mathbf{a}_3$	$\mathbf{a}_4(\mathbf{e}_1)$	$\mathbf{a}_5(\mathbf{e}_2)$
$\mathbf{a}_1$	5	34	1	$\frac{8}{3}$	0	$\frac{1}{3}$	$\frac{2}{3}$
$\mathbf{a}_3$	2	2	0	$\frac{1}{3}$	1	$-\frac{1}{3}$	$\frac{1}{3}$
$z_j - c_j$		174	0	12	0	1	4

As $\mathbf{a}_3$ is the 2nd basis vector, $c_3 = 3$ will replace $c_{B2} = c_4 = 0$ to form new $\mathbf{c}_B$ in the third table. Now verify the result of $z_j - c_j$ row using the formula $z_j - c_j = \mathbf{c}_B \mathbf{y}_j - c_j$ $[i = 1, 2, \cdots, 5]$ and $z_0 = \mathbf{c}_B \mathbf{x}_B$. If we use the first method of calculating the $z_j - c_j$ row or the third row we even need not write $\mathbf{c}_B$ from the second table and onwards.

We have already traced that all $z_i - c_j \geq 0$ and hence the solution in optimal. Final basis is $B = (\mathbf{a}_1, \mathbf{a}_3)$. Hence the optimal value of z, $\max z = z_0 = 174$ corresponding to the B.F.S. $\mathbf{x}_B = [x_{B1}, x_{B2}] = [x_1, x_3] = [34, 2]$ be at $x_1 = 34$, $x_3 = 2$, $x_2 = 0$ (non-basic). The slack variables are not present in the B.F.S. But they may present at the optimal stage.

Check formula at each stage: If the initial basis be a unit basis and initial unit vectors be slack vectors, then there exists a check formula at each stage which is almost perfect but not with hundred percent certainty.

Procedure: The components of **b** are b_1 and b_2, where $b_1 = 30$, $b_2 = 36$. $\mathbf{a}_4(\mathbf{e}_1)$ and $\mathbf{a}_5(\mathbf{e}_2)$ form the initial unit basis which are slack vectors; $z_4 - c_4 = 0$, $z_5 - c_5 = 0$.

$(z_4 - c_4)b_1 + (z_5 - c_5)b_2 = 0 \times 30 + 0 \times 36 = 0$ which is equal to z_0 obtained from the initial table which verifies the correctness of the procedure. In the second table, $z_4 - c_4 = 5$, $z_5 - c_5 = 0$, and $(z_4 - c_4)b_1 + (z_5 - c_5)b_2 = 5 \times 30 + 0 \times 36 = 150$ which is equal to z_0 obtained from the second table and in the final table $z_4 - c_4 = 1$, $z_5 - c_5 = 4$, $(z_4 - c_4)b_1 + (z_5 - c_5)b_2 = 1 \times 30 + 4 \times 36 = 174$ which is equal to the optimal value of the objective function. Thus we may assume that the problem solved, is correct.

Note: (1) As the final basis $B = (\mathbf{a}_1, \mathbf{a}_3)$ the basic variables at the final stage are $x_{B1} = x_1 = 34$, $x_{B2} = x_3 = 2$.

(2) The vectors $\mathbf{a}_1, \mathbf{a}_2, \cdots, \mathbf{a}_5$ and $\mathbf{b}$ do not change in any iteration; only the column vectors $\mathbf{y}_j$ under the heading $\mathbf{a}_j$ and $\mathbf{x}_B = \mathbf{y}_0$ under the heading $\mathbf{b}$ change in each iteration due to the change of basis in each stage.

(3) During computation in each table, the same notation x_{Bi} is used to denote the ith component of the basic variable. Similarly the same notation y_{ij} is used to represent the element corresponding to ith row and jth column. Practically they differ from table to table.

(4) It is important to note that during each iteration, the column vectors $\mathbf{y}_j$ under the heading $\mathbf{a}_j$ (basis vectors) are always unit vectors and they can be arranged to form a unit basis and $z_j - c_j$ under the heading $\mathbf{a}_j$ (basis vectors) are zero.

(5) It is interesting to note that at each stage $\mathbf{y}_4$ and $\mathbf{y}_5$ under the slack vectors $\mathbf{a}_4(\mathbf{e}_1), \mathbf{a}_5(\mathbf{e}_2)$ constitute the basis inverse in each iteration. (Proof of which is given in theorem 8.5.1).

For example, the final basis

$$B = (\mathbf{a}_1, \mathbf{a}_3) = \begin{bmatrix} 1 & -2 \\ 1 & 1 \end{bmatrix} \quad \text{and} \quad B^{-1} = \begin{bmatrix} 1/3 & 2/3 \\ -1/3 & 1/3 \end{bmatrix}$$

which has been reflected under the vectors $\mathbf{a}_4$ and $\mathbf{a}_5$ in the third or final table.

(6) During computation of the second table and onwards, $\mathbf{c}$ and $\mathbf{c}_B$ are not essential.

▶ **Example 8.10.2** *Solve the following L.P.P. by simplex method.*

$$\textit{Maximize,} \quad z = 4x_1 + 7x_2$$

subject to

$$\begin{aligned} 2x_1 + x_2 &\le 1000 \\ x_1 + x_2 &\le 600 \\ -x_1 - 2x_2 &\ge -1000, \quad x_1, x_2 \ge 0. \end{aligned}$$

[C.U.(P)'80]

Solution. This is a maximization problem.

Multiplying the third constraint by (-1) we get $x_1 + 2x_2 \le 1000$.

Hence all $b_i \ge 0$ and all constraints are attached with sign "$\le$" type. Introducing three slack variables x_3, x_4, x_5 one to each constraint we get the following converted equations

$$\begin{aligned} 2x_1 + x_2 + x_3 &= 1000 \\ x_1 + x_2 + x_4 &= 600 \\ x_1 + 2x_2 + x_5 &= 1000 \end{aligned}$$

The adjusted objective function z is given by

$$z = 4x_1 + 7x_2 + 0 \cdot x_3 + 0 \cdot x_4 + 0 \cdot x_5, \quad \mathbf{c} = (4, 7, 0, 0, 0).$$

$$\mathbf{a}_1 = \begin{bmatrix} 2 \\ 1 \\ 1 \end{bmatrix}, \quad \mathbf{a}_2 = \begin{bmatrix} 1 \\ 1 \\ 2 \end{bmatrix}, \quad \mathbf{a}_3 = \begin{bmatrix} 1 \\ 0 \\ 0 \end{bmatrix},$$

$$\mathbf{a}_4 = \begin{bmatrix} 0 \\ 1 \\ 0 \end{bmatrix}, \quad \mathbf{a}_5 = \begin{bmatrix} 0 \\ 0 \\ 1 \end{bmatrix} \quad \text{and} \quad \mathbf{b} = \begin{bmatrix} 1000 \\ 600 \\ 1000 \end{bmatrix} \geq \mathbf{0}.$$

Here the slack vectors $\mathbf{a}_3, \mathbf{a}_4, \mathbf{a}_5$ are unit vectors $\mathbf{e}_1, \mathbf{e}_2$ and $\mathbf{e}_3$ which together constitute the unit basis $B = (\mathbf{a}_3, \mathbf{a}_4, \mathbf{a}_5) = I_3$ and taking this as initial basis we have

Initial B.S. $\mathbf{x}_B = B^{-1}\mathbf{b} = I_3^{-1}\mathbf{b} = I_3\mathbf{b} = \mathbf{b} \geq \mathbf{0}$ which is feasible.

Then B is the admissible basis and

$$\begin{aligned}
\mathbf{x}_B &= [x_{B1}, x_{B2}, x_{B3}] = [x_3, x_4, x_5] = [b_1, b_2, b_3] = [1000, 600, 1000], \\
\mathbf{c}_B &= (c_{B1}, c_{B2}, c_{B3}) = (c_3, c_4, c_5) = (0, 0, 0) = \mathbf{0}, \\
z_0 &= \mathbf{c}_B\mathbf{x}_B = \mathbf{0}\mathbf{x}_B = 0, \\
\mathbf{y}_j &= B^{-1}\mathbf{a}_j = I_3\mathbf{a}_j = \mathbf{a}_j = [j = 1, 2, 3, 4, 5], \\
z_j - c_j &= \mathbf{c}_B\mathbf{y}_j - c_j = \mathbf{0}\mathbf{y}_j - c_j = -c_j[j = 1, 2, 3, 4, 5].
\end{aligned}$$

With the data given above, we can construct the initial simplex table.

Initial simplex table:

		$\mathbf{c}$	4	7	0	0	0	
Basis	$\mathbf{c}_B$	$\mathbf{b}$	$\mathbf{a}_1$	$\mathbf{a}_2$	$\mathbf{a}_3$	$\mathbf{a}_4$	$\mathbf{a}_5$	Min. ratio
$\mathbf{a}_3$	0	1000	2	1	1	0	0	$\frac{1000}{1} = 1000$
$\mathbf{a}_4$	0	600	1	1	0	1	0	$\frac{600}{1} = 600$
$\mathbf{a}_5^*$	0	1000	1	2*	0	0	1	$\frac{1000}{2} = 500^*$
$z_j - c_j$		$z_0 = 0$	−4	−7*	0	0	0	

Construction of the 2nd table from the first table:

Fourth row ($z_j - c_j$ row):

$$z_0 = \frac{2 \times 0 - (-7 \times 1000)}{2} = 3500,$$

$$y_{41} = z_1 - c_1 = \frac{2 \times (-4) - (-7 \times 1)}{2} = -\frac{1}{2},$$

$$y_{42} = z_2 - c_2 = 0,$$

$$y_{43} = z_3 - c_3 = 0,$$

$$y_{44} = z_4 - c_4 = 0,$$

$$y_{45} = z_5 - c_5 = \frac{2 \times 0 - (-7 \times 1)}{2} = \frac{7}{2}.$$

Third row: Divide all elements of the key row (third row) of the first table by 2 to get all elements of the third row of the second table.

First row:

$$x_{B1} = y_{10} = \frac{2 \times 1000 - 1 \times 1000}{2} = 500,$$

$$y_{11} = \frac{2 \times 2 - 1 \times 1}{2} = \frac{3}{2},$$

$$y_{12} = 0,$$

$$y_{13} = \frac{2 \times 1 - 1 \times 0}{2} = 1,$$

$$y_{14} = 0,$$

$$y_{15} = \frac{2 \times 0 - 1 \times 1}{2} = -\frac{1}{2}.$$

Second row:

$$x_{B2} = y_{20} = \frac{2 \times 600 - 1 \times 1000}{2} = 100,$$

$$y_{21} = \frac{2 \times 1 - 1 \times 1}{2} = \frac{1}{2},$$

$$y_{22} = 0,$$

$$y_{23} = 0,$$

$$y_{24} = \frac{2 \times 1 - 1 \times 0}{2} = 1,$$

$$y_{25} = \frac{2 \times 0 - 1 \times 1}{2} = -\frac{1}{2}.$$

All $z_j - c_j (j = 1, 2, \cdots, 5)$ are not greater than or equal to zero. $z_1 - c_1 = -\frac{1}{2} < 0$ and all at least one of $y_{i1} \geq 0 (i = 1, 2, 3)$. Then we shall have to construct the next table.

Second simplex table:

$\mathbf{a}_2$ is the vector to enter in the next basis and $\mathbf{a}_5$ will leave the basis and will be replaced by $\mathbf{a}_2$ and $c_{B3} = c_5$ will be replaced by $c_2 = 7$.

		c	4	7	0	0	0	
Basis	$\mathbf{c}_B$	b	$\mathbf{a}_1$	$\mathbf{a}_2$	$\mathbf{a}_3$	$\mathbf{a}_4$	$\mathbf{a}_5$	Min. ratio
$\mathbf{a}_3$	0	500	$\frac{3}{2}$	0	1	0	$-\frac{1}{2}$	$500/\frac{3}{2} = \frac{1000}{3}$
$\mathbf{a}_4^*$	0	100	$\frac{1}{2}^*$	0	0	1	$-\frac{1}{2}$	$100/\frac{1}{2} = 200^*$
$\mathbf{a}_2$	7	500	$\frac{1}{2}$	1	0	0	$\frac{1}{2}$	$500/\frac{1}{2} = 1000$
$z_j - c_j$		3500	$-\frac{1}{2}^*$	0	0	0	$\frac{7}{2}$	

Construction of the third table from the second table:

Fourth row ($z_j - c_j$ row),

$$z_0 = \frac{\frac{1}{2} \times 3500 - (-\frac{1}{2} \times 100)}{\frac{1}{2}} = 3600, \quad \left[\frac{1}{2} = \text{key element}\right]$$

$$y_{41} = z_1 - c_1 = 0,$$

$$y_{42} = z_2 - c_2 = 0,$$

$$y_{43} = z_3 - c_3 = 0,$$

$$y_{44} = z_4 - c_4 = \frac{\frac{1}{2} \times 0 - (-\frac{1}{2} \times 1)}{\frac{1}{2}} = 1,$$

$$y_{45} = z_5 - c_5 = \frac{\frac{1}{2} \times \frac{7}{2} - (-\frac{1}{2} \times -\frac{1}{2})}{\frac{1}{2}} = 3.$$

Here all $z_j - c_i \geq 0$. Thus we reach at the optimal stage. We need not complete the table except to calculate x_{B1}, x_{B2} and x_{B3}. But here also we make the table complete.

Second row: Divide all elements of the second row (key row) of the second table to get the second row of the third table.

First row:

$$x_{B1} = y_{10} = \frac{\frac{1}{2} \times 500 - 100 \times \frac{3}{2}}{\frac{1}{2}} = 200,$$

$$y_{11} = 0,$$

$$y_{12} = 0,$$

$$y_{13} = \frac{\frac{1}{2} \times 1 - \frac{3}{2} \times 0}{\frac{1}{2}} = 1,$$

$$y_{14} = \frac{\frac{1}{2} \times 0 - \frac{3}{2} \times 1}{\frac{1}{2}} = -3,$$

$$y_{15} = \frac{\frac{1}{2} \times (-\frac{1}{2}) - \frac{3}{2} \times (-\frac{1}{2})}{\frac{1}{2}} = 1.$$

Third row:

$$x_{B3} = y_{30} = \frac{\frac{1}{2} \times 500 - 1 \times 100}{\frac{1}{2}} = 400,$$

$$y_{31} = 0,$$

$$y_{32} = 1,$$

$$y_{33} = 0,$$

$$y_{34} = \frac{\frac{1}{2} \times 0 - \frac{1}{2} \times 1}{\frac{1}{2}} = -1,$$

$$y_{35} = \frac{\frac{1}{2} \times \frac{1}{2} - \frac{1}{2} \times (-\frac{1}{2})}{\frac{1}{2}} = 1.$$

Third simplex table:

$\mathbf{a}_1$ is the vector to enter in the next basis by replacing $\mathbf{a}_4$ of the previous basis and $c_1 = 4$ will replace $c_{B2} = c_4 = 0$ from the previous $\mathbf{c}_B$.

Basis	c_B	b	$\mathbf{a}_1$	$\mathbf{a}_2$	$\mathbf{a}_3$	$\mathbf{a}_4$	$\mathbf{a}_5$
$\mathbf{a}_3$	0	200	0	0	1	−3	1
$\mathbf{a}_1$	4	200	1	0	0	2	−1
$\mathbf{a}_2$	7	400	0	1	0	−1	1
$z_j - c_j$		3600	0	0	0	1	3

All $z_j - c_j \geq 0$, then the solution is optimal. Hence the optimum value of the objective function z corresponding to the final basis $(\mathbf{a}_3, \mathbf{a}_1, \mathbf{a}_2)$, is given by $\max z = 3600$ at B.F.S.

$$\mathbf{x}_B = [x_3, x_1, x_2] = [200, 200, 400],$$

i.e., for $x_1 = 200$, $x_2 = 400$, the objective function of the original problem attains it maximum [$x_3 = 200$ is a slack variable, we need not consider it].

Check formula in each stage:

The initial basis is a unit basis and basis vectors are slack vectors.

Procedure: The components of **b** are $b_1 = 1000$, $b_2 = 600$, $b_3 = 1000$.
In the first table $z_3 - c_3 = 0$, $z_4 - c_4 = 0$, $z_5 - c_5 = 0$,

$$(z_3 - c_3)b_1 + (z_4 - c_4)b_2 + (z_5 - c_5)b_3 = 0\cdot$$

which is equal to z_0 in the first table.

In the second table $z_3 - c_3 = 0$, $z_4 - c_4 = 0$, $z_5 - c_5 = \frac{7}{2}$,

$$(z_3 - c_3)b_1 + (z_4 - c_4)b_2 + (z_5 - c_5)b_3 = \frac{7}{2} \times 1000 = 3500$$

which is equal to z_0 in the second table.

In the third table $z_3 - c_3 = 0$, $z_4 - c_4 = 1$, $z_5 - c_5 = 3$,

$$(z_3 - c_3)b_1 + (z_4 - c_4)b_2 + (z_5 - c_5)b_3 = 1 \times 600 + 3 \times 1000 = 3600$$

which is z_0 of third table. We may assume that the problem is solved correctly.

▶ **Example 8.10.3** *Solve by simplex method.*

$$\textit{Maximize,}\quad z = x_1 - x_2 + 3x_3$$

subject to

$$
\begin{aligned}
x_1 + x_2 + x_3 &\leq 10 \\
2x_1 \quad - x_3 &\leq 2 \\
2x_1 - 2x_2 + 3x_3 &\leq 0, \quad x_1, x_2, x_3 \geq 0.
\end{aligned}
$$

[C.U.(P)'88]

Solution. This is a maximization problem.

$b_i \geq 0$ for all i and the constraints are involved with the sign "$\leq$". Introducing three slack variables x_4, x_5, x_6 one in each constraint, we get the following converted equations

$$
\begin{aligned}
x_1 + x_2 + x_3 + x_4 \qquad &= 10 \\
2x_1 \quad - x_3 \quad + x_5 \quad &= 2 \\
2x_1 - 2x_2 + 3x_3 \qquad + x_6 &= 0
\end{aligned}
$$

The adjusted objective function z is given by

$$z = x_1 - x_2 + 3x_3 + 0 \cdot x_4 + 0 \cdot x_5 + 0 \cdot x_6.$$

Here all slack vectors together constitute a unit basis matrix $B = I_3$ and then as

$$\mathbf{b} = [10, 2, 0], \quad \mathbf{x}_B = B^{-1}\mathbf{b} \geq \mathbf{0}$$

which gives a feasible solution.

Thus initial B.F.S. $= \mathbf{x}_B = B^{-1}\mathbf{b} = \mathbf{b} = [x_{B1}, x_{B2}, x_{B3}] = [x_4, x_5, x_6]$

$$= [b_1, b_2, b_3] = [10, 2, 0].$$

Here the solution is degenerate.

[This problem can be solved by usual method; though Degeneracy occurs at the initial stage].

$$
\begin{aligned}
\mathbf{c} &= (1, -1, 3, 0, 0, 0), \quad \mathbf{c}_B = (c_4, c_5, c_6) = (0, 0, 0) = \mathbf{0} \\
\mathbf{y}_j &= B^{-1}\mathbf{a}_j = I_3^{-1}\mathbf{a}_j = \mathbf{a}_j [j = 1, 2, \cdots, 6] \\
z_B &= \text{Value of the objective function} = \mathbf{c}_B \mathbf{x}_B = 0 \\
z_j - c_j &= \mathbf{c}_B \mathbf{y}_j - c_j = \mathbf{0}\mathbf{y}_j - c_j = -c_j.
\end{aligned}
$$

With these data we shall construct the initial table.

Now without going details we shall solve the problem in a compact form.

Simplex tables:

		c	1	−1	3	0	0	0	
Basis	$\mathbf{c}_B$	b	$\mathbf{a}_1$	$\mathbf{a}_2$	$\mathbf{a}_3$	$\mathbf{a}_4(\mathbf{e}_1)$	$\mathbf{a}_5(\mathbf{e}_2)$	$\mathbf{a}_6(\mathbf{e}_3)$	Min. ratio
$\mathbf{a}_4$	0	10	1	1	1	1	0	0	$\frac{10}{1} = 10$
$\mathbf{a}_5$	0	2	2	0	−1	0	1	0	……
$\mathbf{a}_6^*$	0	0	2	−2	3*	0	0	1	$\frac{0}{3} = 0^*$
$z_j - c_j$		0	−1	1	−3*	0	0	0	
$\mathbf{a}_4^*$	0	10	$\frac{1}{3}$	$\frac{5}{3}^*$	0	1	0	$-\frac{1}{3}$	$10/\frac{5}{3} = 6^*$
$\mathbf{a}_5$	0	2	$\frac{8}{3}$	$-\frac{2}{3}$	0	0	1	$\frac{1}{3}$	……
$\mathbf{a}_3$	3	0	$\frac{2}{3}$	$-\frac{2}{3}$	1	0	0	$\frac{1}{3}$	……
$z_j - c_j$		0	1	−1*	0	0	0	1	
$\mathbf{a}_2$		6							
$\mathbf{a}_5$		6							
$\mathbf{a}_3$		4							
$z_j - c_j$		6	$\frac{6}{5}$	0	0	$\frac{3}{5}$	0	$\frac{4}{5}$	

As all $z_j - c_j \geq 0$ $[j = 1, 2, \cdots, 6]$ in the third table, then the third table is the optimal table and we need not complete the third table. We require to calculate only the column, under the vector **b** which gives the optimal B.F.S. and the optimal value of z. Thus $\max z = 6$ at $x_2 = 6$, $x_5 = 6$, $x_3 = 4$, i.e., for $x_1 = 0$ (non-basic), $x_2 = 6$, $x_3 = 4$ the original problem attains its maximum.

Note. (1) In the first table $\frac{x_{B2}}{y_{23}}$ is not calculated as $y_{23} < 0$. Similarly in the second table $\frac{x_{B2}}{y_{22}}$ and $\frac{x_{B3}}{y_{32}}$ are not calculated as $y_{22}, y_{32} \leq 0$.

Calculation of z_B, elements of $z_j - c_j$ row and $\mathbf{x}_B$

$$y_{40} = z_B = \frac{\frac{5}{3} \times 0 - 10 \times (-1)}{\frac{5}{3}} = 6,$$

$$y_{41} = z_1 - c_1 = \frac{\frac{5}{3} \times 1 - \frac{1}{3} \times (-1)}{\frac{5}{3}} = \frac{6}{5},$$

$$y_{42} = z_2 - c_2 = 0,$$

$$y_{43} = z_3 - c_3 = 0,$$

$$y_{44} = z_4 - c_4 = \frac{\frac{5}{3} \times 0 - 1 \times (-1)}{\frac{5}{3}} = \frac{3}{5},$$

$$y_{45} = z_5 - c_5 = 0,$$

$$y_{46} = z_6 - c_6 = \frac{\frac{5}{3} \times 1 - (-\frac{1}{3}) \times (-1)}{\frac{5}{3}} = \frac{4}{5},$$

$$x_{B1} = \frac{10}{\frac{5}{3}} = 6,$$

$$x_{B2} = \frac{\frac{5}{3} \times 2 - 10 \times (-\frac{2}{3})}{\frac{5}{3}} = 6,$$

$$x_{B3} = \frac{\frac{5}{3} \times 0 - 10 \times (-\frac{2}{3})}{\frac{5}{3}} = 4.$$

We now solve a problem in a compact form.

▶ **Example 8.10.4** *Solve the L.P.P. by simplex method.*

$$\textit{Maximize,} \quad z = 4x_1 + 3x_2$$

subject to

$$\begin{aligned} 3x_1 + \ \ x_2 &\le 15 \\ 3x_1 + 4x_2 &\le 24, \quad x_1, x_2 \ge 0, \end{aligned}$$

Adding slack variables one to each constraint, the converted equations are

$$\begin{aligned} 3x_1 + \ \ x_2 + x_3 \qquad &= 15 \\ 3x_1 + 4x_2 \qquad + x_4 &= 24, \quad x_1, x_2, x_3, x_4 \ge 0. \end{aligned}$$

Simplex tables:

		c	4	3	0	0	
Basic	$\mathbf{c}_B$	$\mathbf{b}$	$\mathbf{a}_1$	$\mathbf{a}_2$	$\mathbf{a}_3(e_1)$	$\mathbf{a}_4(e_2)$	Min. ratio
$\mathbf{a}_3^*$	0	15	3*	1	1	0	$\frac{15}{3} = 5^*$
$\mathbf{a}_4$	0	24	3	4	0	1	$\frac{24}{3} = 8$
$z_j - c_j$		0	−4*	−3	0	0	
$\mathbf{a}_1$	4	5	1	$\frac{1}{3}$	$\frac{1}{3}$	0	$\frac{5}{1/3} = 15$
$\mathbf{a}_4^*$	0	9	0	3*	−1	1	$\frac{9}{3} = 3^*$
$z_j - c_j$		20	0	$-\frac{5}{3}^*$	$\frac{4}{3}$	0	
$\mathbf{a}_1$	4	4					
$\mathbf{a}_2$	3	3					
$z_j - c_j$		25	0	0	$\frac{7}{9}$	$\frac{5}{9}$	

Final basis $B = (\mathbf{a}_1, \mathbf{a}_2)$ optimal value of $z = \max z = 25$ at B.F.S. $\mathbf{x}_B = [x_{B1}, x_{B2}] = [x_1, x_2] = [4, 3]$, i.e., at $x_1 = 4$, and $x_2 = 3$.

▶ **Example 8.10.5** *Solve the L.P.P.*

$$\textit{Minimize,} \quad z = -2x_1 + 3x_2$$

subject to

$$\begin{aligned} 2x_1 - 5x_2 &\le 7 \\ 4x_1 + \ \ x_2 &\le 8 \\ 7x_1 + 2x_2 &\le 16, \quad x_1 \ge 0, x_2 \ge 0. \end{aligned}$$

Solution. The problem is a problem of minimization.

Let $z' = -z$; then $\min z = -\max(-z) = -\max z'$. Hence the problem is a problem of maximization of $z' = -z = -(-2x_1 + 3x_2) = 2x_1 - 3x_2$ and finally $\min z = -\max(z')$ with the same solution set. $b_i \geq 0$ for all i and constraints are associated with the sign "$\leq$".

Introducing three slack variables x_3, x_4 and x_5 (one in each inequation) we get the following converted equations

$$\begin{aligned} 2x_1 - 5x_2 + x_3 \quad\quad\quad &= 7 \\ 4x_1 + x_2 \quad\quad + x_4 \quad &= 8 \\ 7x_1 + 2x_2 \quad\quad\quad + x_5 &= 16. \end{aligned}$$

The adjusted objective function is $z' = 2x_1 - 3x_2 + 0.x_3 + 0.x_4 + 0.x_5$.

Here all the slack vectors are unit vectors which produce a unit basis. Initial

$$\text{B.F.S.} = \mathbf{x}_B = [x_{B1}, x_{B2}, x_{B3}] = [x_3, x_4, x_5] = [7, 8, 16]$$
$$\mathbf{c}_B = (c_{B1}, c_{B2}, c_{B3}) = (c_3, c_4, c_5) = (0, 0, 0)$$
$$z = \mathbf{c}_B\mathbf{x}_B = 0 \text{ and } \mathbf{y}_j = B^{-1}\mathbf{a}_j = \mathbf{a}_j$$
$$z_j - c_j = \mathbf{c}_B\mathbf{y}_j - c_j = \mathbf{0}\mathbf{y}_j - c_j = -c_j$$

Now with the values of $z_j - c_j$ etc. we construct the initial table and solve accordingly.

Simplex tables:

		c	2	−3	0	0	0	
Basis	$\mathbf{c}_B$	**b**	$\mathbf{a}_1$	$\mathbf{a}_2$	$\mathbf{a}_3(\mathbf{e}_1)$	$\mathbf{a}_4(\mathbf{e}_2)$	$\mathbf{a}_5(\mathbf{e}_3)$	Min. ratio
$\mathbf{a}_3$	0	7	2	−5	1	0	0	$\frac{7}{2}$
$\mathbf{a}_4^*$	0	8	4*	1	0	1	0	$\frac{8}{4} = 2^*$
$\mathbf{a}_5$	0	16	7	2	0	0	1	$\frac{16}{7}$
$z_j - c_j$		0	−2*	3	0	0	0	
$\mathbf{a}_3$	0	3	0	$-\frac{11}{2}$	1	$-\frac{1}{2}$	0	
$\mathbf{a}_1$	2	2	1	$\frac{1}{4}$	0	$\frac{1}{4}$	0	
$\mathbf{a}_5$	0	2	0	$\frac{1}{4}$	0	$-\frac{7}{4}$	1	
$z_j - c_j$		4	0	$\frac{7}{2}$	0	$\frac{1}{2}$	0	

As none of $z_j - c_j < 0$, therefore the solution is optimal.

Hence $\max z' = 4$.

Now $\min z = -\max z' = -4$. Hence the minimum value of z is -4 corresponding to the optimal basic feasible solution.

$\mathbf{x}_B = [x_3, x_1, x_5] = [3, 2, 2]$, i.e., for $x_1 = 2$, $x_2 = 0$, the objective function of the original problem attains its minimum [x_2 is a non-basic variable.].

▶ **Example 8.10.6** *Solve the L.P.P.*

$$\textit{Maximize,} \quad z = x_1 + x_2 + 3x_3$$

subject to

$$\begin{aligned} x_1 + 2x_2 - \ x_3 &\le 10 \\ 3x_2 + 2x_3 &\le \ 8 \\ x_2 + 3x_3 &\le 15\ , \quad x_1 \ge 0, x_2 \ge 0 \text{ and } x_3 \ge 0. \end{aligned}$$

Solution. $b_i \ge 0$ for all i.

Introducing three slack variables x_4, x_5 and x_6, one to each constraint we get the following equations

$$\begin{aligned} x_1 + 2x_2 - \ x_3 + x_4 \qquad\qquad &= 10 \\ 0 \cdot x_1 + 3x_2 + 2x_3 \qquad + x_5 \qquad &= \ 8 \\ 0 \cdot x_1 + \ x_2 + 3x_3 \qquad\qquad + x_6 &= 15 \end{aligned}$$

The adjusted objective function z is given by

$$z = x_1 + x_2 + 3x_3 + 0 \cdot x_4 + 0 \cdot x_5 + 0 \cdot x_6.$$

The slack vectors constitute a basis matrix which is a unit matrix. But in this problem the column vector $\mathbf{a}_1$, associated with the variable x_1 is also a unit vector $(\mathbf{e}_1)$. Hence in this problem unit basis matrix is not unique. But as the problem is a problem of maximization and the coefficient of x_1 in the objective function is a positive quantity, the initial basis matrix may be selected in such a way, that x_1 is a basic variable, i.e., the column vector $\mathbf{a}_1(\mathbf{e}_1)$ associated with x_1 be included in the initial unit basis. And due to this selection the problem may be solved quickly. Unit vector $\mathbf{e}_1$, associated with the variable x_4 be kept outside the basis matrix, i.e., x_4 is to be considered as a non-basic variable.

Therefore, initial B.F.S.

$$\mathbf{x}_B = [x_1, x_5, x_6] = [10, 8, 15]$$

$$\mathbf{c}_B = (c_1, c_5, c_6) = (1, 0, 0), \mathbf{y}_j = \mathbf{a}_j$$

$$z = \mathbf{c}_B \mathbf{x}_B = 1 \times 10 + 0 \times 8 + 0 \times 15 = 10$$

$$z_1 - c_1 = z_5 - c_5 = z_6 - c_6 = 0$$

$$z_2 - c_2 = (1, 0, 0)[2, 3, 1] - 1 = 1$$

$$z_3 - c_3 = (1, 0, 0)[-1, 2, 3] - 3 = -4$$

$$z_4 - c_4 = (1, 0, 0)[1, 0, 0,] - 0 = 1.$$

Simplex tables:

		c	1	1	3	0	0	0	
Basis	$\mathbf{c}_B$	**b**	$\mathbf{a}_1(\mathbf{e}_1)$	$\mathbf{a}_2$	$\mathbf{a}_3$	$\mathbf{a}_4(\mathbf{e}_1)$	$\mathbf{a}_5(\mathbf{e}_2)$	$\mathbf{a}_6(\mathbf{e}_3)$	Min. ratio
$\mathbf{a}_1$	1	10	1	2	−1	1	0	0	......
$\mathbf{a}_5^*$	0	8	0	3	2*	0	1	0	$\frac{8}{2} = 4^*$
$\mathbf{a}_6$	0	15	0	1	3	0	0	1	$\frac{15}{3} = 5$
$z_j - c_j$		10	0	1	−4*	1	0	0	
$\mathbf{a}_1$	1	14	1	$\frac{7}{2}$	0	1	$\frac{1}{2}$	0	
$\mathbf{a}_3$	3	4	0	$\frac{3}{2}$	1	0	$\frac{1}{2}$	0	
$\mathbf{a}_6$	0	3	0	$-\frac{7}{2}$	0	0	$-\frac{3}{2}$	1	
$z_j - c_j$		26	0	7	0	1	2	0	

As none of $z_j - c_j < 0$, therefore the solution set is optimal and the optimal value of z is 26 for the B.F.S. $\mathbf{x}_B = [x_1, x_3, x_6] = [14, 4, 3]$, i.e., for $x_1 = 14$, $x_2 = 0$ and $x_3 = 4$ the original objective function attains its maximum [x_2 is a non-basic variable.].

Now we solve a problem and observe how much the method be able to save time and labour.

▶ **Example 8.10.7** *Solve the L.P. problem by simplex method.*

$$\textit{Maximize,} \quad z = 3x_1 + 5x_2 + 4x_3$$

subject to

$$2x_1 + 3x_2 \leq 8$$

$$2x_2 + 5x_3 \leq 10$$

$$3x_1 + 2x_2 + 4x_3 \leq 15\,, \quad x_1, x_2, x_3 \geq 0.$$

[Meerut M.Sc.(Math)'84]

Solution. $\mathbf{b} = [8, 10, 15] \geq \mathbf{0}$. Thus introducing three slack variables, x_4, x_5 and x_6, one to each constraint and taking initial basis $B = (\mathbf{a}_4, \mathbf{a}_5, \mathbf{a}_6) = I_3$, we can start the initial simplex table and then solve in a compact table as shown below.

		c	3	5	4	0	0	0	
Basis	$\mathbf{c}_B$	**b**	$\mathbf{a}_1$	$\mathbf{a}_2$	$\mathbf{a}_3$	$\mathbf{a}_4(\mathbf{e}_1)$	$\mathbf{a}_5(\mathbf{e}_2)$	$\mathbf{a}_6(\mathbf{e}_3)$	Min. ratio
$\mathbf{a}_4^*$	0	8	2	3*	0	1	0	0	$\frac{8}{3}^*$
$\mathbf{a}_5$	0	10	0	2	5	0	1	0	$\frac{10}{2}=5$
$\mathbf{a}_6$	0	15	3	2	4	0	0	1	$\frac{15}{2}$
$z_j - c_j$		0	-3	-5^*	-4	0	0	0	
$\mathbf{a}_2$		$\frac{8}{3}$	$\frac{2}{3}$	1	0	$\frac{1}{3}$	0	0	......
$\mathbf{a}_5^*$		$\frac{14}{3}$	$-\frac{4}{3}$	0	5*	$-\frac{2}{3}$	1	0	$\frac{14}{3}/5=\frac{14}{15}^*$
$\mathbf{a}_6$		$\frac{29}{3}$	$\frac{5}{3}$	0	4	$-\frac{2}{3}$	0	1	$\frac{29}{3}/4=\frac{29}{12}$
$z_j - c_j$		$\frac{40}{3}$	$\frac{1}{3}$	0	-4^*	$\frac{5}{3}$	0	0	
$\mathbf{a}_2$		$\frac{8}{3}$	$\frac{2}{3}$	1	0	$\frac{1}{3}$	0	0	$\frac{8}{3}/\frac{2}{3}=4$
$\mathbf{a}_3$		$\frac{14}{15}$	$-\frac{4}{15}$	0	1	$-\frac{2}{15}$	$\frac{1}{5}$	0	......
$\mathbf{a}_6^*$		$\frac{89}{15}$	$\frac{41}{15}^*$	0	0	$-\frac{2}{25}$	$-\frac{4}{5}$	1	$\frac{89}{15}/\frac{41}{15}=\frac{89}{41}^*$
$z_j - c_j$		$\frac{256}{15}$	$-\frac{11}{15}^*$	0	0	$\frac{17}{15}$	$\frac{4}{5}$	0	
$\mathbf{a}_2$		$\frac{50}{41}$							
$\mathbf{a}_3$		$\frac{62}{41}$							
$\mathbf{a}_1$		$\frac{89}{41}$							
$z_j - c_j$		$\frac{765}{41}$	0	0	0	$\frac{45}{41}$	$\frac{24}{41}$	$\frac{11}{41}$	

In the fourth table, all $z_j - c_j \geq 0$. Thus the fourth table is the optimal table. We now only calculate the elements under the column vector **b** which gives B.F.S. and the value of the objective function. We need not complete the table.

$$\max z = \frac{765}{41} \quad \text{at} \quad x_1 = \frac{89}{41}, \quad x_2 = \frac{50}{41} \quad \text{and} \quad x_3 = \frac{62}{41}.$$

This method is extremely helpful for three or more than three constraints and for fractional cost coefficients. But here the final basis inverse will not be available. Of course, the value of the objective function needs not to be calculated in each table. It may be calculated at the optimal table only by using the formula $z_B = \mathbf{c}_B \mathbf{x}_B$.

Problem having Multiple Optimal Solutions

▶ **Example 8.10.8** *Use simplex method to solve the following L.P.P.*

$$\textit{Maximize}, \quad z = 5x_1 + 2x_2$$

subject to

$$6x_1 + 10x_2 \leq 30$$
$$10x_1 + 4x_2 \leq 20\,, \quad x_1, x_2 \geq 0.$$

[C.U. M.Com.'85]

Is the solution unique? If not, write down the convex combination of the alternative optima.

Solution: The constraints, after the addition of slack variables x_3 and x_4, one to each, are

$$\begin{aligned} 6x_1 + 10x_2 + x_3 \qquad &= 30 \\ 10x_1 + \ 4x_2 \qquad + x_4 &= 20 \ , \quad x_j \geq 0, j = 1, 2, \cdots, 4. \end{aligned}$$

The adjusted objective function $z = 5x_1 + 2x_2 + 0 \cdot x_3 + 0 \cdot x_4$.

$$\mathbf{b} = \begin{bmatrix} 30 \\ 20 \end{bmatrix} \geq \mathbf{0} \quad \text{and} \quad B = (\mathbf{a}_3, \mathbf{a}_4) = I_2 \text{ is a unit matrix.}$$

$$\mathbf{x}_B = B^{-1}\mathbf{b} = \mathbf{b} \geq \mathbf{0}.$$

Thus with the initial basis B we can start the problem

$$\mathbf{y}_j = B^{-1}\mathbf{a}_j = \mathbf{a}_j, \mathbf{c}_B = (0, 0).$$

		c	5	2	0	0	
$\mathbf{c}_B$	Basis	$\mathbf{b}$	$\mathbf{a}_1$	$\mathbf{a}_2$	$\mathbf{a}_3$	$\mathbf{a}_4$	Min. ratio
0	$\mathbf{a}_3$	30	6	10	1	0	$\frac{30}{6} = 5$
0	$\mathbf{a}_4^*$	20	10*	4	0	1	$\frac{20}{10} = 2^*$
	$z_j - c_j$	0	-5^*	-2	0	0	
0	$\mathbf{a}_3^*$	18	0	$\frac{38}{5}^*$	1	$-\frac{3}{5}$	$\frac{18}{38/5} = \frac{90}{38}^*$
5	$\mathbf{a}_1$	2	1	$\frac{2}{5}$	0	$\frac{1}{10}$	$\frac{2}{2/5} = 5$
	$z_j - c_j$	10	0	0*	0	$\frac{1}{2}$	
2	$\mathbf{a}_2$	$\frac{45}{19}$	0	1	$\frac{5}{38}$	$-\frac{3}{38}$	
5	$\mathbf{a}_1$	$\frac{20}{19}$	1	0	$-\frac{2}{19}$	$\frac{5}{38}$	
	$z_j - c_j$	10	0	0	0	$\frac{1}{2}$	

Here in the second table all $z_j - c_j \geq 0$. Hence the optimal solution has been obtained $\max z = 10$ at $x_3 = 18, x_1 = 2$, i.e., for $x_1 = 2, x_2 = 0$ (non-basic), the problem attains its maximum. But here $z_2 - c_2 = 0$ corresponding to a non-basic vector $\mathbf{a}_2$. Thus the solution is not unique. Using $\mathbf{a}_2$ as a vector to enter in the next basis we have $\max z = 10$ remains same but the optimal solution will change which has been shown from the table 3. Other optimal basic solution is $x_1 = \frac{20}{19}, x_2 = \frac{45}{19}$. We know that if there are more than one optimal solution then there exist infinite optimal solutions which will be obtained from the convex combination of the optimal solutions $\mathbf{x}_1 = [2, 0]$, $\mathbf{x}_2 = [\frac{20}{19}, \frac{45}{19}]$. Any optimal solution $\mathbf{x}$ is given by [Alternative optima]

$$\mathbf{x} = \lambda\mathbf{x}_1 + (1-\lambda)\mathbf{x}_2, 0 \le \lambda \le 1$$
$$= \lambda[2,0] + (1-\lambda)\left[\frac{20}{19}, \frac{45}{19}\right]$$

For example, if we take $\lambda = \frac{1}{2}$ then $\mathbf{x} = [1\frac{10}{19}, \frac{45}{38}]$ which is also an alternative optimal solution.

Note. $\mathbf{x}_1 = [x_1 = 2, x_2 = 0]$

Problem having an Unbounded Solution

▶ **Example 8.10.9** *Use the simplex method to solve the L.P.P.*

$$\textit{Maximize,} \quad 2x_2 + x_3$$

subject to

$$x_1 + x_2 - 2x_3 \le 7$$
$$-3x_1 + x_2 + 2x_3 \le 3, \quad x_1, x_2 \textit{ and } x_3 \ge 0 \qquad \text{[C.U.(H)'89]}$$

Solution: Adding two slack variables x_4 and x_5, one to each constraint, the constraints are

$$x_1 + x_2 - 2x_3 + x_4 = 7$$
$$-3x_1 + x_2 + 2x_3 + x_5 = 3, \quad x_1 \ge 0, j = 1, \cdots, 5.$$

and the objective function is $0x_1 + 2x_2 + x_3 + 0 \cdot x_4 + 0x_5$, $\mathbf{b} = [7, 3] \ge \mathbf{0}$ and $B = [\mathbf{a}_4, \mathbf{a}_5] = I_2$ will be the initial unit basis.

Simplex tables

		c	0	2	1	0	0	
Basis	$\mathbf{c}_B$	b	$\mathbf{a}_1$	$\mathbf{a}_2$	$\mathbf{a}_3$	$\mathbf{a}_4$	$\mathbf{a}_5$	Min. ratio
$\mathbf{a}_4$	0	7	1	1	−2	1	0	$\frac{7}{1} = 7$
$\mathbf{a}_5^*$	0	3	−3	1*	2	0	1	$\frac{3}{1} = 3^*$
$z_j - c_j$		0	0	−2*	−1	0	0	
$\mathbf{a}_4^*$	0	4	4*	0	−4	1	−1	$\frac{4}{4} = 1^*$
$\mathbf{a}_2$	2	3	−3	1	2	0	1	......
$z_j - c_j$		6	−6*	0	3	0	2	
$\mathbf{a}_1$	0	1	1	0	−1	$\frac{1}{4}$	$-\frac{1}{4}$	
$\mathbf{a}_2$	2	6	0	1	−1	$\frac{3}{4}$	$\frac{1}{4}$	
$z_j - c_j$		12	0	0	−3	$\frac{3}{2}$	$\frac{1}{2}$	

Note. (1) After completing the $z_j - c_j$ of the third table, we have seen that $z_3 - c_3 = -3 < 0$. Thus we require to complete the table. After completing, we have seen that $y_{i3} \leq 0$ for $i = 1, 2$ for which $z_3 - c_3$ is negative. Then the only conclusion is that the problem has no finite optimal value and the problem is said to have unbounded solution.

(2) For a problem having unbounded solution we cannot trace it before completing the final table.

▶ **Example 8.10.10** *Solve the L.P.P. by simplex method and prove that alternative optimal solutions exist. Find two optimal solutions.*

$$\text{Maximize,} \quad z = 2x_1 - x_2 + 3x_3 + x_4$$

subject to

$$\begin{aligned} 2x_1 + x_2 + 3x_3 + 5x_4 &\leq 12 \\ 3x_1 + 2x_2 + x_3 + 4x_4 &\leq 15 , \quad x_1, x_2, x_3 \text{ and } x_4 \geq 0. \end{aligned}$$

Simplex table

		c	2	−1	3	1	0	0	
Basis	$\mathbf{c}_B$	b	$\mathbf{a}_1$	$\mathbf{a}_2$	$\mathbf{a}_3$	$\mathbf{a}_4$	$\mathbf{a}_5(\mathbf{e}_1)$	$\mathbf{a}_6(\mathbf{e}_2)$	Min. ratio
$\mathbf{a}_5^*$	0	12	2	1	3*	5	1	0	$\frac{12}{3} = 4^*$
$\mathbf{a}_6$	0	15	3	2	1	4	0	1	$\frac{15}{1} = 15$
$z_j - c_j$		0	−2	1	−3*	−1	0	0	......
$\mathbf{a}_3$	3	4	$\frac{2}{3}$	$\frac{1}{3}$	1	$\frac{5}{3}$	$\frac{1}{3}$	0	$\frac{4}{2/3} = 6$
$\mathbf{a}_6^*$	0	11	$\frac{7}{3}^*$	$\frac{5}{3}$	0	$\frac{7}{3}$	$-\frac{1}{3}$	1	$\frac{11}{7/3} = \frac{33}{7}^*$
$z_j - c_j$		12	0*	2	0	4	1	0	
$\mathbf{a}_3$	3	$\frac{6}{7}$							
$\mathbf{a}_1$	2	$\frac{33}{7}$							
$z_j - c_j$		12	0	2	0	4	1	0	

In the second table, all $z_j - c_j \geq 0$. Therefore, we reach at the optimal stage. Then $\max z = 12$ at $x_1 = 0, x_2 = 0, x_3 = 4$ and $x_4 = 0$. Now in the table $z_1 - c_1 = 0$ corresponding to a non-basis vector $\mathbf{a}_1$. Hence alternative optimal solutions exist. Thus taking $\mathbf{a}_1$ to vector enter in the basis we get another optimal solution which is $x_1 = \frac{33}{7}$, $x_2 = 0$, $x_3 = \frac{6}{7}$, $x_4 = 0$ and $\max z = 12$.

▶ **Example 8.10.11** *Solve the L.P.P. by simplex method*

$$\text{Maximize,} \quad z = 2x_1 - 3x_2 - 2x_3 + 6x_4$$

subject to

$$\begin{aligned} 5x_1 - x_2 + 2x_3 + 6x_4 &\leq 20 \\ 2x_1 + 3x_2 + 4x_3 - 5x_4 &\leq 16 \\ x_1 + 2x_2 - 3x_3 + x_4 &\leq 2 , \quad x_1, x_2, x_3 \text{ and } x_4 \geq 0. \end{aligned}$$

Simplex tables

Basis	c_B	c / b	2 / a_1	−3 / a_2	−2 / a_3	6 / a_4	0 / a_5	0 / a_6	0 / a_7	Min. ratio
a_5	0	20	5	−1	2	6	1	0	0	$\frac{20}{6}=\frac{10}{3}$
a_6	0	16	2	3	4	−5	0	1	0	……
a_7^*	0	2	1	2	−3	1*	0	0	1	$\frac{2}{1}=2^*$
z_j-c_j		0	−2	3	2	−6*	0	0	0	
a_5^*	0	8	−1	−13	20*	0	1	0	−6	$\frac{8}{20}=\frac{2}{5}^*$
a_6	0	26	7	13	−11	0	0	1	5	……
a_4	6	2	1	2	−3	1	0	0	1	……
z_j-c_j		12	4	15	−16*	0	0	0	6	
a_3	−2	$\frac{2}{5}$								
a_6	0	$\frac{152}{5}$								
a_4	6	$\frac{16}{5}$								
z_j-c_j		$\frac{92}{5}$	$\frac{16}{5}$	$\frac{23}{5}$	0	0	$\frac{4}{5}$	0	$\frac{6}{5}$	

All $z_j - c_j \geq 0$ in the third table. Thus we reach at the optimal stage. Then

$$\max z = \frac{92}{5} \quad \text{at} \quad x_1 = 0, \quad x_2 = 0, \quad x_3 = \frac{2}{5} \quad \text{and} \quad x_4 = \frac{16}{5},$$

Verification of the result.

$$\max z = -2 \times \frac{2}{5} + 0 \times \frac{152}{5} + 6 \times \frac{16}{5} = \frac{92}{5}.$$

By using Duality theory

$$\max z = 20 \times \frac{4}{5} + 0 \times 16 + 2 \times \frac{6}{5} = \frac{92}{5}.$$

Thus the correctness of the solution has been verified.

8.11 Minimization Problems, Solved as Minimization Problems

► **Example 8.11.1** *Solve the L.P.P. by simplex method*

$$\textit{Minimize,} \quad z = x_1 - 3x_2 + 2x_3$$

subject to

$$\begin{aligned} 3x_1 - \ \ x_2 + 2x_3 &\leq 7 \\ -2x_1 + 4x_2 \qquad &\leq 12 \\ -4x_1 + 3x_2 + 8x_3 &\leq 10, \quad x_1, x_2, x_3 \geq 0. \end{aligned}$$

[C.U.(P)'81,'96,'99]

Simplex table

		c	1	−3	2	0	0	0	
Basis	$\mathbf{c}_B$	**b**	$\mathbf{a}_1$	$\mathbf{a}_2$	$\mathbf{a}_3$	$\mathbf{a}_4(\mathbf{e}_1)$	$\mathbf{a}_5(\mathbf{e}_2)$	$\mathbf{a}_6(\mathbf{e}_3)$	Min. ratio
$\mathbf{a}_4$	0	7	3	−1	2	1	0	0	......
$\mathbf{a}_5^*$	0	12	−2	4*	0	0	1	0	$\frac{12}{4} = 3^*$
$\mathbf{a}_6$	0	10	−4	3	8	0	0	1	$\frac{10}{3}$
$z_j - c_j$		0	−1	3*	−2	0	0	0	
$\mathbf{a}_4^*$	0	10	$\frac{5}{2}^*$	0	2	1	$\frac{1}{4}$	0	$\frac{10}{5/2} = 4^*$
$\mathbf{a}_2$	−3	3	$-\frac{1}{2}$	1	0	0	$\frac{1}{4}$	0	......
$\mathbf{a}_6$	0	1	$-\frac{5}{2}$	0	8	0	$-\frac{3}{4}$	1	......
$z_j - c_j$		−9	$\frac{1}{2}^*$	0	−2	0	$-\frac{3}{4}$	0	
$\mathbf{a}_1$	1	4							
$\mathbf{a}_2$	−3	5							
$\mathbf{a}_6$	0	11							
$z_j - c_j$		−11	0	0	$-\frac{12}{5}$	$-\frac{1}{5}$	$-\frac{4}{5}$	0	

All $z_j - c_j \leq 0$. Thus we reach at the optimal stage.

$$\min(z) = -11 \text{ at } x_1 = 4, \ \ x_2 = 5, \ \ x_3 = 0 \quad \text{(non-basic)}.$$

Remark. Here, to select the vector to enter in the next basis, we shall have to select kth vector in the manner,

$$z_k - c_k = \max_{j=1,\cdots,6} (z_j - c_j, z_j - c_j > 0).$$

Here, that is why we select the second vector to enter in the next basis and at the optimal stage all $z_j - c_j \leq 0$.

Exercise 8

1. Solve the L.P.P. by simplex method:

(a) Maximize, $z = 3x_1 + 2x_2$

$$\text{subject to } 2x_1 + \ \ x_2 \leq 12$$
$$6x_1 + 5x_2 \leq 40\,, \quad x_1 \geq 0, x_2 \geq 0.$$

(b) Maximize, $z = 4x_1 + x_2$

$$\text{subject to } \begin{aligned} x_1 + 2x_2 &\le 10 \\ 4x_1 + 3x_2 &\le 24, \quad x_1 \ge 0, x_2 \ge 0. \end{aligned}$$

(c) Maximize, $z = 5x_1 + 3x_2$

$$\text{subject to } \begin{aligned} 3x_1 + 5x_2 &\le 15 \\ 5x_1 + 2x_2 &\le 10, \quad x_1 \ge 0, x_2 \ge 0. \end{aligned}$$ [C.U.(P)'93]

(d) Maximize, $z = 2x_1 + 3x_2$

$$\text{subject to } \begin{aligned} 7x_1 + 4x_2 &\le 28 \\ 7x_1 + 12x_2 &\le 52, \quad x_1 \ge 0, x_2 \ge 0. \end{aligned}$$

(e) Maximize, $z = 5x_1 + 4x_2 + x_3$

$$\text{subject to } \begin{aligned} 6x_1 + x_2 + 2x_3 &\le 12 \\ 8x_1 + 2x_2 + x_3 &\le 30 \\ 4x_1 + x_2 - 2x_3 &\le 16, \quad x_1 \ge 0, x_2 \ge 0 \text{ and } x_3 \ge 0. \end{aligned}$$

(f) Maximize, $z = 10x_1 + x_2 + 2x_3$

$$\text{subject to } \begin{aligned} x_1 + x_2 - 2x_3 &\le 10 \\ 4x_1 + x_2 + x_3 &\le 20, \quad x_1, x_2, \text{ and } x_3 \ge 0. \end{aligned}$$

[C.U.(P)'87]

2. (a) Maximize, $z = 3x_1 + 2x_2$

$$\text{subject to } \begin{aligned} 5x_1 + x_2 &\le 10 \\ 4x_1 + 5x_2 &\le 60, \quad x_1 \ge 0, x_2 \ge 0. \end{aligned}$$

(b) Solve the following L.P.P. with the help of simplex method:
Maximize, $z = 2x_1 + 2x_2$

$$\text{subject to } \begin{aligned} x_1 - x_2 &\ge -1 \\ -0.5x_1 + x_2 &\le 2, \quad x_1 \ge 0, x_2 \ge 0. \end{aligned}$$

[C.U.M.Com.'86]

(c) Given the following linear programme:
Maximize, $z = 3x_1 + 4x_2$

$$\text{subject to } \begin{aligned} 2x_1 + 3x_2 &\le 120 \\ 3x_1 + 2x_2 &\le 105, \quad x_1 \ge 0, x_2 \ge 0. \end{aligned}$$

[C.U.M.Com.'88]

i. Find an optimum solution to this programme by applying simplex technique.
ii. Is the solution unique? Give reason for your answer.
iii. Is the solution degenerate? Give reason for your answer.

(d) Solve the L.P.P. by using simplex algorithm.
Maximize, $z = 2x_1 - x_2 + \frac{8}{3}x_3$

$$\begin{aligned} \text{subject to } 3x_1 + x_2 - 2x_3 &\leq 6 \\ 2x_1 + 5x_2 + x_3 &\leq 14 \\ x_1 + 4x_2 + 2x_3 &\leq 8, \quad x_1, x_2, \text{ and } x_3 \geq 0. \end{aligned}$$

Mention the optimal basis. From the optimal simplex table write down the optimal basis inverse. Verify that, by using ordinary matrix inverse method.

(e) Solve the problem:
Maximize, $z = 2x_1 + x_2 - 4x_3 + x_4$

$$\begin{aligned} \text{subject to } 2x_1 - 3x_2 + 4x_3 + x_4 &\leq 6 \\ x_1 + 2x_2 + 3x_3 + 4x_4 &\leq 8 \\ 2x_1 + x_2 + 4x_3 + 5x_4 &\leq 10, \quad x_1, x_2, x_3 \text{ and } x_4 \geq 0. \end{aligned}$$

Prove that the problem has multiple optimal solutions and prove that two of them are $x_1 = \frac{9}{2}, x_2 = 1, x_3 = 0, x_4 = 0$ and $x_1 = 4$, $x_2 = 2$, $x_3 = 0$, $x_4 = 0$ and $\max z = 10$.

3. Maximize, $z = 20x_1 + 6x_2 + 8x_3$

$$\begin{aligned} \text{subject to } 8x_1 + 2x_2 + 3x_3 &\leq 250 \\ 4x_1 + 3x_2 &\leq 150 \\ 2x_1 + x_3 &\leq 50, \quad x_1 \geq 0, x_2 \geq 0, \text{ and } x_3 \geq 0. \end{aligned}$$

4. Maximize, $z = 5x_1 + 3x_2$

$$\begin{aligned} \text{subject to } x_1 + x_2 &\leq 2 \\ 5x_1 + 2x_2 &\leq 15 \\ 3x_1 + 8x_2 &\leq 12, \quad x_1 \geq 0, x_2 \geq 0. \end{aligned}$$

5. (a) Maximize, $z = 3x_1 + 2x_2 + x_3$

$$\begin{aligned} \text{subject to } x_1 + 2x_2 - 3x_3 &\leq 8 \\ 3x_2 + x_3 &\leq 6, \quad x_1 \geq 0, x_2 \geq 0 \text{ and } x_3 \geq 0. \end{aligned}$$

(b) Maximize, $z = 3x_1 + 6x_2 + 2x_3$

$$\begin{aligned} \text{subject to } 3x_1 + 4x_2 + x_3 &\leq 2 \\ x_1 + 2x_2 + 3x_3 &\leq 1, \quad x_1, x_2 \text{ and } x_3 \geq 0. \end{aligned}$$

[C.U.(P)'87,'90]

6. Maximize, $z = 3x_1 + 5x_2 + 4x_3$

$$\begin{aligned} \text{subject to } 2x_1 + 3x_3 &\leq 18 \\ 2x_2 + 5x_3 &\leq 18 \\ 3x_1 + 2x_2 + 4x_3 &\leq 25, \quad x_1, x_2 \text{ and } x_3 \geq 0. \end{aligned}$$

7. (a) Maximize, $z = x_1 + 3x_2 + 2x_3$

$$\begin{aligned} \text{subject to } x_1 + 2x_2 \quad & \le 10 \\ 2x_1 \quad + x_3 & \le 8 \\ 2x_2 + x_3 & \le 6, \quad x_1, x_2 \text{ and } x_3 \ge 0. \end{aligned}$$

Also prove that the problem has multiple optimal solutions. Find two optimal B.F.S.

(b) Maximize, $z = 4x_1 + 10x_2$

$$\begin{aligned} \text{subject to } 2x_1 + x_2 & \le 50 \\ 2x_1 + 5x_2 & \le 100 \\ 2x_1 + 3x_2 & \le 90, \quad x_1, x_2 \ge 0. \end{aligned}$$ [Madras B.Sc.'84]

(c) Given the following L.P.P. Maximize $z = 2.5x_1 + x_2$

$$\begin{aligned} \text{subject to } 3x_1 + 5x_2 & \le 15 \\ 5x_1 + 2x_2 & \le 10, \quad x_1, x_2 \ge 0. \end{aligned}$$ [C.U.M.Com.'87]

i. Show that there exist multiple optimal solutions to the problem.

ii. Suppose the first constraint is changed to $1.5x_1 + 2.5x_2 \le 7.5$ and the remaining part of the problem remain unchanged. Will that give us a unique solution? Give reasons for your answer.

[*Hints:* The change of the constraint will not give unique solution as the constraint $1.5x_1 + 2.5x_2 \le 7.5$ is same as the constraint $3x_1 + 5x_2 \le 15$. Only the scale is being changed which does not change the nature of the solution etc.]

8. (a) Minimize, $z = 3x_1 - 2x_2$

$$\begin{aligned} \text{subject to } 4x_1 + x_2 & \le 8 \\ 2x_1 + 4x_2 & \le 20, \quad x_1, x_2 \ge 0. \end{aligned}$$

(b) Minimize, $z = -3x_1 + 4x_2$

$$\begin{aligned} \text{subject to } x_1 + x_2 & \le 8 \\ 2x_1 + 5x_2 & \le 22 \\ x_2 & \le 4, \quad x_1, x_2 \ge 0. \end{aligned}$$

9. Minimize, $z = x_1 - 3x_2 + 2x_3$

$$\begin{aligned} \text{subject to } \quad 3x_1 - x_2 + 2x_3 & \le 7 \\ -2x_1 + 4x_2 \quad & \le 12 \\ -4x_1 + 3x_2 + 8x_3 & \le 10, \quad x_1, x_2 \text{ and } x_3 \ge 0. \end{aligned}$$

[C.U.(P)'81,'96,'99]

10. Maximize, $z = 10x_1 + x_2 + 2x_3$

$$\begin{aligned} \text{subject to } \quad x_1 + x_2 - 2x_3 & \le 10 \\ 4x_1 + x_2 + x_3 & \le 20, \quad x_j \ge 0, j = 1, 2, 3. \end{aligned}$$

[C.U.(P)'86, M.Sc.(Appl.Math.)'77]

11. Maximize, $z = 2x_1 + 4x_2 + x_3 + x_4$

$$\text{subject to } \begin{aligned} x_1 + 3x_2 \quad\quad + x_4 &\le 4 \\ 2x_1 + x_2 \quad\quad\quad &\le 3 \\ x_2 + x_3 + x_4 &\le 3, \quad x_1, x_2, x_3, x_4 \ge 0. \end{aligned}$$

12. Maximize, $z = x_1 + x_2$

$$\text{subject to } \begin{aligned} 2x_1 + 3x_2 &\le 22 \\ 2x_1 + x_2 &\le 14 \\ x_1 - x_2 &\le 4 \\ 3x_1 - 2x_2 &\ge -6, \quad x_1, x_2 \ge 0. \end{aligned}$$

[C.U.(P)'94]

Verify the results by using graphical method.

13. Maximize, $z = 4x_1 + 3x_2 - 14x_3$

$$\text{subject to } \begin{aligned} x_1 + 2x_2 - 3x_3 &\le 4 \\ 4x_1 + x_2 + 2x_3 &\le 10 \\ 3x_1 + 5x_2 + 2x_3 &\le 12, \quad x_1, x_2 \text{ and } x_3 \ge 0. \end{aligned}$$

14. (a) Maximize, $z = 3x_1 + x_2 + 3x_3$

$$\text{subject to } \begin{aligned} 2x_1 + x_2 + x_3 &\le 2 \\ x_1 + 2x_2 + 3x_3 &\le 5 \\ 2x_1 + 2x_2 + x_3 &\le 6, \quad x_1, x_2 \text{ and } x_3 \ge 0. \end{aligned}$$

[B.U.(H)'82]

(b) Maximize, $z = 3x_1 + 2x_2 + 5x_3$

$$\text{subject to } \begin{aligned} x_1 + 2x_2 + x_3 &\le 430 \\ 3x_1 \quad\quad + 2x_3 &\le 460 \\ x_1 + 4x_2 \quad\quad &\le 420, \quad x_1, x_2 \text{ and } x_3 \ge 0. \end{aligned}$$

[C.U.(P)'83, V.U.(H)'93]

(c) Maximize, $z = 3x_1 + 2x_2 + 5x_3$

$$\text{subject to } \begin{aligned} x_1 + 2x_2 + x_3 &\le 430 \\ 3x_1 + 2x_2 \quad\quad &\le 460 \\ x_2 + 4x_3 &\le 420, \quad x_1, x_2 \text{ and } x_3 \ge 0. \end{aligned}$$

15. Solve the problem 4 given in (2.6).

16. (a) Solve the problem 5 given in (2.6).

(b) Solve the problem 13 by simplex method given in (2.6).

[C.U.M.Com.'89]

17. A factory is engaged in manufacturing three products A, B and C which involve lathe work, grinding and assembling. The cutting, grinding and assembling times required for one unit of A, are 2, 1 and 1 hours, respectively. Similarly they are 3, 1, 3 hours for one unit B and 1, 3, 1 hours for one unit of C. The profits on A, B and C are Rs. 2, Rs. 2 and Rs. 4 per unit, respectively. Assuming that there are available 300 hours of lathe time, 300 hours of grinding time and 240 hours of assemble time, how many units of each product should be produced to maximize profits [solve by simplex method].

18. A company produces two types of leather belts, say type A and B. Belt A is superior quality and belt B is of lower quality. Profits on the two types belts are 40 are 30 paise per belt, respectively. Each belt of A type requires twice as much time as required by a belt of B type. If all belts were type B, the company could produce 1,000 belts per day. But the supply of leather is sufficient only for 800 belts per day. Belt A requires a fancy buckle and only 400 fancy buckles are available for this per day. For belt of type B, only 700 buckles are available per day. How should the company manufacture the two types of belt in order to have a maximum overall profit? [C.U.(H)'95]

Answers

1. (a) Max $z = 19$ at $x_1 = 5, x_2 = 2$.
   (b) Max $z = 24$ at $x_1 = 6, x_2 = 0$.
   (c) Max $z = \frac{235}{19}$ at $x_1 = \frac{20}{19}, x_2 = \frac{45}{19}$.
   (d) Max $z = 13\frac{4}{7}$ at $x_1 = \frac{16}{7}, x_2 = 3$.
   (e) Max $z = 48$ for $x_1 = 0, x_2 = 12$ and $x_3 = 0$.
   (f) Max $z = 50$ for $x_1 = 5, x_2 = 0, x_3 = 0$.

2. (a) Max $z = 20$ for $x_1 = 0, x_2 = 10$.
   (b) The problem is said to have an unbounded solution.
   (c) Max $z = 165$ at $x_1 = 15$ and $x_2 = 30$. The optimal solution is unique as $z_3 - c_3 = \frac{6}{5} \neq 0$ and $z_4 - c_4 = \frac{1}{5} \neq 0$ corresponding to non-basis vectors. The optimal solution is not degenerate as the optimal basic variable is $x_1 = 15 \neq 0$ and $x_2 = 30 \neq 0$.
   (d) Max $z = 13$ at $x_1 = \frac{7}{2}$, $x_2 = 0$ and $x_3 = \frac{9}{4}$, optimal basis

$$B(\mathbf{a}_1, \mathbf{a}_5, \mathbf{a}_3) = \begin{bmatrix} 3 & 0 & -2 \\ 2 & 1 & 1 \\ 1 & 0 & 2 \end{bmatrix}, \quad B^{-1} = \begin{bmatrix} 1/4 & 0 & 1/4 \\ -3/8 & 1 & -7/8 \\ -1/8 & 0 & 3/8 \end{bmatrix}$$

   which can be obtained from $(\mathbf{y}_3, \mathbf{y}_4, \mathbf{y}_5)$ under the initial basis vectors $\mathbf{a}_3$, $\mathbf{a}_4$ and $\mathbf{a}_5$ (in the final table).

3. Max $z = 700$ for $x_1 = 0, x_2 = 50$ and $x_3 = 50$.

4. Max $z = 10$ for $x_1 = 2, x_2 = 0$.

5. (a) Max $z = 84$ for $x_1 = 26, x_2 = 0$ and $x_3 = 6$.
   (b) Max $z = 3$ at $x_1 = 0$ (basic), $x_2 = \frac{1}{2}$ and $x_3 = 0$, degeneracy occurs.

6. Max $z = 52$ for $x_1 = \frac{7}{3}, x_2 = 9$ and $x_3 = 0$.

7. (a) Max $z = 13$; $[x_1 = 4, x_2 = 3, x_3 = 0$ (basic)$]$, $[x_1 = 4, x_2 = 3, x_4 = 0$ (basic)$]$. The solutions are degenerate.
   (b) Max $z = 200$ at $x_1 = \frac{75}{4}, x_2 = \frac{25}{2}$, alternative optimal solutions exist.

8. (a) Min $z = -10$ at $x_1 = 0, x_2 = 5$;
   (b) Min $z = -24$ for $x_1 = 8, x_2 = 0$.

9. Min $z = -11$ for $x_1 = 4, x_2 = 5, x_3 = 0$.

10. Max $z = 50$ for $x_1 = 5, x_2 = 0, x_3 = 0$.

11. Max $z = 8$ for $x_1 = 1, x_2 = 1, x_3 = 2, x_4 = 0$.

12. Max $z = 9$ for $x_1 = 5, x_2 = 4$.

13. Max $z = \frac{82}{7}$ for $x_1 = \frac{16}{7}, x_2 = \frac{6}{7}, x_3 = 0$.

14. (a) Max $z = \frac{27}{5}$ for $x_1 = \frac{1}{5}, x_2 = 0, x_3 = \frac{8}{5}$.
    (b) Max $z = 1350$ for $x_1 = 0, x_2 = 100, x_3 = 230$.
    (c) Max $z = 985$ at $x_1 = \frac{460}{3}$, $x_2 = 0$, $x_3 = 105$.

15. Min cost = Rs. 13.00, liquid product 1 unit, dry product 5 units.

16. (a) Min cost = Rs. 2.05 for 15 units of X food and $\frac{5}{4}$ units of Y food.
    (b) Max. rev. = Rs. 2,550 for selling 150 red and 250 blue pen.

17. Product A, 120 product B, 0 and product C, 60. Maximum profit = Rs. 480.

18. Max profit = Rs. 260.00 at A type 200 and B type 600.

Chapter 9

Simplex Method or Simplex Algorithm (II)

9.1 Solution to the problem when some artificial variables are added to get a unit basis out of the coefficient matrix

9.1.1 Artificial variable technique

Before using the simplex method, all inequations are to be converted into equations by introducing either slack or surplus variables. Next problem is to get an initial B.F.S. which will be obtained smoothly and easily if the coefficient matrix contains a unit basis and $\mathbf{b} \geq \mathbf{0}$. We have explained previously how some artificial variables are added to the left hand side of the converted equations to get a unit basis matrix. This technique is used only to get the initial B.F.S. of the problem easily. But it is interesting to note that an equation is being kept an equation even after the addition of a new variable (artificial variable) only in one side of an equation which is not correct from the mathematical point of view. It is only true if the values of all artificial variables be equal to zero. So in solving the problem of this type by using simplex method, one must be sure at the optimal stage, that all artificial variables are at zero level. If it is not possible to bring all artificial variables at zero level at the optimal stage, we conclude that the problem has no feasible solution. In an attempt to solve the problem, by using simplex method, the following three cases may arise when the optimality conditions are satisfied (at the optimal stage).

1. All artificial vectors are not present in the basis which indicates that all artificial variables are at zero level at the optimal stage. Thus the solution obtained is a B.F.S.

2. Some artificial vectors are present in the basis and some artificial variables are at positive level at the optimal stage. In that case there exists no feasible solution to the problem.

3. All artificial variables are at zero level but at least one artificial vector is present in the basis at the optimal stage. Here the solution under test is an optimal solution. Here the converted equations are consistent but some of the

constraints may be redundant. By redundancy we mean that the system has more than enough constraints. There are two methods of solving problems of this type:

(a) *Charnes* method of penalties or Big M-method.

(b) Two phase method due to Dantzig, Orden and Wolfe.

We consider only Big-M method.

9.2 Charnes method of penalties or Big M-method

In this method, initially minimum number of artificial variables are to be inserted in the equations to get a unit basis matrix from the coefficient matrix. To each of the artificial variable, a very high negative price or cost, say — M(M is +ve, very large) is attached in the objective function of the problem. M is so large that the sign of α determines the sign of the expression $\alpha M + \beta$. Due to very high negative prices, the objective function cannot be improved in the presence of artificial variables. With the negative high prices, we proceed to solve the problem in usual method and following cases may arise.

1. At any stage, all artificial vectors may be driven out from the basis. In that case, all artificial variables are at zero level at that stage. Now if the optimality conditions are satisfied at that stage, the problem is solved and the problem has an optimal solution. If the optimality conditions are not satisfied we may proceed further to get an optimal solution omitting all column vectors corresponding to the artificial variables. The artificial vectors are used only as agents to get a unit basis.

2. Some artificial variables are at positive level, though the optimality conditions are satisfied. In that case the problem has no feasible solution at all. Here some artificial vectors must be present in the basis at the final stage.

3. All artificial variables are at zero level and at least one artificial vector is present in the basis. If at this stage, the optimality conditions are satisfied the solution obtained is an optimal solution. Here one or more constraints may be redundant. If the optimality conditions are not satisfied, proceed further as in the previous cases to get an optimal solution.

In this connection it is important to note that once an artificial vector leaves the basis, we forget all about it for ever and never consider it as a vector to enter into the basis at any next iteration. The rule of construction of the tables are same as given in the previous cases.

► **Example 9.2.1** *Solve the L.P.P.*

$$\textit{Minimize,} \quad z = 4x_1 + 3x_2$$

subject to

$$\begin{aligned} x_1 + 2x_2 &\geq 8 \\ 3x_1 + 2x_2 &\geq 12, \quad x_1 \geq 0, x_2 \geq 0. \end{aligned}$$

by Charnes Big M-method. [C.U.(P)'89,'93; V.U.(P)'90]

Solution: The problem here is to minimize z. The components of $\mathbf{b}$ are already +ve.

Let $z' = -z$; then $\min z = -\max(-z) = -\max z'$.

Hence the problem is to maximize $z' = -4x_1 - 3x_2$.

Introducing two surplus variables x_3 and x_4 we get the following converted equations

$$\begin{aligned} x_1 + 2x_2 - x_3 \quad\quad &= 8 \\ 3x_1 + 2x_2 \quad\quad - x_4 &= 12 \end{aligned}$$

The coefficient matrix does not contain a unit basis matrix. To get a unit basis matrix, two artificial variables x_5 and x_6 are added, one in each equation and the equations are

$$\begin{aligned} x_1 + 2x_2 - x_3 \quad\quad + x_5 \quad\quad &= 8 \\ 3x_1 + 2x_2 \quad\quad - x_4 \quad\quad + x_6 &= 12 \end{aligned}$$

Now the coefficient matrix does contain a unit basis.

The adjusted objective function z' is given by

$$z' = -4x_1 - 3x_2 + 0.x_3 + 0.x_4 - Mx_5 - Mx_6$$

[assigning a very large negative price to each of the artificial variables x_5 and x_6, M is positive].

Here vectors

$$\mathbf{a}_5 = \begin{bmatrix} 1 \\ 0 \end{bmatrix} \quad \text{and} \quad \mathbf{a}_6 = \begin{bmatrix} 0 \\ 1 \end{bmatrix}$$

are artificial vectors and the initial basis

$$B = (\mathbf{a}_5, \mathbf{a}_6) = \begin{bmatrix} 1 & 0 \\ 0 & 1 \end{bmatrix}$$

and hence initial (artificial) B.F.S.

$$\begin{aligned} \mathbf{x}_B &= [x_5, x_6] = [8, 12] \\ \mathbf{c}_B &= (c_5, c_6) = (-M, -M), \\ \mathbf{y}_j &= B^{-1}\mathbf{a}_j = \mathbf{a}_j. \end{aligned}$$

$$\begin{aligned} z &= \mathbf{c}_B\mathbf{x}_B = -20M \\ z_1 - c_1 &= (-M, -M)[1, 3] + 4 = -4M + 4 \\ z_2 - c_2 &= (-M, -M)[2, 2] + 3 = -4M + 3 \\ z_3 - c_3 &= (-M, -M)[-1, 0] = M \\ z_4 - c_4 &= (-M, -M)[0, -1] = M \quad \text{and} \\ z_5 - c_5 &= z_6 - c_6 = 0 \end{aligned}$$

Initial Simplex table

		c	-4	-3	0	0	$-M$	$-M$	
Basis	$\mathbf{c}_B$	**b**	$\mathbf{a}_1$	$\mathbf{a}_2$	$\mathbf{a}_3$	$\mathbf{a}_4$	$\mathbf{a}_5(\mathbf{e}_1)$	$\mathbf{a}_6(\mathbf{e}_2)$	**Min. ratio**
$\mathbf{a}_5^*$	$-M$	8	1	2^*	-1	0	1	0	$\frac{8}{2} = 4^*$
$\mathbf{a}_6$	$-M$	12	3	2	0	-1	0	1	$\frac{12}{2} = 6$
$z_j - c_j$			$-4M + 4$	$-4M + 3^*$	M	M	0	0	

$\mathbf{a}_2$ is the incoming vector as $z_2 - c_2$ is the $-$ve minimum and $\mathbf{a}_5$ is the outgoing vector. We shall leave the column corresponding to the artificial vector $\mathbf{a}_5$ for ever.

Second Simplex table

		c	-4	-3	0	0	$-M$	
Basis	$\mathbf{c}_B$	**b**	$\mathbf{a}_1$	$\mathbf{a}_2$	$\mathbf{a}_3$	$\mathbf{a}_4$	$\mathbf{a}_6$	**Min. ratio**
$\mathbf{a}_2$	-3	4	$\frac{1}{2}$	1	$-\frac{1}{2}$	0	0	$4/\frac{1}{2} = 8$
$\mathbf{a}_6^*$	$-M$	4	2^*	0	1	-1	1	$\frac{4}{2} = 2^*$
$z_j - c_j$			$-2M + \frac{5}{2}^*$	0	$-M + \frac{3}{2}$	M	0	

$\mathbf{a}_1$ is incoming vector and $\mathbf{a}_6$ is the outgoing vector. Remove the column corresponding to $\mathbf{a}_6$ for ever.

Third simplex table

		c	-4	-3	0	0
Basis	$\mathbf{c}_B$	**b**	$\mathbf{a}_1$	$\mathbf{a}_2$	$\mathbf{a}_3$	$\mathbf{a}_4$
$\mathbf{a}_2$	-3	3	0	1	$-\frac{3}{4}$	$\frac{1}{4}$
$\mathbf{a}_1$	-4	2	1	0	$\frac{1}{2}$	$-\frac{1}{2}$
$z_j - c_j$		-17	0	0	$\frac{1}{4}$	$\frac{5}{4}$

None of $z_j - c_j < 0$; the solution obtained is an optimal solution. No artificial vector is present in the final basis. Therefore, all artificial variables are at zero level at the final stage. Hence the optimal solution obtained is a B.F.S. and the maximum value of z' is -17 for $x_1 = 2$ and $x_2 = 3$.

Therefore, $\min z = -\max(z') = -(-17) = 17$ for $x_1 = 2$ and $x_2 = 3$.

► **Example 9.2.2** *Solve the L.P.P.*

$$\textit{Maximize,} \quad z = 2x_1 - 3x_2$$

subject to

$$\begin{aligned} -x_1 + \ \ x_2 &\geq -2 \\ 5x_1 + 4x_2 &\leq \ 46 \\ 7x_1 + 2x_2 &\geq \ 32, \quad x_1 \geq 0, x_2 \geq 0. \end{aligned}$$

[C.U.(H)'85,'87,'92; (P)'95, 99]

Solution: 1st constraint is $x_1 - x_2 \le 2$ [making component of $\mathbf{b}$, +ve].

Introducing slack variables x_3, x_4 and surplus variable x_5 one to each of the constraints respectively we get the following converted equations

$$\begin{aligned} x_1 - x_2 + x_3 &= 2 \\ 5x_1 + 4x_2 + x_4 &= 46 \\ 7x_1 + 2x_2 - x_5 &= 32 \end{aligned}$$

The coefficient matrix does not contain a unit basis matrix. To get a unit basis, one artificial variable x_6 is added to the left hand side of the third equation and the set of equations are

$$\begin{aligned} x_1 - x_2 + x_3 &= 2 \\ 5x_1 + 4x_2 + x_4 &= 46 \\ 7x_1 + 2x_2 - x_5 + x_6 &= 32 \ , x_j \ge 0, \ j = (1, 2, \cdots, 6) \end{aligned}$$

The adjusted objective function z is given by

$$z = 2x_1 - 3x_2 + 0.x_3 + 0.x_4 + 0.x_5 - Mx_6$$

[assigning very large negative price to the artificial variable x_6.]

Here

$$\mathbf{a}_3 = \begin{bmatrix} 1 \\ 0 \\ 0 \end{bmatrix}, \quad \mathbf{a}_4 = \begin{bmatrix} 0 \\ 1 \\ 0 \end{bmatrix}, \quad \mathbf{a}_6 = \begin{bmatrix} 0 \\ 0 \\ 1 \end{bmatrix}, \quad \mathbf{b} = \begin{bmatrix} 2 \\ 46 \\ 32 \end{bmatrix} \ge \mathbf{0}$$

and the initial unit basis $B = (\mathbf{a}_3, \mathbf{a}_4, \mathbf{a}_6)$. Hence initial B.F.S. (artificial)

$$\begin{aligned} \mathbf{x}_B &= [x_3, x_4, x_6] = [2, 46, 32] \\ \mathbf{c}_B &= (c_3, c_4, c_6) = (0, 0, -M) \\ z &= -32M \\ \mathbf{y}_j &= B^{-1}\mathbf{a}_j = \mathbf{a}_j \end{aligned}$$

$\mathbf{a}_6$ is an artificial vector.

Simplex tables

		c	2	−3	0	0	0	$-M$	
Basis	$\mathbf{c}_B$	**b**	$\mathbf{a}_1$	$\mathbf{a}_2$	$\mathbf{a}_3(\mathbf{e}_1)$	$\mathbf{a}_4(\mathbf{e}_2)$	$\mathbf{a}_5$	$\mathbf{a}_6(\mathbf{e}_3)$	**Min. ratio**
$\mathbf{a}_3^*$	0	2	1*	−1	1	0	0	0	$\frac{2}{1}=2^*$
$\mathbf{a}_4$	0	46	5	4	0	1	0	0	$\frac{46}{5}=9\frac{1}{5}$
$\mathbf{a}_6$	$-M$	32	7	2	0	0	−1	1	$\frac{32}{7}=4\frac{4}{7}$
z_j-c_j			$-7M-2^*$	$-2M+3$	0	0	M	0	
$\mathbf{a}_1$	2	2	1	−1	1	0	0	0	⋯
$\mathbf{a}_4$	0	36	0	9	−5	1	0	0	$\frac{36}{9}=4$
$\mathbf{a}_6^*$	$-M$	18	0	9*	−7	0	−1	1	$\frac{18}{9}=2^*$
z_j-c_j			0	$-9M+1^*$	$7M+2$	0	M	0	
$\mathbf{a}_1$	2	4	1	0	$\frac{2}{9}$	0	$-\frac{1}{9}$		
$\mathbf{a}_4$	0	18	0	0	2	1	1		
$\mathbf{a}_2$	−3	2	0	1	$-\frac{7}{9}$	0	$-\frac{1}{9}$		
z_j-c_j		2	0	0	$\frac{25}{9}$	0	$\frac{1}{9}$		

As none of $z_j - c_j < 0$, therefore, the optimality conditions are satisfied. The artificial vector $\mathbf{a}_6$ is not present in the final basis. Therefore, the artificial variable x_6 is zero at the final stage. Hence the optimal solution obtained is a B.F.S. and the maximum value of z is 2 for $x_1 = 4$ and $x_2 = 2$.

Note: We even need not compute the mid-body of the third table except the column under **b**.

▶ **Example 9.2.3** *Solve the L.P.P.*

$$\text{Maximize,} \quad z = x_1 + 2x_2$$

subject to

$$\begin{aligned} x_1 - 5x_2 &\le 10 \\ 2x_1 - x_2 &\ge 2 \\ x_1 + x_2 &= 10, \quad x_1 \ge 0, x_2 \ge 0. \end{aligned}$$

[C.U.(P)'91,96]

Solution: Introducing slack variable x_3 and surplus variable x_4 to the first and second constraints respectively, we get the following equations

$$\begin{aligned} x_1 - 5x_2 + x_3 \qquad &= 10 \\ 2x_1 - x_2 \qquad - x_4 &= 2 \\ x_1 + x_2 \qquad &= 10 \end{aligned}$$

The coefficient matrix does not contain a unit basis. In order to get a unit basis two artificial variables x_5 and x_6 are added one to each of second and third equation respectively and the new equations are

$$\begin{aligned} x_1 - 5x_2 + x_3 \quad\quad &= 10 \\ 2x_1 - x_2 \quad - x_4 + x_5 &= 2 \\ x_1 + x_2 \quad\quad + x_6 &= 10 ,\; x_j \geq 0, [j = 1, 2, \cdots, 6] \end{aligned}$$

One slack vector, corresponding to the slack variable x_3 and the two artificial vectors corresponding to the artificial variables x_5 and x_6 constitute a unit basis.

The adjusted objective function z is given by

$$z = x_1 + 2x_2 + 0.x_3 + 0.x_4 - Mx_5 - Mx_6$$

[assigning very large negative price $-M$ to each of the artificial variables x_5 and x_6].

Initial (artificial) B.F.S.

$$\begin{aligned} \mathbf{x} &= [x_3, x_5, x_6] = [10, 2, 10] \\ z &= \mathbf{c}_B \mathbf{x}_B = 0 \times 10 - M \times 2 - M \times 10 = -12M \\ \mathbf{y}_j &= \mathbf{a}_j \end{aligned}$$

Simplex tables

		c	1	2	0	0	$-M$	$-M$	
Basis	$\mathbf{c}_B$	b	$\mathbf{a}_1$	$\mathbf{a}_2$	$\mathbf{a}_3(\mathbf{e}_1)$	$\mathbf{a}_4$	$\mathbf{a}_5(\mathbf{e}_2)$	$\mathbf{a}_6(\mathbf{e}_3)$	Min. ratio
$\mathbf{a}_3$	0	10	1	-5	1	0	0	0	$\frac{10}{1} = 10$
$\mathbf{a}_5^*$	$-M$	2	2^*	-1	0	-1	1	0	$\frac{2}{2} = 1^*$
$\mathbf{a}_6$	$-M$	10	1	1	0	0	0	1	$\frac{10}{1} = 10$
$z_j - c_j$			$-3M - 1^*$	-2	0	M	0	0	$\cdots$
$\mathbf{a}_3$	0	9	0	$-\frac{9}{2}$	1	$\frac{1}{2}$		0	$\cdots$
$\mathbf{a}_1$	1	1	1	$-\frac{1}{2}$	0	$-\frac{1}{2}$		0	$\cdots$
$\mathbf{a}_6^*$	$-M$	9	0	$\frac{3}{2}^*$	0	$\frac{1}{2}$		1	$\frac{9}{3/2} = 6^*$
$z_j - c_j$			0	$-\frac{3}{2}M - \frac{5}{2}^*$	0	$-M/2 - \frac{1}{2}$		0	
$\mathbf{a}_3$	0	36	0	0	1	2			
$\mathbf{a}_1$	1	4	1	0	0	$-\frac{1}{3}$			
$\mathbf{a}_2$	2	6	0	1	0	$\frac{1}{3}$			
$z_j - c_j$		16	0	0	0	$\frac{1}{3}$			

None of $z_j - c_j < 0$. Therefore, the optimality conditions are satisfied. The artificial vectors $\mathbf{a}_5$ and $\mathbf{a}_6$ are driven out from the final basis. Hence the artificial variables are at zero level. Thus the solution obtained is a B.F.S. The maximum value of the objective function of the original problem is 16 for $x_1 = 4$ and $x_2 = 6$.

► **Example 9.2.4** *Solve the L.P.P.*

$$\textit{Minimize,} \quad z = 4x_1 + 8x_2 + 3x_3$$

subject to

$$\begin{aligned} x_1 + x_2 \quad &\geq 2 \\ 2x_1 \quad + x_3 &\geq 0\,, \quad x_1 \geq 0, x_2 \geq 0, x_3 \geq 0. \end{aligned}$$

[C.U.(H)'90]

Hints. No artificial vector is required to insert in the coefficient matrix to get a unit basis matrix. A unit matrix $\mathbf{a}_2 = \mathbf{e}_1$ and $\mathbf{a}_3 = \mathbf{e}_2$ is already present there.

Let $z' = -z = -4x_1 - 8x_2 - 3x_3$

Now $\min z = -\max(z')$. Solve the problem without using artificial vector.

[*Ans.* $\min z = -\max z' = -(-10) = 10$ for $x_1 = \frac{5}{2}$, $x_2 = 0$, $x_3 = 0$. [x_2 and x_3 are non-basic variables.]

Problem having an unbounded solution

► **Example 9.2.5** *Solve the L.P.P.*

$$\textit{Maximize,} \quad z = 3x_1 - x_2$$

subject to

$$\begin{aligned} -x_1 + \ x_2 &\geq 2 \\ 5x_1 - 2x_2 &\geq 2\,, \quad x_1 \geq 0, x_2 \geq 0. \end{aligned}$$

Introducing surplus variables x_3 and x_4, one to each of the constraints, we get the following equations

$$\begin{aligned} -x_1 + \ x_2 - x_3 \qquad &\geq 2 \\ 5x_1 - 2x_2 \qquad - x_4 &\geq 2 \end{aligned}$$

The coefficient matrix does not contain a unit basis.

To get a unit basis, two artificial variables x_5 and x_6 are added to the left hand side of the equations and then

$$\begin{aligned} -x_1 + x_2 - x_3 \qquad + x_5 \qquad &\geq 2 \\ 5x_1 - 2x_2 \qquad - x_4 \qquad + x_6 &\geq 2 \end{aligned}$$

The adjusted objective function z is given by

$$z = 3x_1 - x_2 + 0.x_3 + 0.x_4 - Mx_5 - Mx_6$$

Initial (artificial) B.F.S. is

$$\begin{aligned} \mathbf{x}_B &= [x_5, x_6] = [2, 2] \\ \mathbf{c}_B &= [c_5, c_6] = [-M, -M] \\ z &= \mathbf{c}_B \mathbf{x}_B = -4M, \mathbf{y}_j = \mathbf{a}_j \end{aligned}$$

Simplex tables

		c	3	−1	0	0	$-M$	$-M$	
Basis	c_B	**b**	a_1	a_2	a_3	a_4	$a_5(e_1)$	$a_6(e_2)$	**Min. ratio**
a_5	$-M$	2	−1	1	−1	0	1	0	......
a_6^*	$-M$	2	5*	−2	0	−1	0	1	$\frac{2}{5}^*$
$z_j - c_j$			$-4M-3^*$	$M+1$	M	M	0	0	......
a_5^*	$-M$	$\frac{12}{5}$	0	$\frac{3}{5}^*$	−1	$-\frac{1}{5}$	1		$\frac{12}{5}/\frac{3}{5} = 4^*$
a_1	3	$\frac{2}{5}$	1	$-\frac{2}{5}$	0	$-\frac{1}{5}$	0		......
$z_j - c_j$			0	$-\frac{3}{5}M - \frac{1}{5}^*$	M	$\frac{M}{5} - \frac{3}{5}$	0		
a_2	−1	4	0	1	$-\frac{5}{3}$	$-\frac{1}{3}$			
a_1	3	2	1	0	$-\frac{2}{3}$	$-\frac{1}{3}$			
$z_j - c_j$			0	0	$-\frac{1}{3}$	$-\frac{2}{3}$			

All artificial vectors are driven out from the basis. Now in the third column $z_3 - c_3 = -\frac{1}{3} < 0$ with $y_{i3} \leq 0$ for $i = 1$ and 2. Hence the problem has no finite optimal value of the objective function, i.e., the problem has said to have an unbounded solution.

9.2.1 Problem having no F.S.

▶ **Example 9.2.6** *Solve the L.P.P.*

$$\textit{Maximize,} \quad z = 2x_1 - 3x_2$$

subject to

$$\begin{aligned} 2x_1 + \quad x_2 &\leq 8 \\ 10x_1 + 11x_2 &\geq 100\ , \quad x_1 \geq 0, x_2 \geq 0. \end{aligned}$$

Introducing slack variable x_3 and surplus variable x_4 one to each of the constraints respectively we get the following converted equations

$$\begin{aligned} 2x_1 + \quad x_2 + x_3 \qquad &= 8 \\ 10x_1 + 11x_2 \qquad - x_4 &= 100 \end{aligned}$$

The coefficient matrix does not contain a unit basis. To get a unit basis one artificial variable x_5 is to be added to the L.H.S of the 2nd equation and the set of equations are

$$\begin{aligned} 2x_1 + \quad x_2 + x_3 \qquad\qquad &= 8 \\ 10x_1 + 11x_2 \qquad - x_4 + x_5 &= 100 \end{aligned}$$

$z = 2x_1 - 3x_2 + 0.x_3 + 0.x_4 - Mx_5$ [Assigning very large −ve price to the artificial variable x_5]

The vectors $\mathbf{a}_3$ and $\mathbf{a}_5$ constitute a unit basis. Initial solution

$$\mathbf{x}_B = [x_3, x_5] = [8, 100]$$
$$\mathbf{c}_B = (c_3, c_5) = (0, -M) \text{ and } \mathbf{y}_j = B^{-1}\mathbf{a}_j = \mathbf{a}_j$$
$$z = \mathbf{c}_B\mathbf{x}_B = -100M$$

Simplex tables

		$\mathbf{c}$	2	-3	0	0	$-M$	
Basis	$\mathbf{c}_B$	$\mathbf{b}$	$\mathbf{a}_1$	$\mathbf{a}_2$	$\mathbf{a}_3$	$\mathbf{a}_4$	$\mathbf{a}_5$	**Min. ratio**
$\mathbf{a}_3^*$	0	8	2	1*	1	0	0	$\frac{8}{1} = 8^*$
$\mathbf{a}_5$	$-M$	100	10	11	0	-1	1	$\frac{100}{11} = 9\frac{1}{11}$
$z_j - c_j$			$-2-10M$	$-3-11M^*$	0	M	0	
$\mathbf{a}_2$	-3	8	2	1	1	0	0	
$\mathbf{a}_5$	$-M$	12	-12	0	-11	-1	1	
$z_j - c_j$			$12M-8$	0	$11M-3$	M	0	

The optimality conditions have been satisfied. But the artificial vector $\mathbf{a}_5$ is in the basis at positive level. Hence the only conclusion is that the problem has no F.S. in this case. There is no need to calculate the value of the objective function at the final stage.

N.B. Verify the result by geometrical method.

► **Example 9.2.7** *Solving by Big M-method prove that the following L.P.P. has no feasible solution.*

$$\textit{Maximize,} \quad z = 2x_1 - x_2 + 5x_3$$

subject to

$$x_1 + 2x_2 + 2x_3 \le 2$$
$$\tfrac{5}{2}x_1 + 3x_2 + 4x_3 = 12$$
$$4x_1 + 3x_2 + 2x_3 \ge 24, \quad x_1, x_2, x_3 \ge 0.$$

$\mathbf{b} = [2, 12, 24] \ge \mathbf{0}$. In order to get a unit basis we require to add one slack variable x_4, one artificial variable x_5 and one surplus and other artificial variable x_6 and x_7 in the three constraints respectively. Thus $z = 2x_1 - x_2 + 5x_3 + 0.x_4 - Mx_5 + 0.x_6 - Mx_7$ and the constraints are

$$x_1 + 2x_2 + 2x_3 + x_4 = 2$$
$$\tfrac{5}{2}x_1 + 3x_2 + 4x_3 + x_5 = 12$$
$$4x_1 + 3x_2 + 2x_3 - x_6 + x_7 = 24, \quad x_j \ge 0 [j = 1, 2, \cdots, 7].$$

Simplex tables

Basis	c_B	c b	2 a_1	-1 a_2	5 a_3	0 a_4	$-M$ a_5	0 a_6	$-M$ a_7	Min. ratio
a_4^*	0	2	1^*	2	2	1	0	0	0	$\frac{2}{1} = 2^*$
a_5	$-M$	12	$\frac{5}{2}$	3	4	0	1	0	0	$12/\frac{5}{2} = \frac{24}{5}$
a_7	$-M$	24	4	3	2	0	0	-1	1	$\frac{24}{4} = 6$
$z_j - c_j$			$-\frac{13}{2}M^* - 2$	$-6M + 1$	$-6M - 5$	0	0	M	0	
a_1		2								
a_5		7								
a_7		16								
$z_j - c_j$			0	$7M + 5$	$7M - 1$	$\frac{13}{2}M + 2$	0	M	0	

All $z_j - c_j \geq 0$, $j = 1, 2, \cdots, 7$ in the second table. Thus we need not complete the second table. Two artificial variables x_5 and x_7 are present at the positive level in the optimal solution. Then the only conclusion is that the problem has no feasible solution.

Note: $z_j - c_j$ of the second table is calculated using the formula as shown in the examples solved previously.

▶ **Example 9.2.8** *Solve the L.P.P. using artificial variables*

$$\textit{Minimize,} \quad z = -3x_1 + 2x_2$$

subject to

$$\begin{aligned} x_1 - 4x_2 &\leq -14 \\ -3x_1 + 2x_2 &\leq \quad 6, \quad x_1, x_2 \geq 0. \end{aligned}$$

[C.U.(P)'88]

The problem can be written in the maximization standard form

$$\text{maximize,} \quad z' = 3x_1 - 2x_2 + 0.x_3 - M.x_4 + 0.x_5$$

subject to

$$\begin{aligned} -x_1 + 4x_2 - x_3 + x_4 \qquad &= 14 \\ -3x_1 + 2x_2 \qquad\qquad + x_5 &= \ 6 \end{aligned}$$

where $x_1, x_2 \geq 0$ are decision variables, $x_3 \geq 0$ surplus variable, $x_4 \geq 0$ artificial variable, and $x_5 \geq 0$ slack variable.

Simplex tables

		c	3	−2	0	−M	0	
Basis	$\mathbf{c}_B$	**b**	$\mathbf{a}_1$	$\mathbf{a}_2$	$\mathbf{a}_3$	$\mathbf{a}_4$	$\mathbf{a}_5$	**Min. ratio**
$\mathbf{a}_4$	$-M$	14	−1	4	−1	1	0	$\frac{14}{4} = \frac{7}{2}$
$\mathbf{a}_5^*$	0	6	−3	2*	0	0	1	$\frac{6}{2} = 3^*$
$z_j - c_j$			$M - 3$	$-4M + 2^*$	M	0	0	
$\mathbf{a}_4^*$	$-M$	2	5*	0	−1	1	−2	
$\mathbf{a}_2$	−2	3	$-\frac{3}{2}$	1	0	0	$\frac{1}{2}$	
$z_j - c_j$			$-5M^*$	0	M	0	$2M - 1$	
$\mathbf{a}_1$	3	$\frac{2}{5}$	1	0	$-\frac{1}{5}$		$-\frac{2}{5}$	
$\mathbf{a}_2$	−2	$\frac{18}{5}$	0	1	$-\frac{3}{10}$		$-\frac{1}{10}$	
$z_j - c_j$			0	0	0		−1	

In the third table $z_5 - c_5 = -1$ (negative), and both y_{15}, y_{25} are negative. Thus we stop here. The problem has no finite optimal solution and the problem is said to have an unbounded solution.

► **Example 9.2.9** *Solve the following problem by Big M-method*

$$\textit{Minimize,} \quad z = 2x_1 + 3x_2$$

subject to

$$\begin{aligned} 2x_1 + 7x_2 &\geq 22 \\ x_1 + x_2 &\geq 6 \\ 5x_1 + x_2 &\geq 10, \quad x_1, x_2 \geq 0. \end{aligned}$$

Verify it by using graphical method. [C.U.(P)'94,99]

Solution: After the introduction of the surplus and artificial variables to get a unit matrix, the constraints are

$$\begin{aligned} 2x_1 + 7x_2 - x_3 \quad\quad + x_6 \quad\quad &= 22 \\ x_1 + x_2 \quad - x_4 \quad\quad + x_7 \quad &= 6 \\ 5x_1 + x_2 \quad\quad - x_5 \quad\quad + x_8 &= 10, \quad x_j \geq 0[j = 1, 2, \cdots, 8.] \end{aligned}$$

x_6, x_7, x_8 are artificial variables and the objective function,

$$\text{maximize} \quad -2x_1 - 3x_2 + 0.x_3 + 0 \cdot x_4 + 0 \cdot x_5 - Mx_6 - Mx_7 - Mx_8,$$

where $\min z = -\max(-z)$

		c	−2	−3	0	0	0	−M	−M	−M	
Basis	$\mathbf{c}_B$	**b**	$\mathbf{a}_1$	$\mathbf{a}_2$	$\mathbf{a}_3$	$\mathbf{a}_4$	$\mathbf{a}_5$	$\mathbf{a}_6$	$\mathbf{a}_7$	$\mathbf{a}_8$	Min. ratio
$\mathbf{a}_6^*$	$-M$	22	2	7*	−1	0	0	1	0	0	$\frac{22}{7}=\frac{22}{7}^*$
$\mathbf{a}_7$	$-M$	6	1	1	0	−1	0	0	1	0	$\frac{6}{1}=6$
$\mathbf{a}_8$	$-M$	10	5	1	0	0	−1	0	0	1	$\frac{10}{1}=10$
z_j-c_j			$-8M+2$	$-9M+3^*$	M	M	M	0	0	0	
$\mathbf{a}_2$	−3	$\frac{22}{7}$	$\frac{2}{7}$	1	$-\frac{1}{7}$	0	0		0	0	$\frac{22}{7}/\frac{2}{7}=11$
$\mathbf{a}_7$	$-M$	$\frac{20}{7}$	$\frac{5}{7}$	0	$\frac{1}{7}$	−1	0		1	0	$\frac{20}{7}/\frac{5}{7}=4$
$\mathbf{a}_8^*$	$-M$	$\frac{48}{7}$	$\frac{33}{7}^*$	0	$\frac{1}{7}$	0	−1		0	1	$\frac{48}{7}/\frac{33}{7}=\frac{16}{11}^*$
z_j-c_j			$-\frac{38M}{7}^*+2$	0	$-\frac{2M}{7}+\frac{3}{7}$	M	M		0	0	
$\mathbf{a}_2$	−3	$\frac{30}{11}$	0	1	$-\frac{5}{33}$	0	$\frac{2}{33}$		0		$\frac{30}{11}/\frac{2}{33}=45$
$\mathbf{a}_7^*$	$-M$	$\frac{20}{11}$	0	0	$\frac{4}{33}$	−1	$\frac{5}{33}^*$		1		$\frac{20}{11}/\frac{5}{33}=12^*$
$\mathbf{a}_1$	−2	$\frac{16}{11}$	1	0	$\frac{1}{33}$	0	$-\frac{7}{33}$		0		.....
z_j-c_j			0	0	$-\frac{4M}{33}+\frac{13}{33}$	M	$-\frac{5M}{33}+\frac{8}{33}^*$		0		
$\mathbf{a}_2$	−3	2	0	1	$-\frac{1}{5}$	$\frac{2}{5}$	0				
$\mathbf{a}_5$	0	12	0	0	$\frac{4}{5}$	$-\frac{33}{5}$	1				
$\mathbf{a}_1$	−2	4	1	0	$\frac{1}{5}$	$-\frac{7}{5}$	0				
z_j-c_j		−14	0	0	$\frac{1}{5}$	$\frac{8}{5}$	0				

Max $(-z) = -14$ at $x_1 = 4$, $x_2 = 2$, $x_5 = 12$, i.e., Min $z = -(-14) = 14$ at $x_1 = 4, x_2 = 2$. Graphically the problem has been solved in Example 7.1.5.

▶ **Example 9.2.10** *Solve the following L.P. problem by Big M-method and prove that the problem has finite optimal solution and finite value of the objective function.*

$$\textit{Minimize,} \quad z = 3x_1 + 5x_2$$

subject to

$$\begin{aligned} x_1 + 2x_2 &\geq 8 \\ 3x_1 + 2x_2 &\geq 12 \\ 5x_1 + 6x_2 &\leq 60, \quad x_1, x_2 \geq 0. \end{aligned}$$

[B.U.(H)'87]

Solution: After the introduction of slack, surplus and artificial variables so that there exists a unit matrix which will from initial unit basis; the constraints are

$$\begin{aligned} x_1 + 2x_2 - x_3 \qquad\qquad + x_6 \qquad &= 8 \\ 3x_1 + 2x_2 \qquad - x_4 \qquad\qquad + x_7 &= 12 \\ 5x_1 + 6x_2 \qquad\qquad + x_5 \qquad\qquad &= 60, \quad x_1, x_2 \geq 0; x_3, x_4 \geq 0 \end{aligned}$$

surplus variables $x_5 \geq 0$, slack variable; x_6, $x_7 \geq 0$, artificial variables.

The objective function can be converted as

$$\text{Maximize } (-z) = -3x_1 - 5x_2 + 0 \cdot x_3 + 0 \cdot x_4 + 0 \cdot x_5 - Mx_6 - Mx_7,$$

where $\quad \min z = -\max(-z)$.

We now solve the maximization problem.

Simplex tables

Basis	c_B	c / b	-3 / $\mathbf{a}_1$	-5 / $\mathbf{a}_2$	0 / $\mathbf{a}_3$	0 / $\mathbf{a}_4$	0 / $\mathbf{a}_5(\mathbf{e}_3)$	$-M$ / $\mathbf{a}_6(\mathbf{e}_1)$	$-M$ / $\mathbf{a}_7(\mathbf{e}_2)$	Min. ratio
$\mathbf{a}_6$	$-M$	8	1	2	-1	0	0	1	0	$\frac{8}{1}=8$
$\mathbf{a}_7^*$	$-M$	12	3^*	2	0	-1	0	0	1	$\frac{12}{3}=4^*$
$\mathbf{a}_5$	0	60	5	6	0	0	1	0	0	$\frac{60}{5}=12$
z_j-c_j			$-4M+3^*$	$-4M+5$	M	M	0	0	0	
$\mathbf{a}_6^*$	$-M$	4	0	$\frac{4}{3}^*$	-1	$\frac{1}{3}$	0	1		$\frac{4}{4/3}=3^*$
$\mathbf{a}_1$	-3	4	1	$\frac{2}{3}$	0	$-\frac{1}{3}$	0	0		$\frac{4}{2/3}=6$
$\mathbf{a}_5$	0	40	0	$\frac{8}{3}$	0	$\frac{5}{3}$	1	0		$\frac{40}{8/3}=15$
z_j-c_j			0	$-\frac{4}{3}M+3^*$	M	$-\frac{1}{3}M+1$	0	0		
$\mathbf{a}_2$	-5	3	0	1	$-\frac{3}{4}$	$\frac{1}{4}$	0			
$\mathbf{a}_1$	-3	2	1	0	$\frac{1}{2}$	$-\frac{1}{2}$	0			
$\mathbf{a}_5$	0	32	0	0	2	1	1			
z_j-c_j		-21	0	0	$\frac{9}{4}$	$\frac{1}{4}$	0			

All $z_j - c_j \geq 0$ and all artificial variables are driven out from the basis. Then the finite optimal value with finite optimal solution exists. $\max(-z) = -21$ at $x_1 = 2, x_2 = 3$. Therefore, $\min z = -(-21) = 21$ at $x_1 = 2, x_2 = 3$.

Remark. Verify the result by graphical method.

▶ **Example 9.2.11** *Solve the following L.P. Problem by Big M-method and prove that the problem has no feasible solution.*

$$\textit{Maximize,}\quad z = 5x_1 + 11x_2$$

subject to

$$\begin{aligned} 2x_1 + x_2 &\leq 4 \\ 3x_1 + 4x_2 &\geq 24 \\ 2x_1 - 3x_2 &\geq 6, \quad x_1 \geq 0, x_2 \geq 0. \end{aligned}$$

Solution: After the introduction of slack, surplus and artificial variables so that there exists a unit matrix which will form an initial unit basis, the constraints are

$$\begin{aligned} 2x_1 + x_2 + x_3 &= 4 \\ 3x_1 + 4x_2 - x_4 + x_6 &= 24 \\ 2x_1 - 3x_2 - x_5 + x_7 &= 6 \end{aligned}$$

$x_1, x_2 \geq 0, x_3 \geq 0$ slack variable, $x_4, x_5 \geq 0$, surplus variable, $x_6, x_7 \geq 0$ are artificial variables and the adjusted objective function is

$$z = 5x_1 + 11x_2 + 0 \cdot x_3 + 0 \cdot x_4 + 0 \cdot x_5 - Mx_6 - Mx_7$$

Simplex tables

		c	5	11	0	0	0	$-M$	$-M$	
Basis	$\mathbf{c}_B$	**b**	$\mathbf{a}_1$	$\mathbf{a}_2$	$\mathbf{a}_3(\mathbf{e}_1)$	$\mathbf{a}_4$	$\mathbf{a}_5$	$\mathbf{a}_6(\mathbf{e}_2)$	$\mathbf{a}_7(\mathbf{e}_3)$	**Min. ratio**
$\mathbf{a}_3^*$	0	4	2^*	1	1	0	0	0	0	$\frac{4}{2}=2^*$
$\mathbf{a}_6$	$-M$	24	3	4	0	-1	0	1	0	$\frac{24}{4}=6$
$\mathbf{a}_7$	$-M$	6	2	-3	0	0	-1	0	1	......
z_j-c_j			$-5M-5^*$	$-M-11$	0	M	M	0	0	
$\mathbf{a}_1$	5	2	1	$\frac{1}{2}$	$\frac{1}{2}$	0	0	0	0	
$\mathbf{a}_6$	$-M$	18	0	$\frac{5}{2}$	$-\frac{3}{2}$	-1	0	1	0	
$\mathbf{a}_7$	$-M$	2	0	-4	-1	0	-1	0	1	
z_j-c_j			0	$\frac{3}{2}M-\frac{17}{2}$	$\frac{5}{2}M+\frac{5}{2}$	M	M	0	0	

All $z_j - c_j \geq 0$ [M is so large that $\frac{3}{2}M - \frac{17}{2} \geq 0$]. But the artificial variables are present in the basis at positive level. The only conclusion is that the problem has no feasible solution.

Exercise 9

Solve the following L.P.P. by Big-M method:

1. (a) Minimize, $z = x_1 + x_2$

$$\text{subject to } 2x_1 + x_2 \geq 4$$
$$x_1 + 7x_2 \geq 7, \quad x_1, x_2 \geq 0.$$
[C.U.(H)'85]

(b) Minimize, $z = 2x_1 + x_2$

$$\text{subject to } 8x_1 + 5x_2 \geq 50$$
$$x_1 + x_2 \geq 7, \quad x_1, x_2 \geq 0.$$

(c) Minimize, $z = 15x_1 + 24x_2$

$$\text{subject to } 3x_1 + 3x_2 \geq 4$$
$$x_1 + 4x_2 \geq 3, \quad x_1, x_2 \geq 0.$$

(d) Maximize, $z = 10x_1 + 15x_2$

$$\text{subject to } x_1 + x_2 = 2$$
$$3x_1 + 2x_2 \leq 6, \quad x_1, x_2 \geq 0.$$
[C.U.(P)'85]

(e) Maximize, $z = x_1 + 5x_2$

$$\text{subject to } 3x_1 + 4x_2 \leq 6$$
$$x_1 + 3x_2 \geq 3, \quad x_1, x_2 \geq 0.$$
[C.U.(H)'80]

2. (a) Maximize, $z = 5x_1 + 8x_2$

$$\text{subject to } 3x_1 + 2x_2 \geq 3$$
$$x_1 + 4x_2 \geq 4$$
$$x_1 + x_2 \leq 5, \quad x_1, x_2 \geq 0.$$

(b) Minimize, $z = 4x_1 + 2x_2$

$$\begin{aligned} \text{subject to } 3x_1 + x_2 &\geq 27 \\ x_1 + x_2 &\geq 21 \\ x_1 + 2x_2 &\geq 30 \end{aligned}$$

[C.U.(P)'86,'96; M.Sc.(Appl.Math.)'77]

3. Maximize, $z = x_1 + 2x_2 + 3x_3 - x_4$

$$\begin{aligned} \text{subject to } \quad x_1 + 2x_2 + 3x_3 &= 15 \\ 2x_1 + x_2 + 5x_3 &= 20 \\ x_1 + 2x_2 + x_3 + x_4 &= 10, \quad x_1, x_2, x_3 \text{ and } x_4 \geq 0. \end{aligned}$$

[C.U.(P)'90; V.U.(H)'95]

4. Maximize, $z = 2x_1 + x_2 + 3x_3$

$$\begin{aligned} \text{subject to } \quad x_1 + x_2 + 2x_3 &\leq 5 \\ 2x_1 + 3x_2 + 4x_3 &= 12. \end{aligned}$$

[C.U.(P)'90; J.U.(M.Sc)'81]

5. Minimize, $z = 30x_1 + 36x_2$

$$\begin{aligned} \text{subject to } \quad x_1 + x_2 &\geq 5 \\ 2x_1 + 3x_2 &\geq 2 \\ -2x_1 + x_2 &\geq 2, \quad x_1, x_2 \geq 0. \end{aligned}$$

6. Maximize, $z = x_1 - 2x_2 + 3x_3$

$$\begin{aligned} \text{subject to } -2x_1 + x_2 + 3x_3 &= 2 \\ 2x_1 + 3x_2 + 4x_3 &= 1, \quad x_1, x_2 \text{ and } x_3 \geq 0. \end{aligned}$$

7. Find the B.F.S or solutions of the L.P.P. Maximize, $z = x_1 + 2x_2 + 4x_3$ subject to

$$\begin{aligned} -\ x_1 + x_2 &= 5 \\ -\ 2x_1 + x_3 &= 2, \quad x_1, x_2 \text{ and } x_3 \geq 0. \end{aligned}$$

From the B.F.S. obtained from the problem can we compute directly the optimal value of the of the objective function z? Justify it with proper reason. Verify your statement by using simplex algorithm.

[*Hints:* $[x_1 = 0, x_2 = 5, x_3 = 2]$. No, we cannot compute directly the optimal value of z from the B.F.S. This can only be done if we have previous information that the problem has finite optimal value. That is why it may give erroneous result, if we compute max z directly from here. It can be shown by simplex algorithm that the problem has no finite optimal value of the objective function.]

8. (a) Maximize, $z = 3x_1 - x_2$

$$\begin{aligned} \text{subject to } 2x_1 + x_2 &\geq 2 \\ x_1 + 3x_2 &\geq 3 \\ x_2 &\leq 4, \quad x_1, x_2 \geq 0. \end{aligned}$$

[C.U.(P)'92]

(b) Maximize, $z = -2x_1 + 5x_2$

$$\text{subject to } \begin{aligned} 4x_1 - 5x_2 &\geq -20 \\ x_1 + x_2 &\geq 10 \\ x_2 &\geq 2, \quad x_1, x_2 \geq 0. \end{aligned}$$

9. (a) Maximize, $z = 3x_1 + 6x_2 + 2x_3$

$$\text{subject to } \begin{aligned} 3x_1 + 4x_2 + x_3 &= 2 \\ x_1 + 3x_2 + 2x_3 &= 1, \quad x_1, x_2 \text{ and} x_3 \geq 0. \end{aligned}$$

[C.U.(P)'90]

(b) Minimize, $z = 2x_1 + x_2$

$$\text{subject to } \begin{aligned} 3x_1 + x_2 &\geq 3 \\ 4x_1 + 3x_2 &\geq 6 \\ x_1 + 2x_2 &\geq 2. \end{aligned}$$

10. Solve the following L.P.P. using artificial variables. Minimize, $z = -3x_1 + 2x_2$ subject to

$$\begin{aligned} x_1 - 4x_2 &\leq -14 \\ -3x_1 + 2x_2 &\leq 6, \quad x_1, x_2 \geq 0. \end{aligned}$$

[C.U.(P)'88]

11. Solve the following problem graphically and satisfy that the system is infeasible. Maximize, $z = 2x + 3y$ subject to

$$\begin{aligned} x + y &\leq 1 \\ x + 2y &\geq 3. \end{aligned}$$

[C.U.(M.Com)'89]

Solve it again by simplex method and comment on the identification of the criterion of the infeasibility of the system in the simplex method.

12. Minimize, $z = 2x_1 - 3x_2 + 6x_3$

$$\text{subject to } \begin{aligned} 3x_1 - 4x_2 - 6x_3 &\leq 2 \\ 2x_1 + x_2 + 2x_3 &\geq 11 \\ x_1 + 3x_2 - 2x_3 &\leq 5, \quad x_j \geq 0, [j = 1, 2, 3.] \end{aligned}$$

13. (a) Minimize, $z = 2x_1 + x_2$

$$\text{subject to } \begin{aligned} 3x_1 + x_2 &\geq 3 \\ 4x_1 + 3x_2 &\geq 6 \\ x_1 + 2x_2 &\geq 2, \quad x_j \geq 0, [j = 1, 2.] \end{aligned}$$

(b) Maximize, $z = 2x_1 + 3x_2$

$$\text{subject to } \begin{aligned} x_1 + x_2 &\leq 8 \\ x_1 + 2x_2 &= 5 \\ 2x_1 + x_2 &\leq 8, \quad x_1 \geq 0, x_2 \geq 0. \end{aligned}$$

[C.U.(P)'84,'9

14. Minimize, $z = 4x_1 + x_2$

$$\text{subject to } \begin{aligned} 3x_1 + x_2 &= 3 \\ 4x_1 + 3x_2 &\geq 6 \\ x_1 + 2x_2 &\leq 3, \quad x_1, x_2 \geq 0. \end{aligned}$$ [C.U.(H)'83,'91,'99]

15. Maximize, $z = 2x_1 + 3x_2 - x_3$

$$\text{subject to } \begin{aligned} 2x_1 + 5x_2 - x_3 &\leq 5 \\ x_1 + x_2 + 2x_3 &= 6 \\ 2x_1 - x_2 + 3x_3 &= 7, \quad x_1, x_2 \text{ and } x_3 \geq 0. \end{aligned}$$

16. Prove that, Maximize, $z = 2x_1 + x_2 - x_3 + 4x_4$

$$\text{subject to } \begin{aligned} 3x_1 - 5x_2 + x_3 - 2x_4 &= 7 \\ 6x_1 - 10x_2 - x_3 + 5x_4 &= 11, \quad x_j \geq 0, [j = 1, \cdots, 4.] \end{aligned}$$

has an unbounded solution.

17. Prove that, Minimize, $z = 3x_1 - x_2 + 10x_3$

$$\text{subject to } \begin{aligned} x_1 + 2x_2 + 3x_3 &\leq 6 \\ 4x_1 + x_2 + x_3 &= 32 \\ 2x_1 + x_2 + 2x_3 &\geq 72, \quad x_j \geq 0, [j = 1, 2, 3.] \end{aligned}$$

has no feasible solution.

18. Maximize, $z = x_1 - x_2 + 2x_3 + 3x_4$

$$\text{subject to } \begin{aligned} 2x_1 + x_2 + 3x_3 + 2x_4 &= 11 \\ 3x_1 - 3x_2 + 5x_3 + x_4 &= 17, \quad x_j \geq 0, [j = 1, \cdots, 4.] \end{aligned}$$

19. Minimize, $3x_1 - 2x_2$

$$\text{subject to } \begin{aligned} 3x_1 + 4x_2 &\geq 12 \\ x_1 - 3x_2 &\leq 6 \\ x_1 - 2x_2 &\leq -4, \quad x_1, x_2 \geq 0. \end{aligned}$$

Verify the result by using graphical method.

20. Maximize, $6x_1 - 4x_2 + 4x_3 + x_4$ subject to

$$\begin{aligned} 2x_1 - x_2 + 3x_3 + x_4 &= 6 \\ 4x_1 - 2x_2 - x_3 + 2x_4 &= 10, \quad x_j \geq 0, [j = 1, \cdots, 4.] \end{aligned}$$

21. Solve the following L.P. problem using big-M method.

Minimize, $-3x_1 + x_2 + 3x_3 - x_4$

$$\text{subject to } \begin{aligned} x_1 + 2x_2 - x_3 + x_4 &= 0 \\ 2x_1 - 2x_2 + 3x_3 + 3x_4 &= 9 \\ x_1 - x_2 + 2x_3 - x_4 &= 6, \quad x_j \geq 0, [j = 1, 2, \cdots, 4.] \end{aligned}$$

[C.U.(H)'89]

22. Prove that

$$B = (\mathbf{a}_1, \mathbf{a}_3) = \begin{bmatrix} 2 & 3 \\ 3 & 5 \end{bmatrix}$$

can be considered as an admissible basis to start the simplex procedure of the problem 18 and then solve the problem by applying simplex algorithm without using any artificial variable.

[*Hints:* $\det B = 1 \neq 0, B^{-1} = \begin{bmatrix} 5 & -3 \\ -3 & 2 \end{bmatrix}, B^{-1}\mathbf{b} = \begin{bmatrix} 5 & -3 \\ -3 & 2 \end{bmatrix}\begin{bmatrix} 11 \\ 17 \end{bmatrix} = \begin{bmatrix} 4 \\ 1 \end{bmatrix} \geq \mathbf{0}$. Hence B can be taken as an admissible basis to start the simplex procedure. $\mathbf{c}_B = (1, 2)$. Now calculate $\mathbf{y}_j = B^{-1}\mathbf{a}_j [j = 1, 2, \cdots, 4]$ and $z_j - c_j = \mathbf{c}_B\mathbf{y}_j - c_j$ and now construct the simplex table without using artificial variables and solve the problem by applying simplex algorithm technique.]

23. Prove that the L.P.P. Maximize, $z = 5x_1 + x_2 - 2x_3 + x_4$ subject to

$$\begin{aligned} x_1 + 5x_2 - 8x_3 + 3x_4 &= 6 \\ 3x_1 - x_2 + x_3 + x_4 &= 2, \quad x_j \geq 0, [j = 1, 2, \cdots, 4.] \end{aligned}$$

has an unbounded solution.

Answers

1. (a) Min $z = 31/13$ at $x_1 = 21/13$, $x_2 = 10/13$;
   (b) Min $z = 10$ at $x_1 = 0$, $x_2 = 10$;
   (c) Min $z = 25$ at $x_1 = 7/9$, $x_2 = 5/9$;
   (d) Max $z = 30$ at $x_1 = 0$, $x_2 = 2$;
   (e) Max $z = 15/2$ at $x_1 = 0$, $x_2 = 3/2$.
2. (a) Max $z = 40$ at $x_1 = 0$, $x_2 = 5$;
   (b) Min $z = 48$ at $x_1 = 3$, $x_2 = 18$.
3. Max $z = 15$ at $x_1 = 5/2$, $x_2 = 5/2$, $x_3 = 5/2$, $x_4 = 0$.
4. Max $z = 8$ at $x_1 = 3$, $x_2 = 2$, $x_3 = 0$.
5. Min $z = 174$ at $x_1 = 1$, $x_2 = 4$.
6. No feasible solution.
8. (a) Max $z = 9$ at $x_1 = 3$, $x_2 = 0$; (b) Unbounded solution.
9. (a) $x_1 = 2/5$, $x_2 = 1/5$, $x_3 = 0$ Max $z = 12/5$; (b) Min $z = 12/5$ at $x_1 = 3/5$, $x_2 = 6/5$.
10. Unbounded solution.
11. Solving by Big-M method, at the optimal level the artificial variable x_5 is present at the positive level and $x_5 = 3$ there. Hence the problem is infeasible.
12. Min $z = 9$ at $x_1 = 0$, $x_2 = 4$, $x_3 = 7/2$.
13. (a) Min $z = 12/5$ at $x_1 = 3/5$, $x_2 = 6/5$;
    (b) Max $z = 28/3$ at $x_1 = 11/3$, $x_2 = 2/3$;
    (c) Max $z = -3$ at $x_1 = 0$, $x_2 = 4/3$, $x_3 = 5/3$.
14. Min $z = 18/5$ at $x_1 = 3/5$, $x_2 = 6/5$.
15. Max $z = 3$ at $x_1 = 1$, $x_2 = 1$, $x_3 = 2$.

16. Unbounded solution.

17. No feasible solution.

18. Max $z = 58/7$ at $x_1 = 0, x_2 = 0, x_3 = 23/7, x_4 = 4/7$.

19. Min $z = -12/5$ at $x_1 = 4/5, x_2 = 12/5$.

20. Max $z = 116/7$ at $x_1 = 18/7, x_2 = 0, x_3 = 2/7, x_4 = 0$.

21. Min $z = 7$ at $x_1 = 1, x_2 = 1, x_3 = 3, x_4 = 0$.

22. Initial table

		c	1	−1	2	3
Basis	$\mathbf{c}_B$	b	$\mathbf{a}_1$	$\mathbf{a}_2$	$\mathbf{a}_3$	$\mathbf{a}_4$
$\mathbf{a}_1$	1	4	1	14	0	7
$\mathbf{a}_3$	2	1	0	−9	1	−4
	$z_j - c_j$	6	0	−3	0	−4

Short Answer Type Questions with Answers

1. (a) What are slack and surplus variables?

   (b) Convert the following constraints (inequation) into equations using either slack or surplus variables.

   i. $2x_1 + x_2 + x_3 \leq 7$;
   ii. $x_1 - 2x_2 + x_3 \leq -14$;
   iii. $7x_1 + x_2 + 3x_3 - x_4 \geq 15$;
   iv. $-4x_1 + 2x_2 - x_3 \geq -10$;
   v. $|2x_1 - 3x_2 + x_3| \leq 24$;
   vi. $|3x_1 + x_2 - x_3| \geq 10$.

   [all variables are non-negative]

[**Ans.** (a) A non-negative variable x_{r+1} is added to the left hand side of the inequation $a_{11}x_1 + a_{12}x_2 + \cdots + a_{1r}x_r \leq b_1$ to convert it into an equation $a_{11}x_1 + a_{12}x_2 + \cdots + a_{1r}x_r + x_{r+1} = b_1$. the variable x_{r+1} is called a *slack variable;* similarly a non-negative variable x_{r+1} is subtracted from the left hand side of the constraint (inequation) $a_{11}x_1 + a_{12}x_2 + \cdots + a_{1r}x_r \geq b_r$ to convert it into an equation $a_{11}x_1 + a_{12}x_2 + \cdots + a_{1r}x_r - x_{r+1} = b_r$. This variable x_{r+1} is called a *surplus variable.*

(b) $x_1 - 2x_2 + x_3 + x_4 = -14$. Here x_4 is a slack variable. But if we need to convert −14 to 14 then the equation reduces to $-x_1 + 2x_2 - x_3 - x_4 = 14$ and here the slack variable x_4 changes to a surplus variable.

(e) $|2x_1 - 3x_2 + x_3| \leq 24$ means $2x_1 - 3x_2 + x_3 \leq 24$ and $2x_1 - 3x_2 + x_2 \geq -24$. Then after adding slack and subtracting surplus variables we have $2x_1 - 3x_2 + x_3 + x_4 = 24$ and $2x_1 - 3x_2 + x_3 - x_5 = -24$. Here x_4 is a slack and x_5 is a surplus variable.]

2. Transform the L.P.P.: Maximize, $z = 2x_1 + 3x_2$

$$\begin{aligned} \text{subject to}\quad 2x_1 - 3x_2 &\leq 10 \\ -2x_1 + x_2 &\leq -14 \\ 3x_1 + 4x_2 &= 17, \quad x_1 \geq 0, x_2 \geq 0 \end{aligned}$$

such that all constraints are of equality type (standard form).

[*Ans.* Maximize, $z = 2x_1 + 3x_2 + 0.x_3 + 0.x_4$

$$\begin{aligned} \text{subject to} \quad 2x_1 + 3x_2 + x_3 \quad &= 10 \\ -2x_1 + x_2 \quad + x_4 &= -14 \\ 3x_1 + 4x_2 \quad &= 17, \quad x_j \geq 0[j = 1, 2, \cdots, 4.] \end{aligned}$$

Of course, the second constraint can be written as $2x_1 - x_2 - x_4 = 14$ and here the variable x_4 is a surplus variable.]

3. Transform the L.P.P., minimize, $z = 2x_1 - x_2 + x_3$

$$\begin{aligned} \text{subject to} \quad 3x_1 - 4x_2 + x_3 &\leq 14 \\ -10 \leq x_1 - x_2 + 2x_3 &\leq 10, x_j \geq 0[j = 1, 2, 3.] \end{aligned}$$

such that all constraints are of equality type.

[*Ans.* Minimize, $z = 2x_1 - x_2 + x_3 + 0.x_4 + 0.x_5 + 0.x_6$

$$\begin{aligned} \text{subject to} \; 3x_1 - 4x_2 + x_3 + x_4 \quad &= 14 \\ x_1 - x_2 + 2x_3 \quad + x_5 &= 10 \\ x_1 - x_2 + 2x_3 \quad - x_6 &= -10, \quad x_j \geq 0[j = 1, 2, \cdots, 6.] \end{aligned}$$

x_6 is a surplus variable. But if we write the equation in the form $-x_1 + x_2 - 2x_3 + x_6 = 10$, then x_6 reduces to a slack variable.]

4. Transform the L.P.P.; maximize, $z = 4x_1 + x_2$

$$\begin{aligned} \text{subject to} \; 3x_1 - x_2 &\leq 10 \\ 2x_1 + x_2 &\geq 20 \\ 3x_1 - x_2 &= -14, \quad x_j \geq 0, \; [j = 1, 2] \end{aligned}$$

such that the usual simplex procedure can be applied with the presence of an initial unit basis. Write down the initial basic feasible solution also.

[*Ans.* Maximize, $z = 4x_1 + x_2 + 0.x_3 + 0.x_4 - Mx_5 - Mx_6$

$$\begin{aligned} \text{subject to} \quad 3x_1 - x_2 + x_3 \quad &= 10 \\ 2x_1 + x_2 \quad - x_4 + x_5 &= 20 \\ -3x_1 + x_2 \quad + x_6 &= 14, \quad x_j \geq 0[j = 1, 2, \cdots, 6.] \end{aligned}$$

x_3 is a slack, x_4 is a surplus and x_5 and x_6 are artificial variables.

$\mathbf{a}_3(\mathbf{e}_1), \mathbf{a}_5(\mathbf{e}_2), \mathbf{a}_6(\mathbf{e}_3)$ constitute the initial unit basis.

I.B.F.S. (Artificial) $\mathbf{x}_B = [x_3, x_5, x_6] = [10, 20, 14]$. M is a very large positive quantity.]

5. (a) Transform the L.P.P.; maximize, $z = 3x_1 + x_2 - 3x_3 + x_4$

$$\begin{aligned} \text{subject to} \quad x_1 + x_2 - 3x_3 + x_4 &= 17 \\ -3x_1 \quad + x_3 - x_4 &\leq -27, \quad x_j \geq 0[j = 1, 2, 3, 4.] \end{aligned}$$

such that usual simplex procedure can be applied with the presence of an initial unit basis. Write down the initial B.F.S. also.

[*Ans.* Maximize, $z = 3x_1 + x_2 - 3x_3 + x_4 + 0.x_5 - Mx_6$

$$\begin{aligned} \text{subject to} \quad x_1 + x_2 - 3x_3 + x_4 \quad &= 17 \\ 3x_1 \quad - x_3 + x_4 - x_5 + x_6 &= 27, \quad x_j \geq 0[j = 1, \cdots, 6.] \end{aligned}$$

The vectors $\mathbf{a}_2(\mathbf{e}_1)$, and $\mathbf{a}_6(\mathbf{e}_2)$ constitute the initial basis. Only one artificial vector is required to get a unit basis. $\mathbf{a}_6$ is an artificial vector.

I.B.F.S. (Artificial) $\mathbf{x}_B = [x_2; x_6] = [17, 27]$

(b) Introducing slack and surplus variables, write down the following L.P.P. in the standard maximizing form.

Minimize, $z = 3x_1 - 2x_2 + 4x_3$

$$\text{subject to } x_1 - x_2 + 3x_3 \geq 1$$
$$2x_1 + 3x_2 - 5x_3 \geq -3$$
$$4x_1 + 2x_2 \geq 2, \quad x_1, x_2, x_3 \geq 0. \qquad (1)$$

[C.U.(P)'88]

[*Ans.* The standard maximizing form is maximize, $z^* = -3x_1 + 2x_2 - 4x_3 + 0x_4 + 0x_5 + 0x_6$

$$\text{subject to } x_1 - x_2 + 3x_3 - x_4 = 1$$
$$-2x_1 - 3x_2 + 5x_3 + x_5 = 3$$
$$4x_1 + 2x_2 - x_6 = 2, \quad x_j \geq 0[j = 1, \cdots, 6.] \qquad (2)$$

together with $\min z = -\max z^*$ (3)

x_4 and x_6 are surplus variables and x_5 (slack), all non-negative. [The second constraint is multiplied by (-1) to make component of **b**, positive].

Note that the problem (1) and (2) will have the same optimal solution set (if it exists at all) but the optimal values of the objective functions will be quite different which has been stated in (3). Thus without mentioning (3), (2) cannot be considered as the standard maximizing form of (1).]

6. Find the minimum number of artificial vectors required to get an initial unit basis in the following problem to solve it using the usual simplex procedure. Find the I.B.F.S. also.

Maximize, $z = 2x_1 + x_2 - x_3$

$$\text{subject to } x_1 - 2x_3 \leq 7$$
$$x_1 + x_2 - 2x_3 \geq 7$$
$$x_1 + x_3 = 4, \quad x_1 \geq 0, x_2 \geq 0, x_3 \geq 0.$$

[*Ans.* Minimum number of artificial vectors required is one. The problem can be transformed in the manner, maximize, $z = 2x_1 + x_2 - x_3 + 0.x_4 + 0.x_5 - Mx_6$

$$\text{subject to } x_1 - 2x_3 + x_4 = 7$$
$$x_1 + x_2 - 2x_3 - x_5 = 7$$
$$x_1 + x_3 + x_6 = 4, \quad x_j \geq 0[j = 1, 2, \cdots, 6.]$$

x_4 is a slack variable and x_6 is an artificial variable.

The vectors $\mathbf{a}_4(\mathbf{e}_1), \mathbf{a}_2(\mathbf{e}_2), \mathbf{a}_6(\mathbf{e}_3)$ constitute the initial unit basis.

I.B.F.S. $\mathbf{x}_B = [x_4, x_2, x_6] = [7, 7, 4]$]

7. In a L.P.P.; Maximize, $z = 2x_1 - 3x_2 + x_3 + x_4$

$$\text{subject to } x_1 + 2x_2 + x_4 \geq 5$$
$$3x_2 + x_3 - 2x_4 \geq 3, \quad x_j \geq 0[j = 1, \cdots, 4.]$$

do we need to introduce artificial variable to get an initial unit basis? What are the initial unit vectors? What will be the initial B.F.S? Construct the initial table.

[*Ans.* $\mathbf{b} = [5, 3] \geq \mathbf{0}$. The vectors $\mathbf{a}_1 = \mathbf{e}_1$ and $\mathbf{a}_3 = \mathbf{e}_2$; then $\mathbf{a}_1$ and $\mathbf{a}_3$ will constitute the initial unit basis and there is no need to introduce artificial variable to get a unit basis. Only subtract two surplus variables x_5 and x_6.

$$B = (\mathbf{a}_1, \mathbf{a}_3), B = I_2$$

I.B.S. $\mathbf{x}_B = B^{-1}\mathbf{b} = \mathbf{b} = [x_1, x_3] = [5, 3] \geq \mathbf{0}$. Thus $B = (\mathbf{a}_1, \mathbf{a}_3) = I_2$ can be taken as admissible basis to start a simplex algorithm. $\mathbf{c}_B = (2, 1)$, $\mathbf{y}_j = B^{-1}\mathbf{a}_j = \mathbf{a}_j$, $\mathbf{c} = (2, -3, 1, 1, 0, 0)$, $z = \mathbf{c}_B\mathbf{x}_B = 13$, $z_j - c_j = \mathbf{c}_B B^{-1}\mathbf{a}_j - c_j = \mathbf{c}_B\mathbf{a}_j - c_j = (0, -3, 0, -1, 0, 0)$. With these data initial table can be constructed.]

8. Is it strictly necessary from a theoretical point of view to have an initial basis, a unit basis and the requirement vector, non-negative in the case of solving a L.P.P. by simplex method?

[*Ans.* No, it is not strictly necessary. For detailed discussion see Ex. 11 given below.]

9. Why a unit basis is taken as an initial basis in solving a L.P.P. by simplex method?

Or, What are the advantages of using a unit basis as an initial basis?

[*Ans.* If the initial basis be a unit basis B and $\mathbf{b} \geq \mathbf{0}$, it is possible to compute initial B.F.S. $\mathbf{x}_B$ easily and $\mathbf{x}_B = B^{-1}\mathbf{b} = \mathbf{b} \geq \mathbf{0}$. Moreover we can compute $\mathbf{y}_j$ easily and $\mathbf{y}_j = B^{-1}\mathbf{a}_j = \mathbf{a}_j$. That is why a unit basis is taken as an initial basis.]

10. Write a short note about simplex algorithm.

[*Ans.* See 8.2.]

11. In a problem $\mathbf{a}_1 = \begin{bmatrix} -1 \\ 2 \end{bmatrix}, \mathbf{a}_2 = \begin{bmatrix} 1 \\ 1 \end{bmatrix}, \mathbf{a}_3 = \begin{bmatrix} -2 \\ 1 \end{bmatrix}$ and $\mathbf{b} = \begin{bmatrix} -4 \\ 10 \end{bmatrix}$, calculate the basic solution taking $B = (\mathbf{a}_1, \mathbf{a}_2)$ as basis and prove that the solution is feasible. Calculate also $\mathbf{y}_1, \mathbf{y}_2, \mathbf{y}_3$ with the same basis.

12. During simplex iterations, what are the values of $\mathbf{y}_j$ and $z_j - c_j$ corresponding to the basis vectors $\mathbf{a}_j$?

[*Ans.* The values of $\mathbf{y}_j$ corresponding to all basis vectors are unit vectors and they are such that they constitute a unit matrix I_m after proper arrangement and $z_j - c_j = 0$ corresponding to all basis vectors [for proof see theorem 8.5.1 and the appendix]]

13. What is the nature of the solutions obtained during each simplex iteration?

[*Ans.* In each time the solution obtained is a B.F.S, i.e., an extreme point solution. Of course, the solution may be non-degenerate or degenerate.]

14. What is the criterion for selecting a vector to enter in the next basis and which vector will leave the current basis?

[*Ans.* In a maximization problem, in a table if $z_k - c_k = \min(z_j - c_j)$, $z_j - c_j < 0$ then $\mathbf{a}_k$ is the vector to enter in the next basis. Now if $\mathbf{a}_k$ be the vector to enter in the next basis then compute min. $\left(\frac{x_{Bi}}{y_{ik}}, y_{ik} > 0\right)$. If the minimum value be unique and equal to $\frac{x_{Br}}{y_{rk}}$, then y_{rk} is the key (pivot) element and the vector in the rth position of the basis will leave the current basis which will be replaced by $\mathbf{a}_k$.

If the minimum be not unique, then generally the selection is done arbitrarily corresponding to the minimum ratio and the next B.F.S. will be degenerate.]

15. How can you compute $z_j - c_j$ initially?

[*Ans.* As the initial basis is a unit basis B then $z_j - c_j = \mathbf{c}_B\mathbf{y}_j - c_j = \mathbf{c}_B B^{-1}\mathbf{a}_j - c_j = \mathbf{c}_B\mathbf{a}_j - c_j$. With this formula we can compute $z_j - c_j$ initially.]

16. Write down the transformation formula with the help of which simplex tables are being transformed.

[*Ans.* If we take $x_{Bi} = y_{i0}$, $z_B = y_{m+1.0}$ and $z_j - c_j = y_{m+1.j}$ then the transformation formulae are given by

$$y'_{ij} = y_{ij} - y_{ik}\frac{y_{rj}}{y_{rk}} \quad [j = 0, 1, \cdots, n; i = 1, 2, \cdots, m+1, i \neq r.]$$

$y'_{rj} = \frac{y_{rj}}{y_{rk}}$, where y_{rk} is the key element and y_{ij} and y'_{ij} are the values in the current and next table. The $(m+1)$th row gives the values of z and $z_j - c_j$. [This is a very important property to remember.]

17. Are the vectors $\mathbf{c}$ and $\mathbf{c}_B$ essential for computation of the simplex tables beginning from the second table?

[*Ans.* No. The tables can be computed with the help of the formula given in Q.30 without using $\mathbf{c}$ and $\mathbf{c}_B$ starting from the second table and onwards.]

18. What are the data required to construct the initial simplex table?

[*Ans.* To construct the initial table, we require I.B.F.S. $= \mathbf{x}_B, \mathbf{c}_B$ (the associated cost vector), $\mathbf{y}_j$, $z = \mathbf{c}_B\mathbf{x}_B$ (the value of the objective function corresponding to B.F.S. $\mathbf{x}_B$) and $z_j - c_j = \mathbf{c}_B\mathbf{y}_j - c_j$. Now, if the initial basis be a unit basis B and $\mathbf{b} \geq \mathbf{0}$, then $\mathbf{x}_B = B^{-1}\mathbf{b} = \mathbf{b} \geq \mathbf{0}$, $\mathbf{y}_j = B^{-1}\mathbf{a}_j = \mathbf{a}_j$ and $z_j - c_j = \mathbf{c}_B\mathbf{y}_j - c_j = \mathbf{c}_B\mathbf{a}_j - c_j$ and $z = \mathbf{c}_B\mathbf{x}_B = \mathbf{c}_B\mathbf{b}$.]

19. Why artificial variables are introduced in solving a L.P.P. by simplex method?

[*Ans.* Each artificial vector is a unit vector. That is why artificial variables are introduced to get an initial unit basis. But care should be taken to ensure that minimum number of artificial variables be introduced to get a unit basis.]

20. If z be a homogeneous linear function, then prove that minimum of $z =$ $-$maximum of $(-z)$.

[*Ans.* See the theorem 8.5.1.]

21. How can you solve a minimization problem converting it into a maximization problem?

[*Ans.* First of all, change z to $-z$ and find the maximum value of $-z$ subject to the given constraints and last of all, minimum of $z = -$ maximum of $(-z)$ with the same set of optimal solution of the maximization problem.]

22. If $x_{Bi}[i = 1, \cdots, m]$ be a B.F.S. [not optimal] at any iteration, $\mathbf{a}_k$ be the vector to enter in the next basis and y_{rk} be the key element then what will be the B.F.S. at the next iteration?

[*Ans.* $x'_{Br} = \frac{x_{Br}}{y_{rk}}, x'_{Bi} = x_{Bi} - y_{ik}\frac{x_{Br}}{y_{rk}}\begin{bmatrix} i = 1, \cdots, m \\ i \neq r \end{bmatrix}$.]

23. What is the criterion for optimal solution in solving a L.P.P. by simplex method?

[*Ans.* In the case of a maximization, if at any iteration all $z_j - c_j \geq 0, [j = 1, \cdots, n]$, we can say that we reach at the optimal stage provided no artificial variable is present at the positive level in the final B.F.S.]

24. What is the criterion for the existence of an unbounded solution?

[*Ans.* In a maximization problem, if at any iteration at least one $z_j - c_j < 0[j = 1, \cdots, n]$ and all $y_{ij} \leq 0$ corresponding to that column, we conclude that the problem has no finite optimal solution, i.e., the problem is said to have an unbounded solution.]

25. What is the criterion for the existence of an alternative optimal solution?

[*Ans.* If at the optimal stage at least one $z_j - c_j = 0$ corresponding to non-basic vectors, the problem is said to have alternative optimal solution or solutions.]

26. How can you detect that a L.P.P. has no feasible solution?

[*Ans.* In Charne's M method, if at the optimal stage, at least one artificial variable is present at the positive level, i.e., the value of one artificial variable (basic) is +ve, we conclude that the problem has no feasible solution and we need not compute further.]

27. If $\min\left(\frac{x_{Bi}}{y_{ik}}, y_{ik} > 0\right)$ is not unique [$\mathbf{a}_k$ be the vector to enter in the next basis] what will be the nature of the next solution?

[*Ans.* The next solution will be definitely degenerate.]

28. What is the minimum number of zeros in the set of n values of $z_j - c_j$ if there are m constraints $(m < n)$?

[*Ans.* m zeros.]

29. A simplex table at any intermediate stage of a maximization problem is given below:

		$\mathbf{c}$	2	−5	−1	−3	1	1
Basis	$\mathbf{c}_B$	$\mathbf{b}$	$\mathbf{a}_1$	$\mathbf{a}_2$	$\mathbf{a}_3$	$\mathbf{a}_4$	$\mathbf{a}_5$	$\mathbf{a}_6$
		2	0	1	2	1	4	0
		3	1	0	−2	0	2	0
		3	0	0	1	0	4	1
$z_j - c_j$		3	0	2	−8	0	−5	0

(a) Which are the basis vectors? (b) Write down the associated cost vector $\mathbf{c}_B$. (c) What is the B.F.S. at this stage? (d) Is the solution optimal? (e) What is the value of the objective function at this stage? (f) If the solution be not optimal which vector will enter in the next basis and which vector will leave the current basis?

[*Ans.* (a) $B = (\mathbf{a}_4, \mathbf{a}_1, \mathbf{a}_6)$, (b) $\mathbf{c}_B = (-3, 2, 1)$, (c) $\mathbf{x}_B = [x_4, x_1, x_6] = [2, 3, 3]$ (d) No, solution is not optimal. (e) $z = 3$, (f) $\mathbf{a}_3$ is the entering vector and $\mathbf{a}_4$ will leave the current basis which will be replaced by $\mathbf{a}_3$.]

Note: Though $\mathbf{y}_2$ under $\mathbf{a}_2 = [1, 0, 0]$, still $\mathbf{a}_2$ is not a basis vector because $z_2 - c_2 \neq 0$.

30. A simplex table (maximization problem) is given at any intermediate stage [cost vector $\mathbf{c}$ and $\mathbf{c}_B$ are not given]

Basis	$\mathbf{b}$	$\mathbf{a}_1$	$\mathbf{a}_2$	$\mathbf{a}_3$	$\mathbf{a}_4$	$\mathbf{a}_5$
	1	1	0	1	2	1
	8	4	1	0	−3	0
$z_j - c_j$	10	−3	0	−1	0	0

(a) Which vectors are in the basis? (b) Is the solution optimal? (c) If not, select the vector which will enter in the next basis and the key element. (d) Compute the next table. (e) Will the next solution be optimal? (f) If so, find the maximum value of the objective function and the corresponding optimal basic variables. (g) Is there any alternative optimal solution?

[*Ans.* (a) $B = (\mathbf{a}_5, \mathbf{a}_2)$. (b) Solution is not optimal. (c) $\mathbf{a}_1$ is the vector to enter in the next basis. $y_{11} = 1$ is the key element and $\mathbf{a}_5$ is the vector to leave the current basis which will be replaced by $\mathbf{a}_1$. (d) Next table is given below. (e) Next solution is optimal. (f) Max $z = 13$ at $x_1 = 1, x_2 = 4$ (all other variables are non-basic variables). (g) No.

Basis	$\mathbf{b}$	$\mathbf{a}_1$	$\mathbf{a}_2$	$\mathbf{a}_3$	$\mathbf{a}_4$	$\mathbf{a}_5$
$\mathbf{a}_1$	1	1	0	1	2	1
$\mathbf{a}_2$	4	0	1	−4	−11	−4
$z_j - c_j$	13	0	0	2	6	3

Note. It is very interesting to note that during computation from the second table, $\mathbf{c}_B$ and $\mathbf{c}$ are not essential.

Chapter 10

Duality Theory

Associated with every L.P.P. there exists a corresponding L.P.P. The original L.P.P. is known as the *primal* problem and the corresponding problem is known as the *dual* problem.

10.1 Concept of Duality

Let us consider the following diet problem.

The daily requirement of a patient is 20 and 30 units of vitamins v_1 and v_2 respectively. The food F_1 contains 3 units of v_1 and 4 units of v_2, the food F_2 contains 2 units of v_1 and 3 units of v_2. Find the minimum cost of buying the vitamins if the cost per unit of F_1 and F_2 be Rs.7 and Rs.5 respectively.

10.2 Mathematical Formulation

Chart

Food	$F_1(x_1)$	$F_2(x_2)$	Requirement
v_1	3	2	20 units
v_2	4	3	30 units
Cost	Rs.7	Rs.5	per unit

Let x_1 units of F_1 and x_2 units of F_2 be required to get the minimum amount of vitamins. This is a problem of minimization. The L.P.P. is

$$\text{Minimize,} \quad z = 7x_1 + 5x_2$$

subject to

$$3x_1 + 2x_2 \geq 20$$
$$4x_1 + 3x_2 \geq 30\,, \quad x_1 \geq 0, x_2 \geq 0.$$

Let us now consider the corresponding problem.

A dealer sells the above mentioned vitamins v_1 and v_2 separately. His problem is to fix the cost per unit of v_1 and v_2 in such a way that the price of F_1 and F_2 do not exceed the amount mentioned above. His problem is also to get a maximum amount in selling the vitamins.

Let w_1 and w_2 be the price per unit of v_1 and v_2 respectively.

Therefore the problem is

$$\text{Maximize,} \quad z^* = 20w_1 + 30w_2$$

subject to

$$\begin{aligned} 3w_1 + 4w_2 &\leq 7 \\ 2w_1 + 3w_2 &\leq 5\,, \quad w_1 \geq 0, w_2 \geq 0. \end{aligned}$$

Now if we take

$$A = \begin{bmatrix} 3 & 2 \\ 4 & 3 \end{bmatrix}, \quad \mathbf{b} = \begin{bmatrix} 20 \\ 30 \end{bmatrix}, \quad \mathbf{x} = \begin{bmatrix} x_1 \\ x_2 \end{bmatrix} \quad \text{and} \quad \mathbf{c} = (7, 5)$$

the initial problem can be written as

$$\text{Minimize,} \quad z_x = \mathbf{cx}$$

subject to

$$A\mathbf{x} \geq \mathbf{b}, \quad \mathbf{x} \geq \mathbf{0}.$$

Now the corresponding problem is

$$\text{Maximize,} \quad z_w = \mathbf{b'w}$$

subject to

$$A'\mathbf{w} \leq \mathbf{c}', \quad \mathbf{w} \geq \mathbf{0}.$$

$A', \mathbf{b}'$ and $\mathbf{c}'$ are the transposes of the matrix A, vector $\mathbf{b}$ and $\mathbf{c}$ respectively.

The above is an example of primal-dual problem. In the above problem all constraints are inequations. Generally the initial problem is known as the primal and the corresponding problem is known as the dual problem.

There are three types of primal-dual problems.

1. **Symmetric primal-dual problems :** Here all constraints of the both primal and dual problems are inequations and the variables are non-negative.
2. **Unsymmetric primal-dual problems :** Here all constraints of primal are equations and all primal variables non-negative.
3. **Mixed type problems :** Here some constraints of primal are equations and some variables are unrestricted.

Note: Each and every unsymmetric and mixed type problem can be converted into symmetric type problems by proper adjustment which has been shown later.

10.3 Standard Form of Primal

A L.P.P. is said to be in standard form if

1. All constraints involve the sign '$\leq$' in a problem of maximization, or
2. All constraints involve the sign '$\geq$' in a problem of minimization.

10.4 Standard symmetric primal problem

The following L.P.P.

$$\text{Maximize}, \quad z_x = c_1x_1 + c_2x_2 + \cdots + c_nx_n$$

subject to

$$\left.\begin{array}{l} a_{11}x_1 + a_{12}x_2 + \cdots + a_{1n}x_n \leq \quad b_1 \\ a_{21}x_1 + a_{22}x_2 + \cdots + a_{2n}x_n \leq \quad b_2 \\ \cdots\cdots\cdots\cdots\cdots\cdots\cdots\cdots\cdots\cdots \\ a_{m1}x_1 + a_{m2}x_2 + \cdots + a_{mn}x_n \leq \quad b_m \end{array}\right\} \tag{10.4.1}$$

$x_1 \geq 0, x_2 \geq 0, \cdots, x_n \geq 0.$

is known as a standard symmetric primal.

Here the constants

$$b_i[i = 1, 2, \cdots, m]$$
$$\text{and } c_j[j = 1, 2, \cdots, n]$$

are unrestricted in sign, i.e., b_i and c_j may be positive, negative or zero.

The corresponding dual problem is

$$\text{Minimize}, \quad z_w = b_1w_1 + b_2w_2 + \cdots + b_mw_m$$

subject to

$$\left.\begin{array}{l} a_{11}w_1 + a_{21}w_2 + \cdots + a_{m1}w_m \geq \quad c_1 \\ a_{12}w_1 + a_{22}w_2 + \cdots + a_{m2}w_m \geq \quad c_2 \\ \cdots\cdots\cdots\cdots\cdots\cdots\cdots\cdots\cdots\cdots \\ a_{1n}w_1 + a_{2n}w_2 + \cdots + a_{mn}w_m \geq \quad c_n \end{array}\right\} \tag{10.4.2}$$

$w_1 \geq 0, w_2 \geq 0, \cdots, w_m \geq 0.$

$$\mathbf{w} = \begin{bmatrix} w_1 \\ w_2 \\ \vdots \\ w_m \end{bmatrix}$$

is an m component dual variable vector.

There are n constraints in the dual problem. Putting

$$A = \begin{bmatrix} a_{11} & a_{12} & \cdots & a_{1n} \\ a_{21} & a_{22} & \cdots & a_{2n} \\ \cdots & \cdots & \cdots & \cdots \\ a_{m1} & a_{m2} & \cdots & a_{mn} \end{bmatrix}, \; b = \begin{bmatrix} b_1 \\ b_2 \\ \vdots \\ b_m \end{bmatrix}, \; \mathbf{x} = \begin{bmatrix} x_1 \\ x_2 \\ \vdots \\ x_n \end{bmatrix} \text{ and } \mathbf{c} = (c_1, c_2, \cdots, c_n)$$

(10.4.1) can be written as

$$\text{Maximize}, \quad z_x = \mathbf{cx}$$

subject to

$$A\mathbf{x} \leq \mathbf{b}, \quad \mathbf{x} \geq \mathbf{0}. \tag{10.4.3}$$

and the corresponding dual problem (10.4.2) can be written as

$$\text{Minimize,} \quad z_w = \mathbf{b}'\mathbf{w}$$

subject to

$$A'\mathbf{w} \geq \mathbf{c}', \quad \mathbf{w} \geq \mathbf{0}. \tag{10.4.4}$$

The variable vector $\mathbf{w}$ is an m-component column vector. $A', \mathbf{b}'$ and $\mathbf{c}'$ are the transposes of the matrix A, vector $\mathbf{b}$ and $\mathbf{c}$ respectively. $\mathbf{b}$ is unrestricted in sign here.

The rule of transformation is

1. Co-efficient matrix of the dual is the transpose of the co-efficient matrix of the primal.
2. Interchanging of the cost vector and requirement vector.
3. Changing of the direction of the sign of inequalities.
4. Minimization of the objective function instead of maximization.

10.5 Important Remark

Initially, we cannot consider both maximization and minimization problem as primal problems. We shall have to stick to one of the problems as primal and after establishing the theorem (10.5.1), we may consider both types of problem as primal problems and the corresponding problem as the dual problem. But we cannot do this before establishing that most vital theorem. Before establishing that theorem, Hadley considered maximization problem as primal but Saul I. Gass considered minimization problem as primal and they never confused it any where. In this book, except the problem, given in the concept of duality, we shall initially start taking maximization problem as the primal problem.

Theorem 10.5.1 *Dual of the dual is the primal itself.*

Proof. Let the primal problem be,

$$\text{Maximize,} \quad (z_x = \mathbf{cx})$$

subject to

$$A\mathbf{x} \leq \mathbf{b}, \quad \mathbf{x} \geq \mathbf{0}. \tag{10.5.1}$$

The dual of which is, Minimize, $(z_w = \mathbf{b}'\mathbf{w})$ subject to

$$A'\mathbf{w} \geq \mathbf{c}', \quad \mathbf{w} \geq \mathbf{0}. \tag{10.5.2}$$

The problem (10.5.2) is equivalent to the problem,

$$\text{Maximize,} \quad (-\mathbf{b}'\mathbf{w})$$

subject to

$$-A'\mathbf{w} \leq -\mathbf{c}', \quad \mathbf{w} \geq \mathbf{0} \tag{10.5.3}$$

[multiplying the constraints of (10.5.2) by (−1)]

$$\text{where} \quad \min(\mathbf{b'w}) = -\max(-\mathbf{b'w}). \tag{10.5.4}$$

The dual problem (10.5.3) now "looks like" a primal of type (10.5.1) and hence considering it as a primal, the dual of it can be written as,

$$\text{Minimize,} \quad (-\mathbf{cx})$$

subject to

$$-A\mathbf{x} \geq -\mathbf{b}, \quad \mathbf{x} \geq \mathbf{0}. \tag{10.5.5}$$

Now, (10.5.5) is equivalent to the form,

$$\text{Maximize,} \quad (\mathbf{cx})$$

subject to

$$A\mathbf{x} \leq \mathbf{b}, \quad \mathbf{x} \geq \mathbf{0} \tag{10.5.6}$$

[multiplying the constraints of (10.5.5) by −1]

$$\text{where} \quad \min(-\mathbf{cx}) = -\max(\mathbf{cx}). \tag{10.5.7}$$

Now, due to the presence of two negative signs, one before $\max(-\mathbf{b'w})$ of (10.5.4) and one before $\max(\mathbf{cx})$ of (10.5.7), the maximum value of [if it exists at all] (10.5.6) will also be same as that of (10.5.1).

Hence (10.5.6) is exactly the original problem (10.5.1) and thus proves that 'the dual of the dual is the primal itself.'

From this, we conclude now, that if either problem (10.4.3) or (10.4.4) is considered as a primal then the other will be its dual.

i.e., if we consider the primal be,

$$\text{Minimize,} \quad z = \mathbf{cx}$$

subject to

$$A\mathbf{x} \geq \mathbf{b}, \quad \mathbf{x} \geq \mathbf{0} \tag{10.5.8}$$

then its dual will be,

$$\text{Maximize,} \quad z_w = \mathbf{b'w}$$

subject to

$$A'\mathbf{w} \leq \mathbf{c}', \quad \mathbf{w} \geq \mathbf{0} \tag{10.5.9}$$

and vice-versa.

Note: (1) $\max f(x) =$ Maximum value of $f(x)$.

(2) For verification of the proof, see page 9, *last paragraph*, page 131, 132 and page 222-224 of 'Linear Programming' by Hadley.

▶ **Example 10.5.1** *Write down the dual of the symmetric primal*

$$\textit{maximize,} \quad z = 2x_1 - 3x_2$$

subject to

$$\begin{aligned} x_1 - 4x_2 &\leq 10 \\ -x_1 + x_2 &\leq 3 \\ -x_1 - 3x_2 &\geq 4, \quad x_1 \geq 0, x_2 \geq 0. \end{aligned}$$

The third constraint can be written as

$$x_1 + 3x_2 \leq -4.$$

Hence the problem in the standard form is,

$$\text{maximize,} \quad z_x = (2, -3)[x_1, x_2]$$

subject to

$$\begin{bmatrix} 1 & -4 \\ -1 & 1 \\ 1 & 3 \end{bmatrix} \begin{bmatrix} x_1 \\ x_2 \end{bmatrix} \leq \begin{bmatrix} 10 \\ 3 \\ -4 \end{bmatrix} \quad \text{and } x_1 \geq 0, x_2 \geq 0.$$

Therefore the dual problem is,

$$\text{minimize,} \quad z_w = (10, 3, -4)[w_1, w_2, w_3]$$

subject to

$$\begin{bmatrix} 1 & -1 & 1 \\ -4 & 1 & 3 \end{bmatrix} \begin{bmatrix} w_1 \\ w_2 \\ w_3 \end{bmatrix} \geq \begin{bmatrix} 2 \\ -3 \end{bmatrix} \quad \text{and } w_1 \geq 0, w_2 \geq 0, w_3 \geq 0.$$

or the dual problem is,

$$\text{minimize,} \quad z_w = 10w_1 + 3w_2 - 4w_3$$

subject to

$$\begin{aligned} w_1 - w_2 + w_3 &\geq 2 \\ -4w_1 + w_2 + 3w_3 &\geq -3, \quad w_1 \geq 0, w_2 \geq 0, w_3 \geq 0. \end{aligned}$$

Note: In this problem,

$$A = \begin{bmatrix} 1 & -4 \\ -1 & 1 \\ 1 & 3 \end{bmatrix}, \quad \mathbf{b} = \begin{bmatrix} 10 \\ 3 \\ -4 \end{bmatrix} \quad \text{and } \mathbf{c} = (2, -3)$$

and **b** is unrestricted in sign.

▶ **Example 10.5.2** *Write down the dual of the following problem,*

$$\textit{minimize,} \quad z = 2x_1 + 3x_2 + 4x_3$$

subject to

$$\begin{aligned} x_1 + x_2 + x_3 &\geq 4 \\ -2x_1 + x_2 &\geq 1 \\ 2x_2 - 3x_3 &\leq 2, \quad x_1 \geq 0, x_2 \geq 0, x_3 \geq 0. \end{aligned}$$

Expressing the problem in the standard form, it is given by,

$$\text{minimize,}\quad z_x = (2,3,4)[x_1, x_2, x_3]$$

subject to

$$\begin{bmatrix} 1 & 1 & 1 \\ -2 & 1 & 0 \\ 0 & -2 & 3 \end{bmatrix} \begin{bmatrix} x_1 \\ x_2 \\ x_3 \end{bmatrix} \geq \begin{bmatrix} 4 \\ 1 \\ -2 \end{bmatrix}$$

[multiply both sides of the third constraint by -1, to have "$\geq$" as all constraints must have '$\geq$']

Let us consider this problem as the primal problem. Hence the dual problem is,

$$\text{maximize,}\quad z_w = (4,1,-2)[w_1, w_2, w_3]$$

subject to

$$\begin{bmatrix} 1 & -2 & 0 \\ 1 & 1 & -2 \\ 1 & 0 & 3 \end{bmatrix} \begin{bmatrix} w_1 \\ w_2 \\ w_3 \end{bmatrix} \leq \begin{bmatrix} 2 \\ 3 \\ 4 \end{bmatrix} \text{ and } w_1 \geq 0, w_2 \geq 0, w_3 \geq 0$$

or the problem is,

$$\text{maximize,}\quad z_w = 4w_1 + w_2 - 2w_3$$

subject to

$$\begin{aligned} w_1 - 2w_2 \quad\quad & \leq 2 \\ w_1 + w_2 - 2w_3 & \leq 3 \\ w_1 + \quad\quad 3w_3 & \leq 4, \quad w_1 \geq 0, w_2 \geq 0, w_3 \geq 0. \end{aligned}$$

Note: This transformation is made directly by using the property, established in the theorem (10.5.1) and the result stated in (10.5.8) and (10.5.9).

▶ **Example 10.5.3** *Using the theory "dual of the dual is the primal" verify this in the following problems.*

(i) Maximize $z = 2x_1 + x_2 - x_3$ subject to

$$\begin{aligned} 4x_1 - x_2 + x_3 & \leq 4 \\ x_1 + 3x_2 + 4x_3 & \leq 8, \quad x_1, x_2, x_3 \geq 0. \end{aligned}$$

This is a maximization problem which can be written in the manner,

$$\text{maximize,}\quad z_x = (2,1,-1)[x_1, x_2, x_3]$$

subject to

$$\begin{bmatrix} 4 & -1 & 1 \\ 1 & 3 & 4 \end{bmatrix} \begin{bmatrix} x_1 \\ x_2 \\ x_3 \end{bmatrix} \leq \begin{bmatrix} 4 \\ 8 \end{bmatrix}, \quad x_1, x_2, x_3 \geq 0.$$

The dual of it is,

$$\text{minimize,}\quad z_w = 4w_1 + 8w_2$$

subject to

$$\begin{aligned} 4w_1 + \ \ w_2 &\geq \ \ 2 \\ -w_1 + 3w_2 &\geq \ \ 1 \\ w_1 + 4w_2 &\geq -1\ , \quad w_1, w_2 \geq 0. \end{aligned}$$

Now considering the minimization problem as primal; the dual of it,

$$\text{maximize,}\quad 2v_1 + v_2 - v_3$$

subject to

$$\begin{aligned} 4v_1 - \ \ v_2 + \ \ v_3 &\leq 4 \\ v_1 + 3v_2 + 4v_3 &\leq 8\ , \quad v_1, v_2, v_3 \geq 0 \end{aligned}$$

which is nothing but the primal problem. Hence the theorem 'dual of the dual is the primal' is verified for the above problem.

(ii) Minimize, $z = 3x_1 + 4x_3$ subject to

$$\begin{aligned} 4x_1 + 2x_2 - \ \ x_3 &\geq 12 \\ x_1 + 5x_2 + \ \ x_3 &\geq 18 \\ x_1 - \ \ x_2 + 7x_3 &\geq \ \ 2\ , \quad x_1, x_2, x_3 \geq 0 \end{aligned}$$

Considering the minimization problem as primal, the dual of it is,

$$\text{maximize,}\quad z_w = 12w_1 + 18w_2 + 2w_3$$

subject to

$$\begin{aligned} 4w_1 + \ \ w_2 + \ \ w_3 &\leq 3 \\ 2w_1 + 5w_2 - \ \ w_3 &\leq 0 \\ -w_1 + \ \ w_2 + 7w_3 &\leq 4\ , \quad w_1, w_2, w_3 \geq 0. \end{aligned}$$

Now considering the maximization problem as primal, the dual of it is,

$$\text{minimize,}\quad 3v_1 + 4v_3$$

subject to

$$\begin{aligned} 4v_1 + 2v_2 - \ \ v_3 &\geq 12 \\ v_1 + 5v_2 + \ \ v_3 &\geq 18 \\ v_1 - \ \ v_2 + 7v_3 &\geq \ \ 2\ , \quad v_1, v_2, v_3 \geq 0. \end{aligned}$$

Thus the theorem has been verified.

(iii) Maximize, $z = 3x_1 + 2x_2$ subject to

$$\begin{aligned} 2x_1 + x_2 &\leq 5 \\ x_1 + x_2 &\leq 3\ , \quad x_1, x_2 \geq 0. \end{aligned}$$

[C.U.(H)'86]

Considering the maximization problem as the primal, the dual of it is,

$$\text{minimize,}\quad 5w_1 + 3w_2$$

subject to

$$\begin{aligned} 2w_1 + w_2 &\geq 3 \\ w_1 + w_2 &\geq 2\,, \quad w_1, w_2 \geq 0. \end{aligned}$$

Now considering the minimization problem as primal, the dual of it is,

$$\text{maximize,} \quad 3v_1 + 2v_2$$

subject to

$$\begin{aligned} 2v_1 + v_2 &\leq 5 \\ v_1 + v_2 &\leq 3\,, \quad v_1, v_2 \geq 0. \end{aligned}$$

Hence the theorem is verified.

10.6 Properties of Symmetric Primal-Dual Problems

10.6.1 Weak Duality Theorem

Theorem 10.6.1 *If* $\mathbf{x}_0$ *be any F.S. to the primal, maximize,* $z_x = \mathbf{cx}$ *subject to* $A\mathbf{x} \leq \mathbf{b}$, $\mathbf{x} \geq \mathbf{0}$ *and* $\mathbf{w}_0$ *be any F.S. to its dual problem, minimize,* $z_w = \mathbf{b}'\mathbf{w}$ *subject to* $A'\mathbf{w} \geq \mathbf{c}'$, $\mathbf{w} \geq \mathbf{0}$, *then* $\mathbf{cx}_0 \leq \mathbf{b}'\mathbf{w}_0$.

Proof. We have for any feasible solution $\mathbf{w}_0$ of the dual,

$$A'\mathbf{w}_0 \geq \mathbf{c}', \quad \text{or,} \quad (A'\mathbf{w}_0)' \geq (\mathbf{c}')' \Rightarrow \mathbf{w}_0'A \geq \mathbf{c} \tag{10.6.1}$$

$\mathbf{x}_0$ be any feasible solution of the primal [n-component column vector.]

Post multiplying (10.6.1) by $\mathbf{x}_0$, we have

$$\begin{aligned} & (\mathbf{w}_0'A)\mathbf{x}_0 \geq \mathbf{cx}_0 \quad [\because \quad \mathbf{x}_0 \geq \mathbf{0}] \\ \text{or,} \quad & \mathbf{w}_0'(A\mathbf{x}_0) \geq \mathbf{cx}_0 \\ \text{or,} \quad & \mathbf{w}_0'\mathbf{b} \geq \mathbf{cx}_0 \quad [\because \quad A\mathbf{x}_0 \leq \mathbf{b}] \\ \text{or,} \quad & \mathbf{b}'\mathbf{w}_0 \geq \mathbf{cx}_0 \Rightarrow \mathbf{cx}_0 \leq \mathbf{b}'\mathbf{w}_0 \quad [\because \quad \mathbf{w}_0'\mathbf{b} \text{ is a scalar}] \end{aligned} \tag{10.6.2}$$

Hence the theorem is proved.

Note: (1) If $\mathbf{x}_0$ and $\mathbf{w}_0$ be the optimal feasible solution of the primal and dual respectively then Max $z_x \leq$ Min z_w.

(2) The relation (10.6.2) holds good even if $\mathbf{w}_0$ be not feasible, i.e., if $\mathbf{x}_0$ be a feasible solution of primal and $\mathbf{w}_0$ be any solution of the dual (feasible or infeasible) $\mathbf{cx}_0 \leq \mathbf{b}'\mathbf{w}_0$ which is obvious from the deduction.

(3) This relation is also true for unsymmetric primal problem which will be proved later.

Theorem 10.6.2 *If* $\mathbf{x}^*$ *and* $\mathbf{w}^*$ *be any two feasible solutions of the primal, maximize,* $z = \mathbf{cx}$, *subject to* $A\mathbf{x} \leq \mathbf{b}$, $\mathbf{x} \geq \mathbf{0}$ *and the corresponding dual, minimize,* $z_w = \mathbf{b}'\mathbf{w}$ *subject to* $A'\mathbf{w} \geq \mathbf{c}'$, $\mathbf{w} \geq \mathbf{0}$ *respectively and* $\mathbf{cx}^* = \mathbf{b}'\mathbf{w}^*$, *then* $\mathbf{x}^*$ *and* $\mathbf{w}^*$ *are the optimal feasible solutions of the primal and dual respectively.*

Proof. From theorem 10.6.1, for any two F.S. $\mathbf{x}_0$ and $\mathbf{w}_0$ of the primal and dual

$$\mathbf{c}\mathbf{x}_0 \leq \mathbf{b}'\mathbf{w}_0.$$

Then

$$\mathbf{c}\mathbf{x}_0 \leq \mathbf{b}'\mathbf{w}^* \quad [\text{As } \mathbf{w}^* \text{ is a F.S. of the dual}]$$
$$\text{or,} \quad \mathbf{c}\mathbf{x}_0 \leq \mathbf{b}'\mathbf{w}^* = \mathbf{c}\mathbf{x}^* \Rightarrow \mathbf{c}\mathbf{x}_0 \leq \mathbf{c}\mathbf{x}^*$$

from which we get,

$$\max(\mathbf{c}\mathbf{x}) = \mathbf{c}\mathbf{x}^*, \text{ or, } \mathbf{x}^*$$

is an optimal feasible solution of the primal. In the same way we can prove that

$$\min z_w = \min(\mathbf{b}'\mathbf{w}) = \mathbf{b}'\mathbf{w}^*,$$

i.e., $\mathbf{w}^*$ is an optimal feasible solution of the dual.

10.7 Fundamental Duality Theorem (Symmetric Primal-Dual Problem)

Theorem 10.7.1 (a) *If either the primal, maximize, $z_x = \mathbf{c}\mathbf{x}$, subject to $A\mathbf{x} \leq \mathbf{b}$, $\mathbf{x} \geq \mathbf{0}$ or the dual, minimize $z_w = \mathbf{b}'\mathbf{w}$ subject to $A'\mathbf{w} \geq \mathbf{0}$ has a finite optimal solution, then the other problem will also have a finite optimal solution.*

Furthermore, the optimal values of the objective functions in both the problems will be same, i.e., $\max z_x = \min z_w$.

Proof. We first assume that primal has an optimal feasible solution which has been obtained by simplex method.

Let us convert the constraints of the primal in the following form

$$A\mathbf{x} + I_m\mathbf{x}_s = \mathbf{b}, \quad \mathbf{x} \geq \mathbf{0},\ \mathbf{x}_s \geq \mathbf{0}, \tag{10.7.1}$$

where $\mathbf{x}_s$ is a set of m slack variables and I_m is a unit matrix of order m. It is interesting to note that $\mathbf{b}$ is unrestricted in sign as in the original primal problem. Now imagine that an optimal solution has been found without making each component of the requirement vector non-negative (it is possible to solve the problem even if some components of $\mathbf{b}$ are negative).

Let $\mathbf{x}_B$ be the optimal feasible solution of the primal problem corresponding to the final basis B and let $\mathbf{c}_B$ be the associated cost vector. Therefore,

$$\mathbf{x}_B = B^{-1}\mathbf{b} \tag{10.7.2}$$

and the corresponding optimal value of the objective function

$$\max z_x = \mathbf{c}_B\mathbf{x}_B = \mathbf{c}_B(B^{-1}\mathbf{b}). \tag{10.7.3}$$

Since $\mathbf{x}_B$ is optimal, we have

$$z_j - c_j \geq 0$$

in a maximization problem for all j in the final table. Thus

$$\mathbf{c}_B\mathbf{y}_j - c_j \geq 0 \quad [\mathbf{y}_j \text{ is the } j\text{-th column vector of the final table}]$$

$$\text{or,} \quad \mathbf{c}_B B^{-1}\mathbf{a}_j \geq \mathbf{c}_j$$

$$\text{or,} \quad \mathbf{c}_B B^{-1}(\mathbf{a}_1, \mathbf{a}_2, \cdots, \mathbf{a}_n, \mathbf{e}_1, \mathbf{e}_2, \cdots, \mathbf{e}_m) \geq (c_1, c_2, \cdots, c_n, \overbrace{0, 0, \cdots, 0}^{m \text{ times}})$$

[as $\mathbf{e}_1, \mathbf{e}_2$ etc. are slack vectors and 0, 0 etc. are the cost components corresponding to the slack vectors]

$$\text{or,} \quad \mathbf{c}_B B^{-1}(A, I_m) \geq (\mathbf{c}, \mathbf{0})$$

Equating we get,

$$\mathbf{c}_B B^{-1} A \geq \mathbf{c} \quad \text{and} \quad \mathbf{c}_B B^{-1} I_m \geq \mathbf{0}. \tag{10.7.4}$$

Putting

$$\mathbf{c}_B B^{-1} = \mathbf{w}_0' \geq \mathbf{0},$$

where $\mathbf{w}_0' = (w_1, w_2, \cdots, w_m)$, an m-component row vector, we get from (10.7.4)

$$\mathbf{w}_0' A \geq \mathbf{c}, \tag{10.7.5}$$

$$\text{or,} \quad [\mathbf{w}_0' A]' \geq \mathbf{c}' \quad [\text{taking transpose of both sides}] \tag{10.7.6}$$

$$\text{or,} \quad A'\mathbf{w}_0 \geq \mathbf{c}' \tag{10.7.7}$$

which indicates that $\mathbf{w}_0$ is a feasible solution to the dual problem.

Now we are to prove that $\mathbf{w}_0$ is also an optimal solution to the dual problem.

$$\hat{z}_w = \mathbf{b}'\mathbf{w}_0 = \mathbf{w}_0'\mathbf{b} \quad [\text{as } \mathbf{b}'\mathbf{w}_0 \text{ is a scalar}] \tag{10.7.8}$$

$$= (\mathbf{c}_B B^{-1})\mathbf{b} = \mathbf{c}_B(B^{-1}\mathbf{b}) \tag{10.7.9}$$

$$= \mathbf{c}_B\mathbf{x}_B = \mathbf{c}\mathbf{x}_B = \max z_x \quad [\because \mathbf{c}_B\mathbf{x}_B = \mathbf{c}\mathbf{x}_B, \mathbf{0}]] \tag{10.7.10}$$

Hence from the theorem (10.6.2) we can conclude that $\mathbf{w}_0$ is the optimal solution to the dual problem and

$$\hat{z}_w = \min z_w = \max z_x \tag{10.7.11}$$

Now similarly starting with the finite optimal value to the dual problem, if it exists, we can prove that primal has also an optimal value of the objective function and

$$\max z_x = \min z_w. \tag{10.7.12}$$

Note: (1) In fact, in solving the problem (primal) by usual simplex method, the slack vectors change to surplus vectors, for which the components of the requirement vector are negative initially.

The theorem can be re-stated in the following manner.

Statement: A feasible solution $\mathbf{x}^*$ to a primal maximization problem with objective function $\mathbf{cx}$ will be optimal, if and only if, there exists feasible solution $\mathbf{w}^*$ to the dual minimization problem with the objective function $\mathbf{b}'\mathbf{w}$ such that $\mathbf{cx}^* = \mathbf{b}'\mathbf{w}^*$.

Hints of the proof is given. The proof is almost exactly same as given in the above theorem.

Part I: Let $\mathbf{x}^*$ be an optimal feasible solution of the primal. We are to prove that there exists at least one feasible solution set $\mathbf{w}^*$ of the dual such that $\mathbf{cx}^* = \mathbf{b}'\mathbf{w}^*$.

As $\mathbf{x}^*$ is an optimal feasible solution of the primal, then from the fundamental theorem of L.P.P. we can say that there exists at least one B.F.S. $\mathbf{x}_B$ corresponding to a basis (which is optimal basis) B which will make $\mathbf{cx}$ maximum and $\max(\mathbf{cx}) = \mathbf{cx}^* = \mathbf{c}_B\mathbf{x}_B$ where $\mathbf{c}_B$ is the associated cost vector corresponding to the basis B.

Now using the above theorem we can say that there exists at least one feasible solution $\mathbf{w}^*$ to the dual problem such that

$$\mathbf{c}_B\mathbf{x}_B = \mathbf{b}'\mathbf{w}^* = \mathbf{cx}^*. \tag{10.7.13}$$

Part II. Here we make an assumption that there are two F.S. $\mathbf{x}^*$ and $\mathbf{w}^*$ of the primal and dual problems respectively such that $\mathbf{cx}^* = \mathbf{b}'\mathbf{w}^*$ relation holds. We are to prove that $\mathbf{x}^*$ and $\mathbf{w}^*$ are the optimal feasible solutions of the primal and dual respectively. The theorem (10.7.1) is the exact proof of this part.

From the above two theorems we can derive a very important conclusion.

In the duality theory, both the primal and dual will have finite optimal solutions if and only if both problems have at least one feasible solution set.

(2) This theorem is also valid for unsymmetric primal dual problem. But here the dual variables are unrestricted in sign. The proof of the general theorem is too difficult to understand for the students at this stage. Hence the proof is omitted.

Theorem 10.7.2 (b) *If either of primal or dual has unbounded solution then the other will have no feasible solution.*

Proof. Let the primal be,

$$\text{maximize,} \quad z_x = \mathbf{cx}, \quad A\mathbf{x} \leq \mathbf{b}, \; \mathbf{x} \geq \mathbf{0},$$

has an unbounded solution. Now, if $\mathbf{w}^*$ be any feasible solution of the dual [if there exists at all] we have

$$\max z_x \to \infty \leq \mathbf{b}'\mathbf{w}^*, \quad \text{or,} \quad \mathbf{b}'\mathbf{w}^* \geq \infty$$

which implies that the dual may also have an unbounded solution. Now as there are feasible solutions to the dual, at least one of them, say $\mathbf{w}_1$, will have components, all finite which gives $\mathbf{b}'\mathbf{w}_1$ finite which is a contradiction. Thus the dual actually has no feasible solution and thus the question of unboundness does not arise at all. Similarly if the dual has unbounded solution, then the primal will have no feasible solution [see page 241, Art 8–6 of Linear Programming (Hadley), pages 122-123 of L.P.P. Saul I. Gass].

Note: But the converse statement is not necessarily true. Because if the primal problem has no feasible solution, then the dual will have either no feasible solution or unbounded solution. We cannot say more than that from the duality theory. But if we have previous information that the dual has a feasible solution set then it will have only unbounded solution.

▶ **Example 10.7.1** *Primal problem,*

$$\text{maximize,} \quad z = 2x_1 + 7x_2$$

subject to

$$\begin{aligned} -x_1 + x_2 &\leq 0 \\ x_1 - x_2 &\leq -2\,, \quad x_1, x_2 \geq 0. \end{aligned}$$

Using graphical method, or by simplex method it can be easily verified that the problem has no feasible solution. Now consider its dual problem. The dual of which is,

$$\text{minimize,} \quad z_w = -2w_2$$

subject to

$$\begin{aligned} -w_1 + w_2 &\geq 2 \\ w_1 - w_2 &\geq 7\,, \quad w_1, w_2 \geq 0. \end{aligned}$$

From the graph or by simplex method it can also be shown that the dual has also no feasible solution.

▶ **Example 10.7.2** *Solve the primal problem,*

$$\text{minimize,} \quad z = -2x_1 - 2x_2$$

subject to

$$\begin{aligned} x_1 - 5x_2 &\geq 3 \\ x_1 - 2x_2 &\leq 1\,, \quad x_1, x_2 \geq 0 \end{aligned}$$

by solving its dual problem.

The dual of which is given by,

$$\text{maximize,} \quad z_w = 3w_1 - w_2$$

subject to

$$\begin{aligned} w_1 - w_2 &\leq -2 \\ -5w_1 + 2w_2 &\leq -2 \\ \text{or,} \quad -w_1 + w_2 &\geq 2 \\ 5w_1 - 2w_2 &\geq 2\,, \quad w_1, w_2 \geq 0. \end{aligned}$$

The dual problem is said to have an unbounded solution [Solved in Example 9.2.5]. Then the primal problem will have no feasible solution, which can also be verified by the graphical method.

▶ **Example 10.7.3** *Solve the primal problem,*

$$\text{minimize,} \quad z = 8x_1 - 100x_2$$

subject to

$$\begin{aligned} 2x_1 - 10x_2 &\geq 2 \\ x_1 - 11x_2 &\geq -3\,, \quad x_1, x_2 \geq 0 \end{aligned}$$

by solving its dual problem.

The dual of it is given by,

$$\text{maximize,} \quad z_w = 2w_1 - 3w_2$$

subject to

$$\begin{aligned} 2w_1 + \quad w_2 &\le \quad 8 \\ -10w_1 - 11w_2 &\le -100\,, \quad w_1, w_2 \ge 0. \\ \text{or,} \quad 2w_1 + \quad w_2 &\le \quad 8 \\ 10w_1 + 11w_2 &\ge \quad 100\,, \quad w_1, w_2 \ge 0. \end{aligned}$$

The dual problem is solved in Example 9.2.6. The dual problem has no feasible solution. Thus the primal will have either no F.S. or unbounded solution. We cannot say more than that by using duality theory. Now from the graphical method, it can be shown that the primal problem has a feasible solution set. Hence we can now say definitely that the primal has an unbounded solution.

Theorem 10.7.3 *If the primal problem has feasible solutions and the dual has no feasible solution then the primal problem is said to have unbounded solution and vice versa.*

Proof. Since the primal problem has feasible solutions then the primal objective function must have some value or values corresponding to the feasible solutions and if the values be finite then the primal problem has definite optimal value from which we can conclude that the dual has also finite optimal solution. But dual has no feasible solution and therefore no question of finite optimal value arises. Thus the only conclusion is that the primal has no finite optimal solution, i.e., the primal problem is said to have an unbounded solution.

Below given the results of the observations in a tabular form.

Primal	Dual	Conclusion
Unbounded solution of the primal (dual)	—	No feasible solution of the dual (primal).
No F.S. of the primal (dual).	—	Either unbounded solution or no feasible of the dual (primal).
Bounded optimal solution	—	Dual has also bounded optimal solution and vice versa.
Feasible solution	Feasible solution	Finite optimal values of both exist.
Feasible solution	No feasible solution	Unbounded solution of the primal.
No feasible solution	Feasible solution	Unbounded solution of the dual.
No feasible solution	No feasible solution	No optimal solution of the either problem.

Rules for obtaining the dual optimal solution from the final simplex table of the primal problem and vice versa [for symmetrical primal problem.]

Rule 1. The optimal value of the dual objective function is equal to the optimal value of the primal objective function, i.e., $\min z_w = \max z_x$ assuming that the primal problem is a problem of maximization. The result has been proved in theorem (10.7.1).

Rule 2. The values of $z_j - c_j$ for the columns corresponding to the slack (surplus) vectors in the final simplex table of the primal problem, are the values of the corresponding dual optimal variables and vice versa provided the problem is solved as a maximization problem. The result is very important to find out the optimal dual variables and vice versa.

Proof. $z_j - c_j = z_j \geq 0$ [as the prices or costs of slack (surplus) variables are zero]

$$= \mathbf{c}_B \mathbf{y}_j = \mathbf{c}_B B^{-1} \mathbf{a}_j = \mathbf{c}_B B^{-1} \mathbf{e}_i \ [\mathbf{a}_j = \mathbf{e}_i \text{ for slack vectors}]$$

$z_j - c_j$ for the columns corresponding to slack vectors are

$$= \mathbf{c}_B B^{-1}(\mathbf{e}_1, \mathbf{e}_2, \cdots, \mathbf{e}_m)$$
$$= \mathbf{c}_B B^{-1} I_m = \mathbf{c}_B B^{-1}$$
$$= \mathbf{w}_0' \text{ [by condition (10.7.10) of the theorem (10.7.1)]}$$

General discussions about the method of finding the primal (dual) optimal solution from the final simplex table of the dual (primal) problem.

In the case of symmetric primal dual problem, we have mentioned in Rule 2, that if the primal (dual) problem be solved as a maximisation problem by using usual simplex method, i.e., by making $\mathbf{b} \geq 0$ and taking a unit basis as the initial basis (whether the actual problem is a maximization or minimization problem), the values of $z_j - c_j = z_j$ [as $c_j = 0$ for slack and surplus variables] for the columns of the slack and surplus vectors, in the final simplex table, give the corresponding value of the components of the dual (primal) optimal solution. We have nothing to say if all m vectors be slack vectors, which is evident from theorem (10.7.1) and Rule 2. But why are we getting such a magnificient result if the vectors be surplus vectors? Let us try to explain very carefully the theoretical idea behind this important conclusion. Let the constraints of the primal (maximization) problem be

$$A\mathbf{x} \leq \mathbf{b}, \quad \mathbf{x} \geq \mathbf{0},$$
$$\mathbf{b} = [b_1, b_2, \cdots, b_m],$$

$\mathbf{b}$ is unrestricted and let the constraints, after converting all of them into equations, be

$$A\mathbf{x} + I_m \mathbf{x}_s = \mathbf{b}, \quad \mathbf{x} \geq \mathbf{0}, \quad \mathbf{x}_s \geq \mathbf{0},$$

where $\mathbf{x}_s$ is a set of m slack variables and $\mathbf{b}$ remains unrestricted. For simplicity, we assume that only the last component of $\mathbf{b}$, b_m is negative, i.e., — b_m is positive. Now let us try to solve the problem in two ways:

(i) keeping the requirement vector unrestricted in sign as in the original problem and selecting an arbitrary admissible basis B as an initial basis such that

$$B^{-1}\mathbf{b} \geq \mathbf{0},$$

i.e., initial solution is a basic feasible solution. [It is possible and for that see the definition of admissible basis].

(ii) usual method of making requirement vector non-negative by suitable adjustment and selecting a unit basis as the initial basis.

To solve the problem in the second method, the last constraint

$$a_{m1}x_1 + a_{m2}x_2 + \cdots + a_{mn}x_n + \cdots + x_{n+m} = b_m$$

is to be converted in the following way

$$-a_{m1}x_1 - a_{m2}x_2 - \cdots - a_{mn}x_n - x_{n+m} + x_\alpha = -b_m.$$

Here x_α is an artificial variable and the mth slack vector

$$\mathbf{e}_m = [0, 0, \cdots, 1]$$

in the first case changes to a surplus vector $[0, 0, \cdots, -1]$ in the second case and the requirement vector is then

$$[b_1, b_2 \cdots, -b_m] \geq \mathbf{0}.$$

Of course, we may not require to add the artificial variable x_α at all to get the mth unit vector $\mathbf{e}_m$ of the initial unit basis.

Let B_0 and B^* be the optimal basis for the first and second method respectively. As the optimal solutions remain same [assuming that alternative optimal solution is not present] in each case then the optimal solution

$$\mathbf{x}_0^* = B_0^{-1}[b_1, b_2, \cdots, b_m] = B^{*-1}[b_1, b_2, \cdots, -b_m]$$

from which we can write by assuming $b_1 = b_2 = \cdots = b_{m-1} = 0$ and $b_m = 1$, $B_0^{-1}[0, 0, \cdots, 1] = B^{*-1}[0, 0, \cdots, -1]$. The associated cost vector $\mathbf{c}_B$ remains same in the case of optimal basic feasible solution. Thus $\mathbf{c}_B B^{*-1}[0, 0, \cdots, -1] = \mathbf{c}_B B_0^{-1}[0, 0, \cdots, 1]$, which is the mth dual optimal variable and which can be obtained from the $z_j - c_j = z_j = \mathbf{c}_B B^{*-1}[0, 0, \cdots, -1]$ in the final simplex table corresponding to the surplus vector $[0, 0, \cdots, -1]$ in the second method, i.e., in our usual method. This is the mechanism behind this important conclusion and the artificial vector has nothing contribution in finding the mth dual variable. The same rule is also valid if any minimization problem is solved as a maximization problem.

Rule 3. If either problem has unbounded solution, then the other will have no feasible solution [which we have established in theorem (10.7.1)].

10.8 Importance of the Duality Theory

When the number of constraints are greater than the number of variables, duality theory is very helpful in solving the problem by simplex method. For example, let the number of constraints be five and the number of variables be two in the primal problem. If we try to solve the primal problem by simplex method, then the basis will be a 5×5 square matrix and it requires enough time to compute in each table. But if we convert the problem into its dual we get only two constraints instead of five and the dual problem can be solved easily. Now solving the dual problem we get the optimal value of the objective function of the primal as well as the primal optimal variables. Hence by using duality theory we can solve the problems (sometimes) easily and more quickly. Duality theory has also much economical importance.

▶ **Example 10.8.1** *Solve the problem, minimize, $z = 3x_1 + x_2$ subject to*

$$2x_1 + x_2 \geq 14$$
$$x_1 - x_2 \geq 4, \quad x_1, x_2 \geq 0.$$

by solving its dual problem with the help of simplex method. [C.U.(P)'88]

The dual of the problem is,

$$\text{maximize,} \quad z_w = 14w_1 + 4w_2$$

subject to

$$2w_1 + w_2 \leq 3$$
$$w_1 - w_2 \leq 1, \quad w_1, w_2 \geq 0.$$

Solving the dual problem by usual simplex method, we have the simplex tables given below are

		$\mathbf{c}$	14	4	0	0	
Basis	$\mathbf{c}_B$	$\mathbf{b}$	$\mathbf{a}_1$	$\mathbf{a}_2$	$\mathbf{a}_3(\mathbf{e}_1)$	$\mathbf{a}_4(\mathbf{e}_2)$	Min. ratio
$\mathbf{a}_3$	0	3	2	1	1	0	$\frac{3}{2}$
$\mathbf{a}_4^*$	0	1	1*	−1	0	1	$\frac{1}{1} = 1^*$
$z_j - c_j$		0	−14*	−4	0	0	
$\mathbf{a}_3^*$	0	1	0	3*	1	−2	$\frac{1}{3}$
$\mathbf{a}_1$	14	1	1	−1	0	1	...
$z_j - c_j$		14	0	−18*	0	14	
$\mathbf{a}_2$	4	$\frac{1}{3}$	0	1	$\frac{1}{3}$	$-\frac{2}{3}$	
$\mathbf{a}_1$	14	$\frac{4}{3}$	1	0	$\frac{1}{3}$	$\frac{1}{3}$	
$z_j - c_j$		20	0	0	6	2	

All $z_j - c_j \geq 0$. Then the solution is optimal.

$$\max z_w = 20 \text{ at } w_1 = \tfrac{4}{3} \text{ and } w_2 = \tfrac{1}{3}.$$

Now

$$z_3 - c_3 = 6$$
$$\text{and} \quad z_4 - c_4 = 2$$

corresponding to the unit slack vectors $\mathbf{a}_3(\mathbf{e}_1)$ and $\mathbf{a}_4(\mathbf{e}_2)$ at the optimal stage.

Hence the primal optimal solution is $x_1 = 6$ and $x_2 = 2$, i.e., $\min z = \max z_w = 20$ at $x_1 = 6$, $x_2 = 2$.

Again, we have the final basis

$$B = (\mathbf{a}_2, \mathbf{a}_1) = \begin{bmatrix} 1 & 2 \\ -1 & 1 \end{bmatrix} \text{ and } B^{-1} = \begin{bmatrix} 1/3 & -2/3 \\ 1/3 & 1/3 \end{bmatrix}$$

which are nothing but $\mathbf{y}_3, \mathbf{y}_4$ vectors under the unit vectors $\mathbf{a}_3, \mathbf{a}_4$, in the final simplex table. Hence the primal optimal variables are

$$\mathbf{x}'_B = \mathbf{c}_B B^{-1} = (4, 14) \begin{bmatrix} 1/3 & -2/3 \\ 1/3 & 1/3 \end{bmatrix} = [6 \quad 2].$$

Hence the primal optimal solution is $x_1 = 6$ and $x_2 = 2$, i.e., $\min z = \max z_w = 20$ at $x_1 = 6, x_2 = 2$.

Verification of the correctness of the computation in each table, starting from the second table and onwards.

We can verify the correctness of the computation of each table with almost perfectly [but not exactly] by using duality theory. In the second table $z_3 - c_3 = 0$, $z_4 - c_4 = 14$. The components of the requirement vector [' $\leq$' type constraints] of the dual problem is $[3, 1]$. Now $3 \times 0 + 1 \times 14 = 14$ which is the value of the dual objective function of the second table which is shown under the column $\mathbf{b}$ and in $z_j - c_j$ row. In the third table $z_3 - c_3 = 6$, $z_4 - c_4 = 4$. Now $3 \times 6 + 1 \times 2 = 20$, which is same as shown in the corresponding position in the final table. Thus we may expect a correct calculation [but not with 100% certainty]. But the verification of the result when one or more constraints are equations is not so easy.

► **Example 10.8.2** *Solve the L.P.P.,*

$$\textit{maximize,} \quad z_x = 5x_1 + 4x_2$$

subject to

$$\begin{aligned} 3x_1 + 4x_2 &\leq 24 \\ 3x_1 + 2x_2 &\leq 18 \\ x_2 &\leq 5, \quad x_1 \geq 0, x_2 \geq 0. \end{aligned}$$

by simplex method.

Hence find out the minimum value of the objective function of the dual problem and the corresponding dual optimal solution. Verify the result in solving the dual problem by simplex method.

Primal problem is,

$$\text{maximize,} \quad z_x = 5x_1 + 4x_2$$

subject to

$$\begin{aligned} 3x_1 + 4x_2 &\leq 24 \\ 3x_1 + 2x_2 &\leq 18 \\ x_2 &\leq 5. \end{aligned}$$

Introducing three slack variables x_3, x_4 and x_5 we get the following equations:

$$\begin{aligned} 3x_1 + 4x_2 + x_3 \qquad\qquad &= 24 \\ 3x_1 + 2x_2 \qquad + x_4 \qquad &= 18 \\ x_2 \qquad\qquad + x_5 &= 5. \end{aligned}$$

$$z_x = 5x_1 + 4x_2 + 0 \cdot x_3 + 0 \cdot x_4 + 0 \cdot x_5.$$

Initial B.F.S.

$$\begin{aligned} \mathbf{x}_B &= [x_3, x_4, x_5] = [24, 18, 5] \\ \mathbf{c}_B &= (c_3, c_4, c_5) = (0,0,0), \\ \mathbf{y}_j &= B^{-1}\mathbf{a}_j = \mathbf{a}_j \\ z_B &= \mathbf{c}_B\mathbf{x}_B = 0 \end{aligned}$$

The optimal simplex table is

		c	5	4	0	0	0
Basis	$\mathbf{c}_B$	b	$\mathbf{a}_1$	$\mathbf{a}_2$	$\mathbf{a}_3(\mathbf{e}_1)$	$\mathbf{a}_4(\mathbf{e}_2)$	$\mathbf{a}_5(\mathbf{e}_3)$
$\mathbf{a}_2$	4	3	0	1	$\frac{1}{2}$	$-\frac{1}{2}$	0
$\mathbf{a}_1$	5	4	1	0	$-\frac{1}{3}$	$\frac{2}{3}$	0
$\mathbf{a}_5$	0	2	0	0	$-\frac{1}{2}$	$\frac{1}{2}$	1
$z_j - c_j$		32	0	0	$\frac{1}{3}$	$\frac{4}{3}$	0

$\max z_x = 32$ for $x_1 = 4$ and $x_2 = 3$.

The dual problem is,

$$\text{minimize,} \quad z_w = 24w_1 + 18w_2 + 5w_3$$

subject to

$$\begin{aligned} 3w_1 + 3w_2 + 0 \cdot w_3 &\geq 5 \\ 4w_1 + 2w_2 + \quad w_3 &\geq 4\,, \quad w_1 \geq 0, w_2 \geq 0 \text{ and } w_3 \geq 0. \end{aligned}$$

As $\max z_x = \min z_w$ then $\min z_w = 32$.

10.8.1 Determination of the Dual Optimal Variables

There are three dual variables w_1, w_2 and w_3; x_3, x_4 and x_5 are slack variables in the original primal problem; $\mathbf{a}_3, \mathbf{a}_4$ and $\mathbf{a}_5$ are the unit slack vectors $\mathbf{e}_1, \mathbf{e}_2$ and $\mathbf{e}_3$ respectively.

Hence the dual optimal solution is $w_1 = \frac{1}{3}$, $w_2 = \frac{4}{3}$, $w_3 = 0$ corresponding to $z_j - c_j$ $[j = 3, 4, 5]$ in the final simplex table.

Dual problem: Minimize, $z_w = 24w_1 + 18w_2 + 5w_3$ subject to

$$\begin{aligned} 3w_1 + 3w_2 + 0 \cdot w_3 &\geq 5 \\ 4w_1 + 2w_2 + \quad w_3 &\geq 4\,, \quad w_j \geq 0[j = 1, 2, 3.] \end{aligned}$$

Introducing two surplus variables w_4 and w_5, we get the following equations

$$\begin{aligned} 3w_1 + 3w_2 + 0 \cdot w_3 - w_4 \qquad &= 5 \\ 4w_1 + 2w_2 + \quad w_3 \qquad - w_5 &= 4. \end{aligned}$$

Now to get a unit matrix, only one artificial variable w_6 is to be added in the first equation and then the converted equations are

$$\begin{aligned} 3w_1 + 3w_2 + 0 \cdot w_3 - w_4 \qquad + w_6 &= 5 \\ 4w_1 + 2w_2 + \quad w_3 \qquad - w_5 \qquad &= 4. \end{aligned}$$

This is a problem of minimization.

Let

$$\begin{aligned} z'_w &= -z_w = -24w_1 - 18w_2 - 5w_3 \\ \min(z_w) &= -\max(z'_w). \end{aligned}$$

Now the problem is a problem of maximization,

$$z'_w = -24w_1 - 18w_2 - 5w_3 + 0 \cdot w_4 + 0 \cdot w_5 - Mw_6$$

[assigning very large negative price to the artificial variable w_6].

Initial (artificial) solution $= [w_6, w_3] = [5, 4]$

Simplex tables

Basis $\mathbf{c}_B$	b	$\mathbf{a}_1$	$\mathbf{a}_2$	$\mathbf{a}_3(\mathbf{e}_2)$	$\mathbf{a}_4$	$\mathbf{a}_5$	$\mathbf{a}_6(\mathbf{e}_1)$	Min. ratio
	c	-24	-18	-5	0	0	$-M$	
$\mathbf{a}_6$ $\quad -M$	5	3	3	0	-1	0	1	$\frac{5}{3}$
$\mathbf{a}_3^*$ $\quad -5$	4	4^*	2	1	0	-1	0	$\frac{4}{4} = 1^*$
$z_j - c_j$		$-3M + 4^*$	$-3M + 8$	0	M	5	0	
$\mathbf{a}_6^*$ $\quad -M$	2	0	$\frac{3}{2}^*$	$-\frac{3}{4}$	-1	$\frac{3}{4}$	1	$2/\frac{3}{2} = \frac{4}{3}^*$
$\mathbf{a}_1$ $\quad -24$	1	1	$\frac{1}{2}$	$\frac{1}{2}$	0	$-\frac{1}{4}$	0	$1/\frac{1}{2} = 2$
$z_j - c_j$		0	$-\frac{3}{2}M + 6^*$	$\frac{3}{4}M + 1$	M	$-\frac{3}{4}M + 6$	0	
$\mathbf{a}_2$ $\quad -18$	$\frac{4}{3}$	0	1	$-\frac{1}{2}$	$-\frac{2}{3}$	$\frac{1}{2}$		
$\mathbf{a}_1$ $\quad -24$	$\frac{1}{3}$	1	0	$\frac{1}{2}$	$\frac{1}{3}$	$-\frac{1}{2}$		
$z_j - c_j$	-32	0	0	2	4	3		

As none of $z_j - c_j < 0$, then the solution is optimal. The artificial vector $\mathbf{a}_6$ is driven out from the basis. Hence the solution is a B.F.S. $\max(z'_w) = -32$. Then $\min(z_w) = -\max(z'_w) = 32$ for $w_1 = \frac{1}{3}$, $w_2 = \frac{4}{3}$ and $w_3 = 0$ [non-basic].

Now $\max z_x = \min z_w = 32$. The dual problem is solved as a maximization problem and the values of $z_j - c_j$ corresponding to the surplus vectors $\mathbf{a}_4(-\mathbf{e}_1)$ and $\mathbf{a}_5(-\mathbf{e}_2)$ in the final (third) simplex table are 4 and 3 respectively.

Therefore, $x_1 = 4$, $x_2 = 3$.

Hence the optimal value of primal objective function is $\max z_x = 32$ for $x_1 = 4$, $x_2 = 3$, which is obtained earlier.

► **Example 10.8.3** *Write down the dual of the following problem,*

$$\textit{minimize,} \quad z = 30x_1 + 36x_2$$

subject to

$$\begin{aligned} x_1 + \ x_2 &\geq 5 \\ 2x_1 + 3x_2 &\geq 2 \\ -2x_1 + \ x_2 &\geq 2 , \quad x_1, x_2 \geq 0 \end{aligned}$$

and solving the dual problem find out the optimal solution and the optimal value of the objective function.

The dual problem is, maximize $5w_1 + 2w_2 + 2w_3$ subject to

$$\begin{aligned} w_1 + 2w_2 - 2w_3 &\leq 30 \\ w_1 + 3w_2 + \ w_3 &\leq 36 , \quad w_1, w_2, w_3 \geq 0. \end{aligned}$$

The problem has been solved in Example (8.9.1) and from the final table we get $\max(5w_1 + 2w_2 + 2w_3) = 174$ at $w_1 = 34$, $w_3 = 2$, $w_2 = 0$ and $z_4 - c_4 = 1$, $z_5 - c_5 = 4$. This $\min z = 174$ at $x_1 = 1$, $x_2 = 4$.

▶ **Example 10.8.4** *Solve the primal problem,*

$$\textit{minimize}, \quad z = 10x_1 + 2x_2$$

subject to

$$\begin{aligned} x_1 + 2x_2 + 2x_3 &\geq \ \ 1 \\ x_1 \qquad\quad - 2x_3 &\geq -1 \\ x_1 - \ x_2 + 3x_3 &\geq \ \ 3 , \quad x_1, x_2, x_3 \geq 0 \end{aligned}$$

by solving its dual problem.

The dual of the following problem is

$$\text{maximize}, \quad z_w = w_1 - w_2 + 3w_3$$

subject to

$$\begin{aligned} w_1 + \ w_2 + \ w_3 &\leq 10 \\ 2w_1 \qquad\quad - \ w_3 &\leq \ 2 \\ 2w_1 - 2w_2 + 3w_3 &\leq \ 0 , \quad w_1, w_2, w_3 \geq 0. \end{aligned}$$

The final simplex table of the dual is

		c	1	−1	3	0	0	0
Basis	$\mathbf{c}_B$	**b**	$\mathbf{a}_1$	$\mathbf{a}_2$	$\mathbf{a}_3$	$\mathbf{a}_4(\mathbf{e}_1$	$\mathbf{a}_5(\mathbf{e}_2)$	$\mathbf{a}_6(\mathbf{e}_3)$
$\mathbf{a}_2$	−1	6	$\frac{1}{5}$	1	0	$\frac{3}{5}$	0	$-\frac{1}{5}$
$\mathbf{a}_5$	0	6	$\frac{14}{5}$	0	0	$\frac{2}{5}$	1	$\frac{1}{5}$
$\mathbf{a}_3$	3	4	$\frac{4}{5}$	0	1	$\frac{2}{5}$	0	$\frac{1}{5}$
$z_j - c_j$		6	$\frac{6}{5}$	0	0	$\frac{3}{5}$	0	$\frac{4}{5}$

From the optinial table $\max z = 6$ at $w_1 = 0$, $w_2 = 6$ and $w_3 = 4$. Therefore $\min z_w = 6$. Now $z_3 - c_3 = \frac{3}{5}$, $z_4 - c_4 = 0$ and $z_5 - c_5 = \frac{4}{5}$, thus $\min z_w = 6$, at $x_1 = \frac{3}{5}, x_2 = 0$ and $x_3 = \frac{4}{5}$.

Unsymmetric primal-dual problem and mixed primal-dual problem:

A primal problem expressed in the following form, maximize, $z_x = \mathbf{cx}$ subject to

$$A\mathbf{x} = \mathbf{b}, \quad \mathbf{x} \geq \mathbf{0} \tag{10.8.1}$$

is called an unsymmetric primal problem.

The dual of the problem (10.8.1) is given by, minimize, $z_w = \mathbf{b}'\mathbf{w}$ subject to

$$A'\mathbf{w} \geq \mathbf{c}' \tag{10.8.2}$$

The dual variables $w_1, w_2, \cdots, w_m$ are unrestricted in sign, i.e., the components of $\mathbf{w}$ may be positive or negative or zero. $A', \mathbf{b}'$ and $\mathbf{c}'$ are the transposes of $A, \mathbf{b}$ and $\mathbf{c}$ respectively. We shall now establish the fact that the dual variables will be unrestricted in sign.

10.9 Mixed type problem

When some constraints are equations and some of the variables unrestricted in sign the problem is called as mixed type problem.

10.9.1 Variable unrestricted in sign

The difference of two non-negative variables is a variable which is unrestricted in sign. Let x_1 and x_2 be two non-negative variables. The difference of these two variables is a variable x_3 which is unrestricted in sign. If $x_1 > x_2$, then $x_3 > 0$; if $x_1 < x_2$, then $x_3 < 0$ and if $x_1 = x_2$ then $x_3 = 0$.

10.9.2 Expression of an equation into a set of two inequations

Let

$$\sum_{j=1}^{n} a_{ij}x_j = b_i$$

be a linear equation. It can be expressed as the combination of the two different inequations.

$$\sum_{j=1}^{n} a_{ij}x_j = b_i$$

is equivalent to the intersection of the sets

$$\left.\begin{array}{r} \displaystyle\sum_{j=1}^{n} a_{ij}x_j \leq b_i \\ \text{and} \displaystyle\sum_{j=1}^{n} a_{ij}x_j \geq b_i \end{array}\right\}. \tag{10.9.1}$$

$$\left.\begin{array}{r} \text{or } \sum_{j=1}^{n} a_{ij}x_j \leq b_i \\ \text{and } \sum_{j=1}^{n} (-a_{ij}x_j) \leq -b_i \end{array}\right\}. \tag{10.9.2}$$

Theorem 10.9.1 *If the k-th constraint of a primal be an equation, the k-th dual variable will be unrestricted in sign.*

Before establishing the theorem in general let us first consider a numerical example to give a clear idea about the theorem.

Let the primal problem be, maximize, $z_x = 2x_1 + 3x_2 - x_3$ subject to

$$\begin{array}{l} x_1 + 4x_2 + 2x_3 \leq 7 \\ x_1 - 3x_2 + \ \ x_3 = 4\ , \quad x_1, x_2, x_3 \geq 0. \end{array}$$

Here the second constraint is an equation. We are to establish the fact that the second dual variable w_2 will be unrestricted in sign. The second constraint (equation) can be written by using (10.9.1) and (10.9.2) as follows:

$$\begin{array}{rl} & x_1 - 3x_2 + x_3 \leq \ \ 4 \\ & x_1 - 3x_2 + x_3 \geq \ \ 4 \\ \text{or} & x_1 - 3x_2 + x_3 \leq \ \ 4 \\ & -x_1 + 3x_2 \ \ x_3 \leq -4 \end{array}$$

Hence ultimately there are three inequations expressed in the standard form

$$\begin{array}{r} x_1 + 4x_2 + 2x_3 \leq \ \ 7 \\ x_1 - 3x_2 + \ \ x_3 \leq \ \ 4 \\ -x_1 + 3x_2 - \ \ x_3 \leq -4 \end{array}$$

If three dual variables are $w_1 \geq 0$, $w_2' \geq 0$ and $w_2'' \geq 0$ then by using the rule given in (10.4.2) the dual of the problem is given by,

$$\text{minimize,} \quad z_w = 7w_1 + 4w_2' - 4w_2''$$

subject to

$$\begin{array}{l} w_1 + \ \ w_2' - \ \ w_2'' \geq \ \ 2 \\ 4w_1 - 3w_2' + 3w_2'' \geq \ \ 3 \\ 2w_1 + \ \ w_2' - \ \ w_2'' \geq -1\ , \quad w_1 \geq 0, w_2' \geq 0 \text{ and } w_2'' \geq 0. \end{array}$$

Putting $w_2' - w_2'' = w_2$, the problem can be written as, minimize, $z_w = 7w_1 + 4w_2$

$$\begin{array}{r} w_1 + \ \ w_2 \geq \ \ 2 \\ 4w_1 - 3w_2 \geq \ \ 3 \\ 2w_1 + \ \ w_2 \geq -1 \end{array}$$

and $w_1 \geq 0$, and w_2 is unrestricted in sign as w_2 is the difference of two non-negative variables.

We can also write at once the dual of the primal (unsymmetric) following the rule of transformation given in (10.8.2).

Note: Similarly, if all m constraints of a primal be equations, all m dual variables will be unrestricted in sign.

General proof: As the k-th constraint of the primal is an equation, therefore the primal in the standard form can be written as,

$$\text{maximize,} \quad z_x = c_1x_1 + c_2x_2 + \cdots + c_nx_n$$

subject to

$$\begin{array}{rcccl} a_{11}x_1 + & a_{12}x_2 + \cdots + & a_{1n}x_n & \le & b_1 \\ a_{21}x_1 + & a_{22}x_2 + \cdots + & a_{2n}x_n & \le & b_2 \\ \cdots\cdots & \cdots\cdots & \cdots\cdots & & \cdots \\ a_{k1}x_1 + & a_{k2}x_2 + \cdots + & a_{kn}x_n & \le & b_k \\ -a_{k1}x_1 - & a_{k2}x_2 - \cdots - & a_{kn}x_n & \le & -b_k \\ a_{k+1,1}x_1 + & a_{k+1,2}x_2 + \cdots + & a_{k+1,n}x_n & \le & b_{k+1} \\ \cdots\cdots & \cdots\cdots & \cdots\cdots & & \cdots \\ a_{m1}x_1 + & a_{m2}x_2 + \cdots + & a_{mn}x_n & \le & b_m \end{array}$$

$x_1, x_2, \cdots, x_n \ge 0.$

The dual of the above primal can be written as,

$$\text{minimize, } z_w = b_1w_1 + b_2w_2 + \cdots + b_k(w_k' - w_k'') + b_{k+1}w_{k+1} + \cdots + b_mw_m$$

subject to

$$\begin{array}{l} a_{11}w_1 + a_{21}w_2 + \cdots + a_{k1}(w_k' - w_k'') + a_{k+1,1}w_{k+1} + \cdots + a_{m1}w_m \ge c_1 \\ a_{12}w_1 + a_{22}w_2 + \cdots + a_{k2}(w_k' - w_k'') + a_{k+1,2}w_{k+1} + \cdots + a_{m2}w_m \ge c_2 \\ \cdots\cdots\cdots\cdots\cdots\cdots\cdots\cdots\cdots\cdots\cdots\cdots \\ a_{1n}w_1 + a_{2n}w_2 + \cdots + a_{kn}(w_k' - w_k'') + a_{k+1,n}w_{k+1} + \cdots + a_{mn}w_m \ge c_n. \end{array}$$

Putting $w_k' - w_k'' = w_k$, the dual problem can be written as,

$$\text{minimize, } z_w = b_1w_1 + b_2w_2 + \cdots + b_kw_k + \cdots + b_mw_m$$

subject to

$$\begin{array}{l} a_{11}w_1 + a_{21}w_2 + \cdots + a_{k1}w_k + \cdots + a_{m1}w_m \ge c_1 \\ a_{12}w_1 + a_{22}w_2 + \cdots + a_{k2}w_k + \cdots + a_{m2}w_m \ge c_2 \\ \cdots\cdots\cdots\cdots\cdots\cdots\cdots\cdots\cdots\cdots \\ a_{1n}w_1 + a_{2n}w_2 + \cdots + a_{kn}w_k + \cdots + a_{mn}w_m \ge c_n \end{array}$$

and w_1, w_2, w_{k-1}, w_{k+1}, $\cdots$, $w_m \ge 0$, but w_k is unrestricted in sign as w_k is expressed as the difference of two non-negative variables, w_k' and w_k''. Hence the theorem is proved.

Theorem 10.9.2 *If any variable of the primal problem be unrestricted in sign, the corresponding dual constraint will be strictly an equation.*

Proof. Let the k-th primal variable x_k be unrestricted in sign. Then putting $x_k = x_k' - x_k''$, where x_k' and x_k'' are non-negative variables, the primal problem can be written as,

$$\text{maximize,} \quad z_x = c_1x_1 + c_2x_2 + \cdots + c_k(x_k' - x_k'') + \cdots + c_nx_n$$

subject to

$$\begin{array}{l} a_{11}x_1 + a_{12}x_2 + \cdots + a_{1k}(x'_k - x''_k) + \cdots + a_{1n}x_n \le b_1 \\ a_{21}x_1 + a_{22}x_2 + \cdots + a_{2k}(x'_k - x''_k) + \cdots + a_{2n}x_n \le b_2 \\ \cdots\cdots\cdots\cdots\cdots\cdots\cdots\cdots\cdots\cdots\cdots\cdots \\ a_{m1}x_1 + a_{m2}x_2 + \cdots + a_{mk}(x'_k - x''_k) + \cdots + a_{mn}x_n \le b_m \end{array}$$

and $x_1, x_2, \cdots, x_{k-1}, x'_k, x''_k, \cdots, x_n \ge 0$.

The dual of the above problem is given by,

$$\text{minimize,} \quad z_w = b_1w_1 + b_2w_2 + \cdots + b_mw_m$$

subject to

$$\begin{array}{l} a_{11}w_1 + a_{21}w_2 + \cdots + a_{m1}w_m \ge c_1 \\ a_{12}w_1 + a_{22}w_2 + \cdots + a_{m2}w_m \ge c_2 \\ \cdots\cdots\cdots\cdots\cdots\cdots\cdots\cdots\cdots\cdots \\ \left\{ \begin{array}{l} a_{1k}w_1 + a_{2k}w_2 + \cdots + a_{mk}w_m \ge c_k \\ -a_{1k}w_1 - a_{2k}w_2 - \cdots - a_{mk}w_m \ge -c_k \end{array} \right\} \\ \cdots\cdots\cdots\cdots\cdots\cdots\cdots\cdots\cdots\cdots \\ a_{1n}w_1 + a_{2n}w_2 + \cdots + a_{mn}w_m \ge c_n \end{array}$$

and $w_1, w_2, \cdots, w_m \ge 0$.

The two constraints under the bracket { } are equivalent to an equation

$$a_{1k}w_1 + a_{2k}w_2 + \cdots + a_{mk}w_m = c_k.$$

Hence the k-th dual constraint is an equation which proves the theorem.

▶ **Example 10.9.1** *Find out the dual of the problem, maximize,* $z_x = 2x_1 + 3x_2 - 4x_3$ *subject to*

$$\begin{array}{l} 3x_1 + x_2 + x_3 \le 2 \\ -4x_1 \qquad\quad + 3x_3 \ge 4 \\ x_1 - 5x_2 + x_3 = 5 \end{array}$$

and $x_1 \ge 0$, $x_2 \ge 0$ *and* x_3 *is unrestricted in sign.* [C.U., M.Com.'90] [C.U.(P)'93]

Putting $x_3 = x'_3 - x''_3$ where both x'_3 and x''_3 are non-negative variables, the primal problem can be written in the standard form as,

$$\text{maximize,} \quad z_x = 2x_1 + 3x_2 - 4(x'_3 - x''_3)$$

subject to

$$\begin{array}{l} 3x_1 + x_2 + (x'_3 - x''_3) \le 2 \\ 4x_1 \qquad\quad - 3(x'_3 - x''_3) \le -4 \\ x_1 - 5x_2 + (x'_3 - x''_3) \le 5 \\ -x_1 + 5x_2 - (x'_3 - x''_3) \le -5, \quad x_1 \ge 0, x_2 \ge 0, x'_3 \ge 0, x''_3 \ge 0. \end{array}$$

Remark. Note that the problem is now reduced to a symmetric primal problem.

The dual of above primal problem can be written as,

$$\text{minimize,} \quad z_w = 2w_1 - 4w_2 + 5w'_3 - 5w''_3$$

subject to

$$\begin{aligned} 3w_1 + \ \ 4w_2 + \ \ (w_3' - w_3'') &\geq 2 \\ w_1 - 0 \cdot w_2 - 5(w_3' - w_3'') &\geq 3 \end{aligned}$$

$$\left\{\begin{aligned} w_1 - 3w_2 + (w_3' - w_3'') &\geq -4 \\ -w_1 + 3w_2 - (w_3' - w_3'') &\geq \ \ 4 \end{aligned}\right\}$$

and $w_1, w_2, w_3', w_3'' \geq 0$.

Putting $w_3' - w_3'' = w_3$ and expressing the last two constraints in the form of an equation, the dual problem is,

$$\text{minimize,} \quad z_w = 2w_1 - 4w_2 + 5w_3$$

subject to

$$\begin{aligned} 3w_1 + 4w_2 + \ \ w_3 &\geq \ \ 2 \\ w_1 \qquad\quad - 5w_3 &\geq \ \ 3 \\ w_1 - 3w_2 + \ \ w_3 &= -4 \end{aligned}$$

and $w_1 \geq 0, w_2 \geq 0$ and w_3 is unrestricted is sign.

The dual of the problem can be written at once if we follow the rules established in theorem (10.9.1) and (10.9.2).

(i) As the primal variable x_3 is unrestricted in sign, then the third dual constraint will be an equation {by Theorem (10.9.2)}.

(ii) As the third constraint of the primal is an equation, then the third dual variable will be unrestricted in sign {by Theorem (10.9.1)}.

Now transforming the problem in the form,

$$\text{maximize,} \quad z_w = 2x_1 + 3x_2 - 4x_3$$

subject to

$$\begin{aligned} 3x_1 + \ \ x_2 + \ \ x_3 &\leq \ \ 2 \\ 4x_1 \qquad\quad - 3x_3 &\leq -4 \\ x_1 - 5x_2 + \ \ x_3 &= \ \ 5 \end{aligned} \qquad \text{[all inequations will be of “}\leq\text{” type]}$$

and $x_1 \geq 0, x_2 \geq 0$ and x_3 is unrestricted is sign.

The dual of it can be written as,

$$\text{minimize,} \quad z_w = 2w_1 - 4w_2 + 5w_3$$

subject to

$$\begin{aligned} 3w_1 + 4w_2 + \ \ w_3 &\geq \ \ 2 \\ w_1 \qquad\quad - 5w_3 &\geq \ \ 3 \\ w_1 - 3w_2 + \ \ w_3 &= -4 \end{aligned}$$

$w_1 \geq 0, w_2 \geq 0$ and w_3 is unrestricted in sign.

► **Example 10.9.2** *Find the dual of the following primal problem,*

$$\textit{maximize,} \quad z = x_1 + 4x_2 + 3x_3$$

subject to

$$\begin{aligned} 2x_1 + 3x_2 - 5x_3 &\leq 2 \\ 3x_1 - \ \ x_2 + 6x_3 &\geq 1 \\ x_1 + \ \ x_2 + \ \ x_3 &= 4. \end{aligned}$$

$x_1, x_2 \geq 0$, x_3 *is unrestricted.* [C.U.(P)'92]

Solution. Since x_3 is unrestricted then $x_3 = x_3' - x_3''$, $x_3', x_3'' \geq 0$ and third constraint is equation which can be written in the manner

$$\begin{aligned} x_1 + x_2 + x_3 &\leq 4, \\ x_1 + x_2 + x_3 &\geq 4. \end{aligned}$$

The primal problem after putting the value of x_3, is

$$\text{maximize,} \quad z = x_1 + 4x_2 + 3x_3' - 3x_3''$$

subject to

$$\begin{aligned} 2x_1 + 3x_2 - 5x_3' + 5x_3'' &\leq 2 \\ -3x_1 + x_2 - 6x_3' + 6x_3'' &\leq -1 \\ x_1 + x_2 + x_3' - x_3'' &\leq 4 \\ -x_1 - x_2 - x_3' + x_3'' &\leq -4 \end{aligned}$$

$x_1, x_2, x_3', x_3'' \geq 0$.

The dual problem is,

$$\text{minimize,} \quad z_w = 2w_1 - w_2 + 4(w_3 - w_4)$$

subject to

$$\begin{aligned} 2w_1 - 3w_2 + (w_3 - w_4) &\geq 1 \\ 3w_1 + w_2 + (w_3 - w_4) &\geq 4 \end{aligned}$$

$$\left.\begin{aligned} -5w_1 - 6w_2 + (w_3 - w_4) &\geq 3 \\ 5w_1 + 6w_2 - (w_3 - w_4) &\geq -3 \end{aligned}\right\}$$

$w_1, w_2, w_3, w_4 \geq 0$.

Putting $w_3 - w_4 = w_3^{\circ}$ (which is unrestricted) the dual problem is

$$\text{maximize,} \quad z_w = 2w_1 - w_2 + 4w_3^0$$

subject to

$$\begin{aligned} 2w_1 - 3w_2 + w_3^0 &\geq 1 \\ 3w_1 + w_2 + w_3^0 &\geq 4 \\ -5w_1 - 6w_2 + w_3^0 &= 3\,, \quad w_1 \geq 0, w_2 \geq 0, w_3^0 \text{ is unrestricted.} \end{aligned}$$

Remark: As the 3rd variable x_3 is unrestricted then the 3rd constraint is an equation and as the 3rd constraint of the primal is unrestricted then the 3rd dual variable is unrestricted in sign.

▶ **Example 10.9.3** *Set up the dual of the primal problem,*

$$\textit{minimize,} \quad z = x_1 - 2x_2 + x_3$$

subject to

$$\begin{aligned} 3x_1 - x_2 + 5x_3 &\leq 17 \\ x_1 + 2x_2 - x_3 &= 25 \\ 2x_1 - x_2 + 5x_3 &\geq 57\,, \quad x_1 \geq 0, x_2 \geq 0 \end{aligned}$$

x_3 *is unrestricted in sign.* [C.U.(P)'94]

Solution. Since x_3 is unrestricted then $x_3 = x_3' - x_3''$ and $x_3', x_3'' \geq 0$. Now as the 2nd constraint is equation then we can write the 2nd constraint in the manner

$$x_1 + 2x_2 - x_3 \leq 25,$$
$$x_1 + 2x_2 - x_3 \geq 25.$$

Now putting the value of x_3 and since it is a minimization problem, primal can be written

$$\text{minimize,} \quad z = x_1 - 2x_2 + x_3' - x_3''$$

subject to

$$\begin{aligned} -3x_1 + \ x_2 - 5x_3' + 5x_3'' &\geq -17 \\ -x_1 - 2x_2 + \ x_3' - \ x_3'' &\geq -25 \\ x_1 + 2x_2 - \ x_3' + \ x_3'' &\geq \ 25 \\ 2x_1 - \ x_2 + 5x_3' - 5x_3'' &\geq \ 57, \quad x_1, x_2, x_3', x_3'' \geq 0. \end{aligned}$$

The dual of it is

$$\text{maximize,} \quad z_w = -17w_1 - 25w_2 + 25w_3 + 57w_4$$

subject to

$$\begin{aligned} -3w_1 - \ w_2 + \ w_3 + 2w_4 &\leq \ 1 \\ w_1 - 2w_2 + 2w_3 - \ w_4 &\leq -2 \end{aligned}$$

$$\left.\begin{aligned} -5w_1 + w_2 - w_3 + 5w_4 &\leq \ 1 \\ 5w_1 - w_2 + w_3 - 5w_4 &\leq -1 \end{aligned}\right\}$$

w_1, w_2, w_3, $w_4 \geq 0$, which can be written as

$$\text{maximize,} \quad z_w = -17w_1 - 25(w_2 - w_3) + 57w_4$$

subject to

$$\begin{aligned} -3w_1 - \ (w_2 - w_3) + 2w_4 &\leq \ 1 \\ w_1 - 2(w_2 - w_3) - \ w_4 &\leq -2 \\ -5w_1 + \ (w_2 - w_3) + 5w_4 &= \ 1, \quad w_1, w_2, w_3 w_4 \geq 0. \end{aligned}$$

Putting $w_2 - w_3 = w_2^0$ the problem is

$$\text{maximize,} \quad z_w = -17w_1 - 25w_2^0 + 57w_4$$

subject to

$$\begin{aligned} -3w_1 - \ w_2^0 + 2w_4 &\leq \ 1 \\ w_1 - 2w_2^0 - \ w_4 &\leq -2 \\ -5w_1 + \ w_2^0 - 5w_4 &= \ 1, \quad w_1 \geq 0, w_4 \geq 0 \end{aligned}$$

and w_2^0 is unrestricted in sign.

▶ **Example 10.9.4** *Find out the dual of the problem,*

$$\textit{minimize,} \quad z = 2x_1 - x_2 + 3x_3$$

subject to

$$\begin{aligned} -x_1 + 2x_2 - 4x_3 &\leq 2 \\ 2x_1 - 3x_2 + \ x_3 &= 6 \end{aligned}$$

x_1 *is unrestricted* $x_2 \geq 0, x_3 \geq 0$.

Solution. (1) In this problem first constraint is to be converted in the manner

$$x_1 - 2x_2 + 4x_3 \geq -2$$

[as in the minimization problem, all inequations must be "≥"]. (2) The primal has two constraints with three variables. Thus the dual will contain three constraints with two variables. (3) In the primal problem x_1 is unrestricted. Therefore first dual constraint will be an equation. (4) In the primal second constraint is an equation. Therefore, the second dual variable will be unrestricted in sign. The constraints are

$$\begin{aligned} x_1 - 2x_2 + 4x_3 &\geq -2 \\ 2x_1 - 3x_2 + x_3 &= 6. \end{aligned}$$

The dual problem is thus,

$$\text{maximize}, \quad z_w = -2w_1 + 6w_2$$

subject to

$$\begin{aligned} w_1 + 2w_2 &= 2 \\ -2w_1 - 3w_2 &\leq -1 \\ 4w_1 + w_2 &\leq 3, \quad w_1 \geq 0, w_2 \text{ unrestricted.} \end{aligned}$$

▶ **Example 10.9.5** *Find the dual of the problem,*

$$\textit{maximize}, \quad z = 2x_1 + x_3$$

subject to

$$\begin{aligned} 4x_1 - 5x_2 + x_3 &= 0 \\ x_1 + 2x_2 + 3x_3 &\leq 7, \quad x_1 \geq 0, x_2 \geq 0, \end{aligned}$$

x_3 is unrestricted.

Using the property of the theorems (10.9.1) and (10.9.2) the dual problem can be written at once as given by,

$$\text{minimize}, \quad z_w = 0 \cdot w_1 + 7w_2$$

subject to

$$\begin{aligned} 4w_1 + w_2 &\geq 2 \\ -5w_1 + 2w_2 &\geq 0 \\ w_1 + 3w_2 &= 1 \end{aligned}$$

w_1 is unrestricted, $w_2 \geq 0$.

▶ **Example 10.9.6** *Solve the L.P.P.*

$$\textit{Maximize}, \quad z_x = 4x_1 + 3x_2$$

subject to

$$\begin{aligned} x_1 \qquad &\leq 6 \\ x_2 &\leq 8 \\ x_1 + x_2 &\leq 7 \\ 3x_1 + x_2 &\leq 15 \\ -x_2 &\leq 1, \quad x_1, x_2 \geq 0 \end{aligned}$$

by using duality theory.

We shall first solve the dual problem of it.

The dual of the given problem is,

$$\text{minimize,} \quad z_w = 6w_1 + 8w_2 + 7w_3 + 15w_4 + w_5$$

subject to

$$\begin{aligned} w_1 + 0 \cdot w_2 + w_3 + 3w_4 + 0 \cdot w_5 &\geq 4 \\ 0 \cdot w_1 + w_2 + w_3 + w_4 - w_5 &\geq 3, \quad w_1, w_2, w_3, w_4, w_5 \geq 0. \end{aligned}$$

Let

$$-z_w = z_w^* = -6w_1 - 8w_2 - 7w_3 - 15w_4 - w_5.$$

Introducing the surplus variables w_6 and w_7 one to each of the constraints, the converted equations are

$$\begin{aligned} w_1 + 0 \cdot w_2 + w_3 + 3w_4 + 0 \cdot w_5 - w_6 \quad &= 4 \\ 0 \cdot w_1 + w_2 + w_3 + w_4 - w_5 \quad - w_7 &= 3. \end{aligned}$$

The problem is a problem of maximization of z_w^* and $\min(z_w) = -\max(z_w^*)$.

No artificial vector is required to get a unit basis. $\mathbf{a}_1(\mathbf{e}_1)$ and $\mathbf{a}_2(\mathbf{e}_2)$ constitute the basis. Initial B.F.S. $\mathbf{w}_B = [w_1, w_2] = [4, 3]$

$$\mathbf{c}_B = (c_1, c_2) = (-6, -8)$$
$$z^* = \mathbf{c}_B \mathbf{w}_B = -48.$$

Simplex Tables

		c	−6	−8	−7	−15	−1	0	0	
Basis	$\mathbf{c}_B$	**b**	$\mathbf{a}_1$	$\mathbf{a}_2$	$\mathbf{a}_3$	$\mathbf{a}_4$	$\mathbf{a}_5$	$\mathbf{a}_6$	$\mathbf{a}_7$	Min. ratio
$\mathbf{a}_1^*$	−6	4	1	0	1	3*	0	−1	0	$\frac{4}{3}$*
$\mathbf{a}_2$	−8	3	0	1	1	1	−1	0	−1	$\frac{3}{1}$
$z_j - c_j$		−48	0	0	−7	−11*	9	6	8	
$\mathbf{a}_4$	−15	$\frac{4}{3}$	$\frac{1}{3}$	0	$\frac{1}{3}$	1	0	$-\frac{1}{3}$	0	$\frac{4}{3}/\frac{1}{3}$
$\mathbf{a}_2^*$	−8	$\frac{5}{3}$	$-\frac{1}{3}$	1	$\frac{2}{3}$*	0	−1	$\frac{1}{3}$	−1	$\frac{5}{3}/\frac{2}{3}$*
$z_j - c_j$		$-\frac{100}{3}$	$\frac{11}{3}$	0	$-\frac{10}{3}$*	0	9	$\frac{7}{3}$	8	
$\mathbf{a}_4$	−15	$\frac{1}{2}$	$\frac{1}{2}$	$-\frac{1}{2}$	0	1	$\frac{1}{2}$	$-\frac{1}{2}$	$\frac{1}{2}$	
$\mathbf{a}_3$	−7	$\frac{5}{2}$	$-\frac{1}{2}$	$\frac{3}{2}$	1	0	$-\frac{3}{2}$	$\frac{1}{2}$	$-\frac{3}{2}$	
$z_j - c_j$		−25	2	5	0	0	4	4	3	

As none of $z_j - c_j < 0$ therefore the solution is optimal.

Maximum value of $z_w^* = -25$ for $w_3 = \frac{5}{2}$, $w_4 = \frac{1}{2}$ [other variables are non-basic].

Hence $\min z_w = -\max z_w^* = -(-25)$ for $w_3 = \frac{5}{2}$ and $w_4 = \frac{1}{2}$.

The maximum value of the primal objective function $\max z_x = \min z_w = 25$. The dual problem is solved as a maximization problem.

10.9.3 Determination of primal optimal solution

$\mathbf{a}_6(-\mathbf{e}_1)$ and $\mathbf{a}_7(-\mathbf{e}_2)$ are the surplus vectors.

$z_6 - c_6 = 4$ and $z_7 - c_7 = 3$ in the final dual simplex table.

Therefore $x_1 = 4$ and $x_2 = 3$ corresponding to $-\mathbf{e}_1$ and $-\mathbf{e}_2$ respectively.

Hence $\max z_x = 25$ for $x_1 = 4$ and $x_2 = 3$.

Note: No artificial variable is introduced to get the initial unit basis. But this does not make any hurdle to determine the primal optimal solution.

▶ **Example 10.9.7** *Solve the following L.P.P. by usual simplex method without using any artificial variable.*

$$\textit{Minimize,} \quad z = x_1 + x_2$$

subject to

$$x_1 + 2x_2 \geq 12$$
$$5x_1 + 6x_2 \geq 48\ , \quad x_1 \geq 0, x_2 \geq 0.$$

Solution. In the usual simplex method, generally initial basis is a unit basis. But in this problem, there will be no unit basis in the absence of artificial variables. Let us now write down the dual of the problem and dual is given by,

$$\text{maximize,} \quad z_w = 12w_1 + 48w_2$$

subject to

$$w_1 + 5w_2 \leq 1$$
$$2w_1 + 6w_2 \leq 1\ , \quad w_1 \geq 0, w_2 \geq 0.$$

In this problem we shall get a unit basis after the introduction of two slack variables only and the constraints are thus

$$w_1 + 5w_2 + w_3 \qquad = 1$$
$$2w_1 + 6w_2 \qquad + w_4 = 1\ , \quad w_j \geq 0, [j = 1, 2, \cdots, 4.]$$

Now solving as usual, the simplex tables are

		c	12	48	0	0
Basis	$\mathbf{c}_B$	$\mathbf{b}$	$\mathbf{a}_1$	$\mathbf{a}_2$	$\mathbf{a}_3$	$\mathbf{a}_4$
$\mathbf{a}_3$	0	1	1	5	1	0
$\mathbf{a}_4^*$	0	1	2	6*	0	1
$z_j - c_j$		0	−12	−48*	0	0
$\mathbf{a}_3$	0	$\frac{1}{6}$	$-\frac{2}{3}$	0	1	$-\frac{5}{6}$
$\mathbf{a}_2$	48	$\frac{1}{6}$	$\frac{1}{3}$	1	0	$\frac{1}{6}$
$z_j - c_j$		8	4	0	0	8

From the above table we have $\min z = \max z_w = 8$ at $w_1 = 0$, $w_2 = \frac{1}{6}$ and $x_1 = 0$, $x_2 = 8$. Thus the first problem has an optimal solution at $x_1 = 0$, $x_2 = 8$.

10.9.4 Solution of mixed type primal dual problem

▶ **Example 10.9.8** *Find the optimal solution of the dual of the problem, maximize z where $z = 5x_1 + 12x_2 + 4x_3$, subject to*

$$\begin{aligned} x_1 + 2x_2 + x_3 &\le 5 \\ 2x_1 - x_2 + 3x_3 &= 2\,, \quad x_1, x_2, x_3 \ge 0. \end{aligned}$$

from the simplex table of the primal. [C.U.(P)'90]

Solving the primal problem using the simplex method, the optimal simplex table is

		c	5	12	4	0	$-M$
Basis	$\mathbf{c}_B$	**b**	$\mathbf{a}_1$	$\mathbf{a}_2$	$\mathbf{a}_3$	$\mathbf{a}_4(\mathbf{e}_1)$	$\mathbf{a}_5(\mathbf{e}_2)$
$\mathbf{a}_2$	12	$\frac{8}{5}$	0	1	$-\frac{1}{5}$	$\frac{2}{5}$	$-\frac{1}{5}$
$\mathbf{a}_1$	5	$\frac{9}{5}$	1	0	$\frac{7}{5}$	$\frac{1}{5}$	$\frac{2}{5}$
$z_j - c_j$		$28\frac{1}{5}$	0	0	$\frac{3}{5}$	$\frac{29}{5}$	$M - \frac{2}{5}$

From the optimal table, we have $\max z = 28\frac{2}{5}$ at $x_1 = \frac{9}{5}, x_2 = \frac{8}{5}, x_3 = 0$.

There are two methods of finding the dual optimal solution.

(i) Final basis

$$B = (\mathbf{a}_2, \mathbf{a}_1), B^{-1} = \begin{bmatrix} 2/5 & -1/5 \\ 1/5 & 2/5 \end{bmatrix}, \quad \mathbf{c}_B = (12, 5).$$

The dual optimal solution,

$$\mathbf{c}_B B^{-1} = (12, 5)\begin{bmatrix} 2/5 & -1/5 \\ 1/5 & 2/5 \end{bmatrix} = \begin{bmatrix} 29/5 \\ -2/5 \end{bmatrix}.$$

Thus the dual optimal solution $w_1^* = \frac{29}{5}, w_2^* = -\frac{2}{5}$ and $\min z_w = \max z = 28\frac{1}{5}$.

(ii) Here the column corresponding to the artificial vector $\mathbf{a}_5(\mathbf{e}_2)$ is essential and we shall not remove that column. $\mathbf{a}_4(\mathbf{e}_1)$ and $\mathbf{a}_5(\mathbf{e}_2)$ are the initial unit basis. Then the dual optimal solution is $w_1^* = \frac{29}{5}$ and $w_2^* = -\frac{2}{5}$ [putting $M = 0$ which are $z_4 - c_4$ and $z_5 - c_5$ in the final simplex table]. Here w_2^* is unrestricted (which is $-$ve here).

Note. The dual objective function is

$$z_w = 5w_1 + 2w_2 \text{ and } 5 \times \frac{29}{5} + 2\left(-\frac{2}{5}\right) = \frac{141}{5} = 28\frac{1}{5}$$

and thus the final result is being checked or verified.

Objective and Short Answer Type Questions with Answers

Duality theory

1. Write down the dual of the given problem, maximize, $z = \mathbf{cx}$ subject to $A\mathbf{x} \le \mathbf{b}, \mathbf{x} \ge \mathbf{0}$.

2. Write down the dual of the problems given below:

   (a) Minimize, $z = \mathbf{cx}$ subject to $D\mathbf{x} \geq \mathbf{d}, \mathbf{x} \geq \mathbf{0}$.

   (b) Minimize, $z = \mathbf{cx}$ subject to $D\mathbf{x} \leq \mathbf{b}, \mathbf{x} \geq \mathbf{0}$.

3. Write down the dual of the following problems:

   (a) Maximize, $z = 4x_1 + 2x_2$

$$\begin{aligned} \text{subject to } 3x_1 + 4x_2 &\leq 7 \\ 7x_1 - 2x_2 &\leq 13, \quad x_1 \geq 0, x_2 \geq 0. \end{aligned}$$

   (b) Maximize, $z = 3x_1 - x_2$

$$\begin{aligned} \text{subject to } 3x_1 + x_2 &\leq 14 \\ 2x_1 - 4x_2 &\geq 4, \quad x_1 \geq 0, x_2 \geq 0. \end{aligned}$$

   (c) Minimize, $z_x = 7x_1 - x_2 + x_3$

$$\begin{aligned} \text{subject to } 2x_1 + x_2 + x_3 &\geq 6 \\ x_1 + x_2 + 3x_3 &\geq 16, \quad x_1 \geq 0, x_2 \geq 0, x_3 \geq 0. \end{aligned}$$

   (d) Minimize, $z_x = 7x_1 + x_2$

$$\begin{aligned} \text{subject to } x_1 + 3x_2 &\leq 17 \\ x_1 - 4x_2 &\leq 6, \quad x_1 \geq 0, x_2 \geq 0. \end{aligned}$$

4. Write down the dual of the given problem, minimize, $2x_1 + 3x_2$

$$\begin{aligned} \text{subject to } 4x_1 + 5x_2 &\geq 7 \\ 3x_1 - 2x_2 &\geq 2, \quad x_1 \geq 0, x_2 \geq 0 \end{aligned}$$

and hence establish the fact that no artificial variable is needed to solve the dual problem in usual simplex method.

[*Ans.* The dual is: maximize, $z_w = 7w_1 + 2w_2$

$$\begin{aligned} \text{subject to } 4w_1 + 3w_2 &\leq 2 \\ 5w_1 - 2w_2 &\leq 3, \quad w_1 \geq 0, w_2 \geq 0 \end{aligned}$$

and hence two slack vectors will constitute the initial unit basis of the dual problem, $\mathbf{b} \geq \mathbf{0}$ and no artificial variable is needed to get the initial unit basis.]

5. Justify or correct the following statements:

   (a) The dual of the maximization problem is a minimization problem.

   (b) The dual of a dual problem is the primal problem.

   (c) If the kth constraint of a primal be an equation, the kth dual variable will be non-negative.

   (d) If the kth variable of a primal problem is unrestricted in sign, the kth dual constraint will be strictly an equation.

   (e) If the primal problem is unbounded, the dual may have finite value of the objective function.

(f) If the primal problem has m constraints with n variables the dual will have the same number of m constraints with n variables.

(g) If either of the primal or dual has finite optimal solution then the other has also finite optimal solution.

(h) If the primal problem has no feasible solution, then the dual will also have no feasible solution.

[*Ans.* (a) Correct (b) Correct (c) If the kth constraint of a primal be an equation, the kth dual variable will be unrestricted in sign (d) Correct (e) If the primal problem is unbounded, the dual will have no feasible solution. (f) If the primal problem has m constraints with n variables then dual will have n constraints with m variables. (g) Correct (h) If the primal problem has no feasible solution, then the dual will have either no F.S. or unbounded solution.]

6. Define (i) symmetric primal and dual, (ii) unsymmetric primal and dual (iii) mixed type primal and dual.

[*Ans.* See 10.4 and 10.9.]

7. If $\mathbf{x}^*$ and $\mathbf{w}^*$ be any two feasible solutions of symmetric primal (maximization problem) and dual problem with objective functions $\mathbf{cx}$ and $\mathbf{b'w}$ respectively, then which one of the statements is correct?

(a) $\mathbf{cx}^* \le \mathbf{b'w}^*$, (b) $\mathbf{cx}^* = \mathbf{b'w}^*$, (c) $\mathbf{cx}^* \ge \mathbf{b'w}^*$

[*Ans.* $\mathbf{cx}^* \le \mathbf{b'w}^*$.]

8. What is the relation between the optimal values of primal and dual problems [assume that both exist]?

[*Ans.* If the primal be (maximization) with objective function z_x and obviously dual (minimization) with objective function z_w then $\max z_x = \min z_w$.]

9. You are given a problem (L.P.) with five constraints with two variables. Which one of the problems, primal or dual will you select to solve? Justify your answer.

[*Ans.* Dual problem will be selected to solve as the dual will have only two constraints with five variables which can be solved easily.]

10. In a primal problem, the 4th constraint is an equation and the 3rd variable is unrestricted in sign. What will be nature of the 4th dual variable and the 3rd dual constraint?

[*Ans.* 4th dual variable will be unrestricted in sign and the 3rd dual constraint will be strictly an equation.]

11. How can you determine the primal optimal solution from the dual simplex table in the case of symmetric problem?

[*Ans.* The values of $z_j - c_j$ for the columns corresponding to the slack and surplus vectors in the final simplex table of dual problem are the values of the corresponding primal optimal variables provided the dual problem is solved as a maximization problem (whether the actual dual problem is a maximization or a minimization problem).]

12. State the necessary and sufficient condition that both the primal and dual will have finite optimal solutions.

(a) If either the primal or dual has finite optimal solution then both the problems will have finite optimal solutions.

Or,

(b) The necessary and sufficient condition that both problems will have finite optimal solutions is the existence of feasible solutions of both problems.

Exercise 10A

Obtain the dual of the given L.P. problems:

1. Maximize, $z = 2x_1 + 3x_2 - 4x_3$

$$\text{subject to } \begin{aligned} 5x_1 - 2x_2 + x_3 &\le 4 \\ x_1 + x_2 - 4x_3 &\le 7, \quad x_1, x_2, x_3 \ge 0. \end{aligned}$$

[C.U.(P)'90 (Old)]

2. (a) Minimize, $z = x_1 - x_2 + 2x_3$

$$\text{subject to } \begin{aligned} x_1 + x_2 + 4x_3 &\ge 7 \\ x_2 - 2x_3 &\ge 10 \\ 3x_1 + x_2 + x_3 &\le 3, \quad x_1, x_2, x_3 \ge 0. \end{aligned}$$

(b) Maximize, $z = x_1 - x_2 + 3x_3$

$$\text{subject to } \begin{aligned} x_1 + x_2 + x_3 &\le 10 \\ 2x_1 - x_3 &\le 2 \\ 2x_1 - 2x_2 + 3x_3 &\le 6, \quad x_1, x_2, x_3 \ge 0. \end{aligned}$$

[C.U.(P)'83]

3. Maximize, $z = -x_1 + 2x_2 + 7x_3$

$$\text{subject to } \begin{aligned} 4x_1 + x_2 + x_3 &\le 2 \\ 7x_1 - x_2 + 5x_3 &\ge 7 \\ 2x_1 + x_2 - x_3 &\le 4, \quad x_1 \ge 0, x_2 \ge 0 \end{aligned}$$

and x_3 is unrestricted in sign.

4. Maximize, $z = x_1 + x_2 + x_3$

$$\text{subject to } \begin{aligned} x_1 - 3x_2 + 4x_3 &= 5 \\ x_1 - 2x_2 &\ge 3 \\ 2x_2 - x_3 &\le 4, \quad x_1, x_2 \ge 0 \end{aligned}$$

and x_3 is unrestricted.

5. Maximize, $z = 3x_1 + x_2 + x_3 - x_4$

$$\begin{aligned} \text{subject to } x_1 + 5x_2 + 3x_3 + 4x_4 &\le 5 \\ x_1 + x_2 &= -1 \\ x_3 - x_4 &\le -5, \quad x_1, x_2, x_3, x_4 \ge 0. \end{aligned}$$

6. Maximize, $z = 3x_1 - 2x_2$

$$\begin{aligned} \text{subject to } x_1 &\le 4 \\ x_2 &\le 6 \\ x_1 + x_2 &\le 5 \\ -x_2 &\le -1, \quad x_1, x_2 \ge 0. \end{aligned}$$

7. Maximize, $z = 2x_1 - 3x_3$

$$\begin{aligned} \text{subject to } x_1 - 2x_2 + x_3 &\ge 0 \\ 2x_1 + 4x_2 + x_3 &\le 27, \quad x_1 \ge 0, x_2 \ge 0, x_3 \ge 0. \end{aligned}$$

8. Minimize, $z = x_1 + x_2 + x_3$

$$\begin{aligned} \text{subject to } 3x_1 - x_2 + x_3 &= 4 \\ 2x_1 + x_2 - x_3 &\le 8 \end{aligned}$$

x_1 is unrestricted in sign, $x_2 \ge 0, x_3 \ge 0$.

9. Maximize, $z = 2x_1 + x_2$

$$\begin{aligned} \text{subject to } 2x_1 + x_2 &\ge 9 \\ 3x_1 &\ge 6 \\ 2x_1 - 3x_2 &= 1 \end{aligned}$$

$x_1 \ge 0$ and x_2 is unrestricted in sign.

10. Minimize, $z = 2x_2 + 3x_3$

$$\begin{aligned} \text{subject to } 2x_1 + x_2 - x_3 &\le 4 \\ -x_1 - x_2 + 4x_3 &\le 16 \end{aligned}$$

x_1 is unrestricted in sign, $x_2 \ge 0, x_3 \ge 0$.

11. Maximize, $z = x_1 - x_2 - x_4$

$$\begin{aligned} \text{subject to } 2x_1 + x_2 - x_3 &= 10 \\ x_2 + x_3 - x_4 &\le 0 \\ x_1 + x_3 + 2x_4 &\ge 6, \quad x_1 \ge 0, x_2 \ge 0, \end{aligned}$$

x_3 is unrestricted in sign $x_4 \ge 0$.

12. (a) Minimize, $z = 2x_1 - x_3 + x_4$

$$\begin{aligned} \text{subject to } x_1 - x_2 - x_3 + x_4 &= 8 \\ 3x_1 + x_3 - 2x_4 &\ge 9, \quad x_1 \ge 0, x_2 \ge 0, x_3 \ge 0, \end{aligned}$$

x_4 is unrestricted in sign.

(b) Maximize, $z = 2x_1 + 3x_2 + x_3$

$$\begin{aligned} \text{subject to } 4x_1 + 3x_2 + x_3 &= 6 \\ x_1 + x_2 + 5x_3 &\le 4, \quad x_j \ge 0 [j = 1, 2, 3.] \end{aligned}$$

[C.U.(H)'80]

13. (a) Construct the dual of the following problem,

minimize, $z = x_1 + 3x_2$

$$\begin{aligned} \text{subject to } \quad x_1 + 2x_2 &\le -6 \\ x_1 - x_2 &\le 2 \\ 2x_1 - x_2 &= 4 \\ -x_1 + x_2 &\ge 7, \quad x_1 \ge 0, x_2 \ge 0. \end{aligned}$$

[C.U., M.Com,'89]

(b) Write down the dual of the following primal,

minimize, $z = x_1 - 2x_2 - x_3 + 2x_4 + x_5$

$$\begin{aligned} \text{subject to } 3x_1 - x_2 + 5x_3 - 7x_4 + x_5 &\le 17 \\ x_1 + 2x_2 - x_3 + 5x_4 + 2x_5 &= 25 \\ 2x_1 - x_2 + 5x_3 - x_4 + 3x_5 &\ge 57, \; x_j \ge 0 [j = 1, 2, \cdots, 5.] \end{aligned}$$

[C.U.(P)'87]

14. Find the dual problem for the following problem,

maximize, $z = 7x_1 + 5x_2 - 2x_3$

$$\begin{aligned} \text{subject to } \quad x_1 + x_2 + x_3 &= 10 \\ 2x_1 - x_2 + 3x_3 &\le 16 \\ 3x_1 + x_2 - 2x_3 &\ge -5, \quad x_1 \ge 0, x_2 \ge 0, \end{aligned}$$

x_3 unrestricted. [C.U., M.Com'89]

15. (a) Obtain the dual of the L.P.P., minimize, $z = x_1 + x_2 + x_3$

subject to the constraints

$$\begin{aligned} x_1 - 3x_2 + 4x_3 &= 5 \\ x_1 - 2x_2 \quad\quad &\le 3 \\ 2x_2 - x_3 &\ge 4, \quad x_1, x_2 \ge 0 \end{aligned}$$

and x_3 is unrestricted in sign. [C.U.(H)'88]

(b) Find the dual of the following programme,

maximize, $z = 2x_1 + 3x_2 + x_3$

$$\begin{aligned} \text{subject to } 4x_1 + 3x_2 + x_3 &= 6 \\ x_1 + 2x_2 + 5x_3 &= 4, \quad x_1, x_2, x_3 \ge 0. \end{aligned}$$

[C.U., M.Com'86(P),'90,'99]

Answers

1. Minimize $z_w = 4w_1 + 7w_2$

$$\begin{aligned} \text{subject to } \quad 5w_1 + w_2 &\geq 2 \\ -2w_1 + w_2 &\geq 3 \\ w_1 - 4w_2 &\geq -4, \quad w_1 \geq 0, w_2 \geq 0. \end{aligned}$$

2. (a) Maximize, $z_w = 7w_1 + 10w_2 - 3w_3$

$$\begin{aligned} \text{subject to } \quad w_1 \qquad - 3w_3 &\leq 1 \\ w_1 + w_2 - w_3 &\leq -1 \\ 4w_1 - 2w_2 - w_3 &\leq 2, \quad w_1, w_2, w_3 \geq 0. \end{aligned}$$

(b) Minimize, $z_w = 10w_1 + 2w_2 + 6w_3$

$$\begin{aligned} \text{subject to } \quad w_1 + 2w_2 + 2w_3 &\geq 1 \\ w_1 \qquad - 2w_3 &\geq -1 \\ w_1 - w_2 + 3w_3 &\geq 3, \quad w_1 \geq 0, w_2 \geq 0, w_3 \geq 0. \end{aligned}$$

3. Minimize, $z_w = 2w_1 - 7w_2 + 4w_3$

$$\begin{aligned} \text{subject to } \quad 4w_1 - 7w_2 + 2w_3 &\geq -1 \\ w_1 + w_2 + w_3 &\geq 2 \\ w_1 - 5w_2 - w_3 &= 7, \quad w_1 \geq 0, w_2 \geq 0, w_3 \geq 0. \end{aligned}$$

4. Minimize $z_w = 5w_1 - 3w_2 + 4w_3$

$$\begin{aligned} \text{subject to } \quad w_1 - w_2 \qquad &\geq 1 \\ -3w_1 + 2w_2 + 2w_3 &\geq 1 \\ 4w_1 \qquad - w_3 &= 1: \end{aligned}$$

w_1 is unrestricted in sign, $w_2 \geq 0$, $w_3 \geq 0$.

5. Minimize, $z_w = 5w_1 - w_2 - 5w_3$

$$\begin{aligned} \text{subject to } \quad w_1 + w_2 \qquad &\geq 3 \\ 5w_1 + w_2 \qquad &\geq 1 \\ 3w_1 \qquad + w_3 &\geq 1 \\ 4w_1 \qquad - w_3 &\geq -1, \quad w_1 \geq 0, w_3 \geq 0, \end{aligned}$$

w_2 is unrestricted in sign.

6. Minimize, $z_w = 4w_1 + 6w_2 + 5w_3 - w_4$

$$\begin{aligned} \text{subject to } \quad w_1 \qquad + w_3 \qquad &\geq 3 \\ w_2 + w_3 - w_4 &\geq -2, \quad w_1 \geq 0, w_2 \geq 0, w_3 \geq 0, w_4 \geq 0. \end{aligned}$$

7. Minimize, $z_w = 27w_2$

$$\begin{aligned} \text{subject to } \quad -w_1 + 2w_2 &\geq 2 \\ 2w_1 + 4w_2 &\geq 0 \\ -w_1 + w_2 &\geq -3, \quad w_1 \geq 0, w_2 \geq 0. \end{aligned}$$

8. Maximize, $z_w = 4w_1 - 8w_2$

$$\begin{aligned} \text{subject to } \quad 3w_1 - 2w_2 &= 1 \\ -w_1 - w_2 &\leq 1 \\ w_1 + w_2 &\leq 1 \end{aligned}$$

w_1 is unrestricted in sign, $w_2 \geq 0$.

9. Minimize, $z_w = -9w_1 - 6w_2 + w_3$

$$\begin{aligned} \text{subject to } \quad -2w_1 - 3w_2 + 2w_3 &\geq 2 \\ -w_1 \qquad - 3w_3 &= 1, \quad w_1 \geq 0, w_2 \geq 0, \end{aligned}$$

w_3 is unrestricted in sign.

10. Maximize, $z_w = -4w_1 - 16w_2$

$$\text{subject to } \begin{aligned} -2w_1 + w_2 &= 0 \\ -w_1 + w_2 &\le 2 \\ w_1 - 4w_2 &\le 3, \quad w_1 \ge 0, w_2 \ge 0. \end{aligned}$$

11. Minimize, $z_w = 10w_1 - 6w_3$

$$\text{subject to } \begin{aligned} 2w_1 \quad\quad - w_3 &\ge 1 \\ w_1 + w_2 \quad\quad &\ge -1 \\ -w_1 + w_2 - w_3 &= 0 \\ - w_2 - 2w_3 &\ge -1, \end{aligned}$$

w_1 is unrestricted in sign $w_2 \ge 0, w_3 \ge 0$.

12. (a) Maximize, $z_w = 8w_1 + 9w_2$

$$\text{subject to } \begin{aligned} w_1 + 3w_2 &\le 2 \\ -w_1 \quad\quad &\le 0 \\ -w_1 + w_2 &\le -1 \\ w_1 - 2w_2 &= 1, \end{aligned}$$

w_1 is unrestricted in sign, $w_2 \ge 0$.

(b) Minimize, $z_w = 6w_1 + 4w_2$

$$\text{subject to } \begin{aligned} 4w_1 + w_2 &\ge 2 \\ 3w_1 + w_2 &\ge 3 \\ w_1 + 5w_2 &\ge 1, \end{aligned}$$

w_1 is unrestricted, $w_2 \ge 0$.

13. (a) Maximize, $z_w = 6w_1 - 2w_2 + 4w_3 + 7w_4$

$$\text{subject to } \begin{aligned} -w_1 - w_2 + 2w_3 - w_4 &\le 1 \\ -2w_1 + w_2 - w_3 + w_4 &\le 3, \quad w_1, w_2, w_4 \ge 0, \end{aligned}$$

w_3 is unrestricted.

(b) Maximize, $z_w = -17w_1 + 25w_2 + 57w_3$

$$\text{subject to } \begin{aligned} -3w_1 + w_2 + 2w_3 &\le 1 \\ w_1 + 2w_2 - w_3 &\le -2 \\ -5w_1 - w_2 + 5w_3 &\le -1 \\ 7w_1 + 5w_2 - w_3 &\le 2 \\ -w_1 + 2w_2 + 3w_3 &\le 1, \quad w_1, w_2 \ge 0, \end{aligned}$$

w_2 is unrestricted.

14. Minimize, $z_w = 10w_1 + 16w_2 + 5w_3$

$$\text{subject to } \begin{aligned} w_1 + 2w_2 - 3w_3 &\ge 7 \\ w_1 - w_2 - w_3 &\ge 5 \\ w_1 + 3w_2 + 2w_3 &= -2, \end{aligned}$$

w_1 unrestricted, $w_2, w_3 \ge 0$.

15. (a) Maximize, $z_w = 5w_1 - 3w_2 + 4w_3$

$$\text{subject to } \begin{aligned} w_1 - w_2 \quad\quad\quad &\le 1 \\ -3w_1 + 2w_2 + 2w_3 &\le 1 \\ 4w_1 \quad\quad\quad - w_3 &= 1, \end{aligned}$$

w_1 unrestricted, $w_2, w_3 \ge 0$.

(b) Minimize, $z_w = 6w_1 + 4w_2$

$$\text{subject to } \begin{aligned} 4w_1 + w_2 &\ge 2 \\ 3w_1 + 2w_2 &\ge 3 \\ w_1 + 5w_2 &\ge 1, \quad w_1, w_2 \text{ unrestricted in sign.} \end{aligned}$$

Exercise 10B

Solving the dual problem obtain the optimum solution, if any to each of the following L.P.P.

1. Minimize, $z = 4x_1 + 3x_2$

$$\text{subject to } \begin{aligned} x_1 + 2x_2 &\geq 8 \\ 3x_1 + 2x_2 &\geq 12, \quad x_1, x_2 \geq 0. \end{aligned}$$

2. Minimize, $z = 15x_1 + 10x_2$

$$\text{subject to } \begin{aligned} 3x_1 + 5x_2 &\geq 5 \\ 5x_1 + 2x_2 &\geq 3, \quad x_1, x_2 \geq 0. \end{aligned}$$

[Cal. Applied Math.'79]

3. Minimize, $z = 3x_1 + x_2$

$$\text{subject to } \begin{aligned} x_1 + x_2 &\geq 1 \\ 2x_1 + 3x_2 &\geq 2, \quad x_1, x_2 \geq 0. \end{aligned}$$

4. Minimize, $z = 2x_1 + 2x_2$

$$\text{subject to } \begin{aligned} 2x_1 + 4x_2 &\geq 1 \\ x_1 + 2x_2 &\geq 1 \\ 2x_1 + x_2 &\geq 1, \quad x_1, x_2 \geq 0. \end{aligned}$$

5. Maximize, $z = 2x_1 + 3x_2$

$$\text{subject to } \begin{aligned} -x_1 + 2x_2 &\leq 4 \\ x_1 + x_2 &\leq 6 \\ x_1 + 3x_2 &\leq 9, \quad x_1, x_2 \geq 0. \end{aligned}$$

6. Solve the primal problem, minimize, $z = -2x_1 + 3x_2$

$$\text{subject to } \begin{aligned} 2x_1 - 5x_2 &\leq 7 \\ 4x_1 + x_2 &\leq 8 \\ 7x_1 + 2x_2 &\leq 16, \quad x_j \geq 0 [j = 1, 2.] \end{aligned}$$

by simplex method and find the dual optimal variables from the final simplex table of the primal problem.

7. Prove that the following two L.P.P.

(a) Maximize, $z_1 = 4x_1 + 3x_2$

$$\text{subject to } \begin{aligned} 3x_1 + x_2 &\leq 15 \\ 3x_1 + 4x_2 &\leq 24, \quad x_1, x_2 \geq 0 \end{aligned}$$

(b) Minimize, $z_2 = 15x_1 + 24x_2$

$$\text{subject to } 3x_1 + 3x_2 \geq 4$$
$$x_1 + 4x_2 \geq 3\,, \quad x_1, x_2 \geq 0$$

have the same optimal value. Give reason for it. Verify it by Graphical Method.

Answers

1. $\min z = 17$ at $x_1 = 2$, $x_2 = 3$. 2. $\min z = \frac{235}{19}$, for $x_1 = \frac{5}{19}$, $x_2 = \frac{16}{19}$. 3. $\min z = 1$ for $x_1 = 0$, $x_2 = 1$. 4. $\min z = \frac{4}{3}$ for $x_1 = \frac{1}{3}$, $x_2 = \frac{1}{3}$. 5. $\max z = \frac{27}{2}$ for $x = \frac{9}{2}$, $x_2 = \frac{3}{2}$. 6. $\min z = -4$ for $x_1 = 2$, $x_2 = 0$, and the dual optimal variables are $w_1 = 0$, $w_2 = \frac{1}{2}$, $w_3 = 0$. 7. $\max z_1 = 25$ at $x_1 = 4$, $x_2 = 3$, $\min z_2 = 25$ at $x_1 = \frac{7}{9}$, $x_2 = \frac{5}{9}$. The optimal values are same as they are dual of each other.

Chapter 11

Transportation Problems and Assignment Problems

11.1 Definition

A transportation problem (simply T.P.) is a particular type linear programming problem. Here, a particular commodity which is stored at different warehouses (origins) is to be transported to different distribution centres (destinations) in such a way that the transportation cost is minimum. Consider a particular example. Let there be m origins $O_1, O_2, \cdots, O_i, \cdots, O_m$ and the quantity available at origin O_i be $a_i[i = 1, 2, \cdots, m]$ and let there be n destinations $D_1, D_2, \cdots, D_j, \cdots, D_n$ and the quantity required, i.e., the demand at D_j be $b_j[j = 1, 2, \cdots, n]$. Let us make an assumption that

$$\sum_{i=1}^{m} a_i = M = \sum_{j=1}^{n} b_j. \tag{11.1.1}$$

This assumption is not restrictive. In a particular problem when $\sum_{i=1}^{m} a_i = \sum_{j=1}^{n} b_j$, i.e., the total available quantity is equal to total demand, it is called as *balanced transportation problem* and when $\sum_{i=1}^{m} a_i \neq \sum_{j=1}^{n} b_j$ it is called as *unbalanced transportation problem*. We shall initially discuss first type problems.

$$\begin{array}{cc|ccccccc|cl}
 & & \multicolumn{7}{c}{\text{Destinations}} & & \\
 & & D_1 & \cdots & D_2 & \cdots & D_j & \cdots & D_n & & \\
\hline
 & O_1 & x_{11} & \cdots & x_{12} & \cdots & x_{1j} & \cdots & x_{1n} & a_1 & \\
 & O_2 & x_{21} & \cdots & x_{22} & \cdots & x_{2j} & \cdots & x_{2n} & a_2 & \\
 & \vdots & \vdots & & \vdots & & \vdots & & \vdots & \vdots & \\
\text{Origins} & O_i & x_{i1} & \cdots & x_{i2} & \cdots & x_{ij} & \cdots & x_{in} & a_i & \text{Capacities} \\
 & \vdots & \vdots & & \vdots & & \vdots & & \vdots & \vdots & \\
 & O_m & x_{m1} & \cdots & x_{m2} & \cdots & x_{mj} & \cdots & x_{mn} & a_m & \\
\hline
 & & b_1 & \cdots & b_2 & \cdots & b_j & \cdots & b_n & & \\
 & & \multicolumn{7}{c}{\text{Demands}} & &
\end{array}$$

$c_{ij}[i = 1, 2, \cdots, m;\ j = 1, 2, \cdots, n]$, the cost of transporting per unit of commodity from the ith origin to jth destination is a known quantity. It is assumed in general that $c_{ij} \geq 0$ for all i and j. But it may be negative under some special conditions. The problem before us is to determine the quantity $x_{ij}[i = 1, 2, \cdots, m;$ $j = 1, 2, \cdots, n]$ which is to be transported from ith origin to jth destination such that the transportation cost is minimum provided the condition (11.1.1) is satisfied. Mathematically, the problem can be written as,

$$\text{minimize,} \quad z = \sum_{i=1}^{m}\sum_{j=1}^{n} c_{ij}x_{ij} \tag{11.1.2}$$

subject to the constraints

$$\sum_{j=1}^{n} x_{ij} = a_i, \ i = 1, 2, \cdots, m \tag{11.1.3}$$

$$\sum_{i=1}^{m} x_{ij} = b_j, \ j = 1, 2, \cdots, n \tag{11.1.4}$$

$$\text{and} \quad \sum_{i=1}^{m} a_i = \sum_{j=1}^{n} b_j.$$

From the above diagram, the constraints (11.1.3) and (11.1.4) can be written easily. The sum of the variables of the ith row is equal to a_i and the sum of the variables of the jth column is equal to b_j.

It is evident that $x_{ij} \geq 0$ for all i and j.

The problem is a minimization problem and

$$z = \sum_{i=1}^{m}\sum_{j=1}^{n} c_{ij}x_{ij}$$

is the *objective function* which is to be minimized.

In this problem, there are $(m+n)$ constraints of which all are equations of mn variables

$$x_{ij}[i = 1, 2, \cdots, m; j = 1, 2, \cdots, n].$$

Since, in general, in a L.P.P. the number of variables are greater than the number of constraints, therefore m and n both must be ≥ 2. Since

$$\sum_{i=1}^{m} a_i = \sum_{j=1}^{n} b_j,$$

all constraint equations are not linearly independent. There are only

$$(m+n) - 1 = (m+n-1)$$

linearly independent equations. This can be easily verified by considering the particular case given in (11.1.5) where the $R(A) = R(Ab)$ of A given in (11.1.7) is 4.

Hence the problem ultimately is a L.P.P. with $(m+n-1)$ independent constraints and mn variables

$$x_{ij}[i = 1, 2, \cdots, m;\ j = 1, 2, \cdots, n].$$

Consider a particular T.P. involving 2 origins and 3 destinations.

Here the problem can be written as

$$\begin{aligned}\text{minimize,}\quad z = {} & c_{11}x_{11} + c_{12}x_{12} + c_{13}x_{13} + c_{21}x_{21} \\ & + c_{22}x_{22} + c_{23}x_{23}\end{aligned}$$

subject to the constraints

$$\left.\begin{array}{lllllll} (1)\ x_{11} + x_{12} + x_{13} & & & & & = a_1 \\ (2) & & x_{21} & + x_{22} & + x_{23} & = a_2 \\ (3)\ x_{11} & & + x_{21} & & & = b_1 \\ (4)\ \quad x_{12} & & & + x_{22} & & = b_2 \\ (2)\ \quad\quad x_{13} & & & & + x_{23} & = b_3 \end{array}\right\} \quad (11.1.5)$$

and $x_{ij} \geq 0$ for all i and j and

$$a_1 + a_2 = b_1 + b_2 + b_3. \quad (11.1.6)$$

In this problem there are $2 \times 3 = 6$ variables and though apparently there are $2 + 3 = 5$ constraints but actually there are only $2 + 3 - 1 = 4$ independent constraints. For example, the constriant (1) can be expressed as,

$$(1) = (3) + (4) + (5) - (2) \quad [\text{as } a_1 + a_2 = b_1 + b_2 + b_3].$$

Hence the first equation is redundant. Similarly it can be shown that any one of the constraints can be reduced to a redundant constraint.

But still if we keep the first constraint intact in its due position, the co-efficient matrix A can be written symmetrically as

$$A = \begin{bmatrix} 1 & 1 & 1 & 0 & 0 & 0 \\ 0 & 0 & 0 & 1 & 1 & 1 \\ 1 & 0 & 0 & 1 & 0 & 0 \\ 0 & 1 & 0 & 0 & 1 & 0 \\ 0 & 0 & 1 & 0 & 0 & 1 \end{bmatrix} \quad (11.1.7)$$

There are six column vectors in the co-efficient matrix A.

A can be expressed as

$$A = (\mathbf{a}_{11}, \mathbf{a}_{12}, \mathbf{a}_{13}, \mathbf{a}_{21}, \mathbf{a}_{22}, \mathbf{a}_{23}) \quad (11.1.8)$$

where $\mathbf{a}_{11}$ is the column vector associated with the variable x_{11} etc. Again each column vector can be expressed as the sum of two unit vectors. For example,

$$\mathbf{a}_{11} = \mathbf{e}_1 + \mathbf{e}_{2+1};\quad \mathbf{a}_{12} = \mathbf{e}_1 + \mathbf{e}_{2+2};\quad \mathbf{a}_{23} = \mathbf{e}_2 + \mathbf{e}_{2+3} \text{ etc.}$$

where all unit vectors are $(2+3)$ component unit column vectors.

In general, it can be written as

$$\mathbf{a}_{ij} = \mathbf{e}_i + \mathbf{e}_{m+j} \quad [i = 1, 2; j = 1, 2, 3]. \tag{11.1.9}$$

The maximum number of vectors which are linearly independent in this set of six vectors is $(2 + 3 - 1) = 4$.

Ultimately this T.P. problem can be written in the matrix form given below:

$$\begin{aligned} \text{minimize,} \quad z &= \mathbf{cx} \text{ subject to} \\ A\mathbf{x} &= \mathbf{b}, \quad \mathbf{x} \geq \mathbf{0} \end{aligned} \tag{11.1.10}$$

where A is the co-efficient matrix given in (11.1.7) or (11.1.8),

$\mathbf{c}$ is an $2 \times 3 = 6$ component row vector,

$\mathbf{x}$ is an 6 component column vector, and

$\mathbf{b} = [a_1, a_2, b_1, b_2, b_3]$ is an $2 + 3 = 5$ component column vector.

The cost vector $\mathbf{c}$ can be written in matrix form

$$\mathbf{c} = \begin{bmatrix} c_{11} & c_{12} & c_{13} \\ c_{21} & c_{22} & c_{23} \end{bmatrix} \tag{11.1.11}$$

which is called as a *cost matrix*.

Similarly a general T.P. involving m origins and n destinations can be written in matrix form given below:

$$\begin{aligned} \text{minimize,} \quad z &= \mathbf{cx} \\ \text{subject to} \quad A\mathbf{x} &= \mathbf{b}, \mathbf{x} \geq \mathbf{0} \end{aligned}$$

where $A = (\mathbf{a}_{11}, \mathbf{a}_{12}, \cdots, \mathbf{a}_{ij}, \cdots, \mathbf{a}_{mn})$ is the co-efficient matrix,

$\mathbf{c} = (c_{11}, c_{12}, \cdots, c_{ij}, \cdots, c_{mn})$ is an mn component row vector (cost vector),

$\mathbf{x} = (x_{11}, x_{12}, \cdots, x_{ij}, \cdots, x_{mn})$ is an mn component column vector (variable vector), and

$\mathbf{b} = [a_1, a_2, \cdots, a_m, b_1, b_2, \cdots, b_n]$ is an $(n + m)$ component column vector.

Each column vector $\mathbf{a}_{ij}$ can be expressed as

$$\mathbf{a}_{ij} = \mathbf{e}_i + \mathbf{e}_{m+j} \tag{11.1.12}$$

where both $\mathbf{e}_i$ and $\mathbf{e}_{m+j}$ are $(m + n)$ component unit column vectors.

As there are only $(m + n - 1)$ linearly independent equations, the maximum number of vectors which will be linearly independent is $m + n - 1$ in the set of mn vectors $\mathbf{a}_{ij} [i = 1, 2, \cdots, m; j = 1, 2, \cdots, n]$.

Any set of $(m + n)$ of more vectors $\mathbf{a}_{ij}$ are always linearly dependent.

Theorem 11.1.1 *There exists a feasible solution in each T.P. which is given by,*

$$x_{ij} = \frac{a_i b_j}{M} \quad [i = 1, 2, \cdots, m;\ j = 1, 2, \cdots, n]$$

where $M = \sum_{i=1}^{m} a_i = \sum_{j=1}^{n} b_j$.

Proof. Since all a_i and b_j are non-negative quantities therefore $x_{ij} \geq 0$ for all i and j.

$$\sum_{j=1}^{n} x_{ij} = \frac{a_i \sum_{j=1}^{n} b_j}{M} = \frac{a_i M}{M} = a_i$$

$$\text{and} \quad \sum_{i=1}^{m} x_{ij} = \frac{b_j \sum_{i=1}^{m} a_i}{M} = \frac{b_j M}{M} = b_j$$

which satisfy the conditions given in (11.1.3) and (11.1.4).

Hence in each T.P. there exists a feasible solution and

$$x_{ij} = \frac{a_i b_j}{M} \quad [i = 1, 2, \cdots, m; j = 1, 2, \cdots, n] \tag{11.1.13}$$

Theorem 11.1.2 *In each T.P. there exists at least one B.F.S. which makes the objective function a minimum.*

Proof. Since there exists a feasible solution, it has at least one B.F.S. (by theorem 6.1.1). As all c_{ij} and x_{ij} are finite quantities, therefore z is a function which is bounded from below and above. Hence the objective function must have finite minimum value. As the objective function has an optimal value, there exists at least one B.F.S. which will be an optimal solution to the problem.

Theorem 11.1.3 *In a balanced T.P. having m origins and n destinations* $(m, n \geq 2)$ *the exact number of basic variables is* $m + n - 1$.

The balanced T.P. is

$$\textit{minimize,} \quad z = \sum_{j=1}^{n} \sum_{i=1}^{m} c_{ij} x_{ij}$$

$$\textit{subject to} \quad \sum_{j=1}^{n} x_{ij} = a_i \quad [i = 1, 2, \cdots, m] \tag{11.1.14}$$

$$\sum_{i=1}^{m} x_{ij} = b_j \quad [j = 1, 2, \cdots, n] \tag{11.1.15}$$

$$\textit{and} \sum_{i=1}^{m} a_i = \sum_{j=1}^{n} b_j$$

There are $m+n$ linear constraints with mn variables and $mn > m+n-1$ ($\because$ $m, n \geq 2$).

From (11.1.14)

$$\sum_{i=1}^{m}\sum_{j=1}^{n} x_{ij} = \sum_{i=1}^{m} a_i = \sum_{j=1}^{n} b_j. \tag{11.1.16}$$

Now summing the first $(n-1)$ constraints of (11.1.15) we get,

$$\sum_{j=1}^{n-1}\sum_{i=1}^{m} x_{ij} = \sum_{j=1}^{n-1} b_j. \tag{11.1.17}$$

Now subtracting (11.1.17) from (11.1.16) we get,

$$\sum_{j=1}^{n}\sum_{i=1}^{m} x_{ij} - \sum_{j=1}^{n-1}\sum_{i=1}^{m} x_{ij} = \sum_{j=1}^{n} b_j - \sum_{j=1}^{n-1} b_j = b_n. \tag{11.1.18}$$

Thus we get,

$$\sum_{i=1}^{m}\left(\sum_{j=1}^{n} x_{ij} - \sum_{j=1}^{n-1} x_{ij}\right) = b_n, \quad \text{or,} \quad \sum_{i=1}^{m} x_{in} = b_n, \tag{11.1.19}$$

which is the last or nth constraint of (11.1.4). Therefore, there are only $(m+n-1)$ linearly independent equations with mn variables $(mn > m+n-1)$. Thus from the definition of the basic solution, we can say that the number of basic variables is exactly $(m+n-1)$.

Remark. All basic variables may not be positive, some of them may be zero. When all basic variables are positive, the solution is called a non-degenerate B.F.S. When at least one basic variable is zero, the solution is called a degenerate B.F.S.

The number of basic cells will be exactly $(m+n-1)$ all of which contain $(m+n-1)$ basic variables which are either all positive basic variables or some variables may be 0 (zero) which has been shown later on.

11.2 Transportation Table

T.P. is a special case of a L.P.P. Therefore a T.P. can be solved by using simplex method. But the method is not suitable in solving a T.P. Here a specially designed table is constructed to solve the problem systematically which is called as a *transportation table.* A specimen of a transportation table with m origins and n destinations is given below.

In this table there are mn squares or rectangles arranged in m rows and n columns. Each square or rectangle is called as a *cell.* The cell which is in the ith row and jth column is called as (i, j)th cell or cell (i, j). Each cost component c_{ij} is displayed at the *south-east* corner of the corresponding cell. A component of a feasible solution x_{ij} (if any) is to be displayed inside a small square situated at the *north-west* corner of the cell (i, j). The different origin capacities and destination demands (requirements) are listed in the outer column and outer row respectively as given in the table (11.1). These quantities are called as *rim requirements.*

Table 11.1: Transportation table

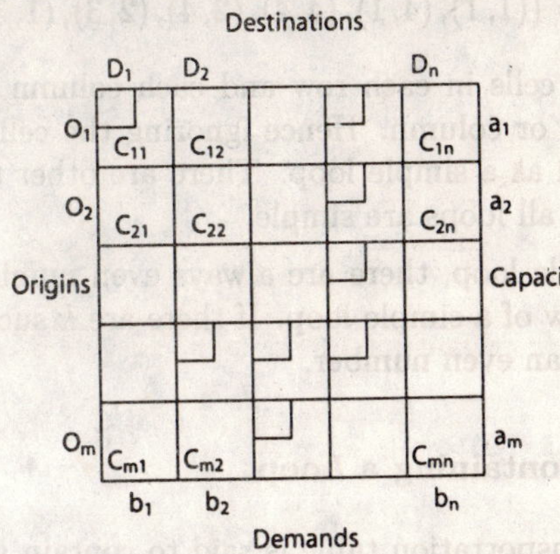

11.2.1 Loops in a Transportation Table and their properties

In a transportation table, an ordered set of four or more cells are said to form a loop (*i*) if and only if two consecutive cells in the ordered set lie either in the same row or in the same column and if (*ii*) the first and the last cell of the set also lie either in the same row or in the same column.

In the following figure one closed circuit is formed in each of the three transportation tables.

Table : 11.2

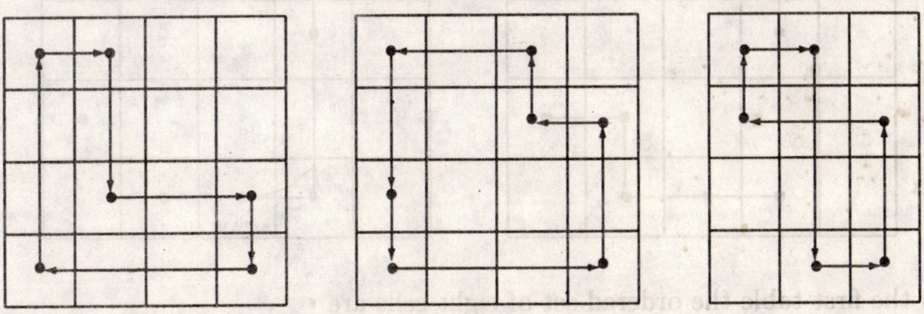

The ordered set of cells in the circuits are

$$(1) \quad L_1 = \{(1,1),(1,2),(3,2),(3,4),(4,4),(4,1)\}$$
$$(2) \quad L_2 = \{(1,1),(3,1),(4,1),(4,4),(2,4),(2,3),(1,3)\}$$
$$(3) \quad L_3 = \{(1,1),(1,2),(4,2),(4,3),(2,3),(2,1)\}$$

[Observe the ordered sets very carefully]

In the first and third table there are only two cells in each row and each column and the first and last cell are in the same row or same column. These loops are called simple loops. In the second table three cells are in the first column. But if we ignore or omit the cell $(3,1)$ and ordered the set of cells in the manner

$$L_2 = \{(1,1),(4,1),(4,4),(2,4),(2,3),(1,3)\}$$

then there are just two cells in each row and each column and the first and last cell are in the same row or column. Hence ignoring the cell $(3,1)$, the 2nd closed circuit is also considered as a simple loop. There are other types of loops. But in transportation problem, all loops are simple.

Note. (1) In a simple loop, there are always even number of cells. There are only two cells in each row of a simple loop. If there are k such rows, the number of cells will be $2k$ which is an even number.

11.2.2 Set of Cells containing a Loop

A set C of cells of a transportation table is said to contain a loop if the cells of C or any subset of C can be ordered so as to form a loop.

11.2.3 Tree and other form of Ordered Set of Cells

In the following two transportation tables with four origins and five destinations, $(4+5-1) = 8$ cells are selected. The cells of the first table are ordered in the following manner:

Table : 11.3

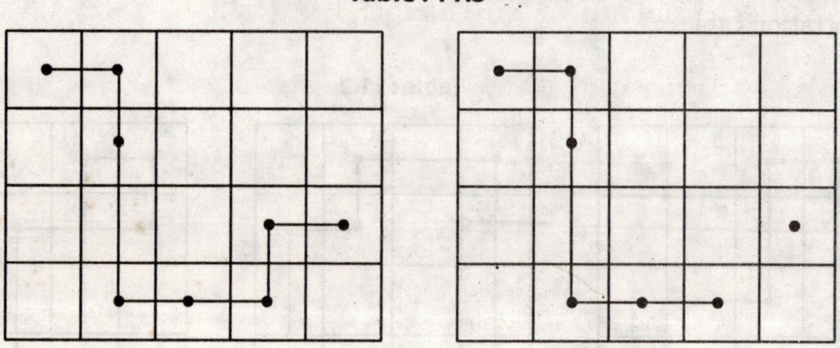

In the first table the ordered set of eight cells are

$$L_1 = \{(1,1),(1,2),(2,2),(4,2),(4,3),(4,4),(3,4),(3,5)\}.$$

The first and the last cells are not placed either in a row or in a column. These set of cells do not contain a loop. But it is possible to connect all cells by line segments obeying the restrictions used in the formation of a loop [except in the case of first and last cell]. This diagram is called as a *tree* and all segments are the branches of the tree.

In the second table, the set of eight cells cannot be ordered in any manner such that even a tree diagram cannot be constructed. But *it is interesting to note that if we select only one more cell out of remaining 12 cells arbitrarily, it is always possible to form a loop connecting either by all cells or by a subset of the cells in both tables* [verify the statement by selecting any cell]. From this geometrical idea we can establish a very important property regarding the linear dependence of a set of column vectors of the co-efficient matrix A of a transportation problem.

We know that the maximum number of column vectors $\mathbf{a}_{ij}$ which will be linearly independent in a T.P. having m origins, n destinations is $(m+n-1)$. Hence for $m=4, n=5$ the maximum number of vectors which will be linearly independent is $(4+5-1)=8$. Any set of nine or more vectors are always linearly dependent for $m=4, n=5$. We have established in the above figures that any set of nine or more cells can be ordered so as to form a loop and there is at least one set of eight cells which cannot be ordered so as to form a loop. From this, we conclude that if a set of four or more cells can be ordered so as to form a loop, the set of corresponding vectors are linearly dependent and there is at least one set of $(m+n-1)$ cells which cannot be ordered so as to form a loop and the set of corresponding vectors are linearly independent.

▶ **Example 11.2.1** *A set X of column vectors of the co-efficient matrix of a T.P. will be linearly dependent if their corresponding cells in the transportation table contain a loop.*

▶ **Example 11.2.2** *A set C of cells of a transportation table contains a loop, if their corresponding column vectors of the co-efficient matrix are linearly dependent (Proof is beyond the scope of the book).*

11.3 Determination of an Initial B.F.S.

So long we have discussed some fundamental properties of a transportation problem which will help in solving a problem. Next problem before us is to determine the initial B.F.S. of the problem and from this we proceed to find another B.F.S. which will improve the value of the objective function. There are various methods of finding an initial B.F.S. *It is interesting to note that in all cases the solution is a B.F.S.*. Of course, the solution may be non-degenerate or degenerate. If the solution is a B.F.S. the *cells to which some allocation is made are called as basic cells.* Obviously the allocated values are the components of the B.F.S. The methods of determining an initial B.F.S. are (1) North-west corner rule, (2) Row minima method, (3) Column minima method, (4) Matrix minima method and (5) Vogel's approximation method (VAM).

11.3.1 North-West Corner Rule

Step 1. Compute $\min(a_1, b_1)$. If $a_1 < b_1 \cdot \min(a_1, b_1) = a_1$ and if $b_1 < a_1 \cdot \min(a_1, b_1) = b_1$. Select $x_{11} = \min(a_1, b_1)$ and allocate the value of x_{11} in the cell $(1, 1)$, i.e., in the cell situated in the *north-west corner* of the transportation table.

Step 2. If $a_1 < b_1$, the capacity of the origin O_1 will be exhausted completely which indicates that all other cells of the first row will remain vacant. But there remain some demand in the destination D_1. Compute $\min(a_2, b_1 - a_1)$. Select

$x_{21} = \min(a_2, b_1 - a_1)$ and allocate the value of x_{21} in the cell $(2, 1)$. Let us now make an assumption that $b_1 - a_1 < a_2$ which indicates that the demand of D_1 is satisfied completely. Of course, this assumption is not essential. With this assumption, the next cell for which some allocation is to be made, is the cell $(2, 2)$ etc.

If $b_1 < a_1$, the demand of the destination D_1 will be satisfied exactly which indicates that all other cells of the first column will remain vacant. But the capacity of origin O_1 will not be exhausted. Compute, $\min(a_1 - b_1, b_2)$. Select $x_{12} = \min(a_1 - b_1, b_2)$ and allocate the value of x_{12} in cell $(1, 2)$. Let us now make an assumption that $a_1 - b_1 < b_2$ which indicates that the capacity of O_1 is exhausted completely. With this assumption the next cell for which some allocation is to be made, is the cell $(2, 2)$ etc. If $a_1 = b_1$, the capacity of the origin O_1 will be exhausted as well as the demand of D_1 will be satisfied simultaneously. In that case, the solution will be *degenerate*. Select either x_{12} or $x_{21} = \min(a_1 - b_1, b_2) = \min(a_2, a_1 - b_1) = 0$ and allocate the value 0 only in one of the two cells $(1, 2)$ or $(2, 1)$. The next cell for which some allocation is to be made is cell $(2, 2)$. In this way proceed step by step until all the rim requirements are satisfied. In general, if an allocation is made in the cell (i, j) in the current step, the next allocation will be made either in cell $(i + 1, j)$ or $(i, j + 1)$. The feasible solution obtained by this method is always a B.F.S. In North-West Corner Rule, a tree diagram can be constructed by connecting all basic cells but no loop will be formed.

▶ **Example 11.3.1** *Determine an initial B.F.S. of the following problems by the method of North-West Corner Rule.*

(i)

	D_1	D_2	D_3	D_4	a_i
O_1	4	6	9	5	16
O_2	2	6	4	1	12
O_3	5	7	2	9	15
b_j	12	14	9	8	43

(ii)

	D_1	D_2	D_3	D_4	a_i
O_1	2	5	4	7	4
O_2	6	1	2	5	6
O_3	4	5	2	4	8
b_j	3	7	6	2	18

Note. Inside the rectangle, the cost matrix is given and outside the rectangle, the rim requirements are given.

Solution: The initial B.F.S. are displayed in the following two tables:

Table 11.4(A)

[12] 4	[4] 6	9	5
2	[10] 6	[2] 4	1
5	7	[7] 2	[8] 9

Table 11.4(B)

[3] 2	[1] 5	4	7
6	[6] 1	[0] 2	5
4	5	[6] 2	[2] 4

Explanation: (i) B.F.S. is displayed in the Table 11.4A : $\min(16, 12) = 12$. Therefore $x_{11} = 12$ and allocate it in the cell $(1, 1)$. The demand of D_1 is satisfied and hence all other cells in first column remain vacant.

As $b_1 = 12 < 16 = a_1$, therefore next allocation will be in cell $(1,2)$ and $x_{12} = \min(16-12, 14) = 4$. Now the capacity of O_1 is exhausted. Next allocation will be in cell $(2,2)$ and $x_{22} = \min(12, 14-4) = 10$. Proceeding similarly we get $x_{23} = 2, x_{33} = 7$ and $x_{34} = 8$ and all the rim requirements are satisfied. The solution obtained is a B.F.S. because the set of cells (which contain components of F.S.) do not contain a loop (number of variables $= 4+3-1 = 6$) and the cost due to this assignment is $4 \times 12 + 6 \times 4 + 6 \times 10 + 4 \times 2 + 2 \times 7 + 9 \times 8 = 226$ units.

(ii) B.F.S. is displayed in Table 11.4B : $\min(4,3) = 3$. Therefore $x_{11} = 3$ and allocate it in the cell $(1,1)$. The demand of D_1 is satisfied and hence all other cells of first column remain vacant. As $b_1 = 3 < 4 = a_1$, therefore next allocation will be in the cell $(1,2)$ and $x_{12} = \min(4-3, 7) = 1$. Now the capacity of O_1 is exhausted. Next allocation will be in cell $(2,2)$ $x_{22} = \min(6, 7-1) = 6$. Now the capacity of O_3 is exhausted and the demand of D_2 is satisfied simultaneously. Therefore either x_{23} or x_{32} will be zero. Let us take $x_{23} = 0$ and proceed similarly until all rim requirements are satisfied. The solution obtained is a degenerate B.F.S. as the number of variables is $4+3-1 = 6$ (one of them being zero) and the set of corresponding cells do not contain a loop.

11.3.2 Row Minima Method

Step 1. Select the smallest cost in the first row. Let it be c_{1j}; compute $\min(a_1, b_j)$. Set $x_{1j} = \min(a_1, b_j)$ and allocate it in the cell $(1,j)$. This is the maximum feasible amount which can be allocated in the cell $(1,j)$. If the smallest cost is not unique, select any one of the minimum cost arbitrarily.

Step 2. If $a_1 < b_j$, the capacity of the origin O_1 will be exhausted. But the demand of destination D_j remains unsatisfied. Cross off[1] the first row and diminish b_j by a_1. Proceeding similarly, allocate the maximum feasible amounts in the cells of the remaining rows starting from the second until all rim requirements are satisfied. If $b_j < a_1$, the total demand of the destination D_j is satisfied but the capacity of the origin O_1 is not exhausted completely. Cross off the jth column and diminish a_1 by b_j. Reconsider the first row and select the next smallest cost of this row. Let it be c_{1k}. Compute, $\min(a_1 - b_j, b_k)$. Set $x_{1k} = \min(a_1 - b_j, b_k)$ and allocate it in the cell $(1,k)$. Let us now make an assumption that $a_1 - b_j < b_k$. Therefore the capacity of O_1 will be exhausted completely [assumption is not restrictive]. Cross off the first row and repeat the above procedure for the second row and so on as in the above method until all rim requirements are satisfied.

Step 3. If $a_1 = b_j; \min(a_1, b_j) = a_1 = b_j$. Set $x_{1j} = a_1 = b_j$ and allocate it in the cell $(1,j)$. Due to this allocation, the capacity of origin O_1 will be exhausted as well the demand of D_j will be satisfied completely. In that case solution will be degenerate. Set $x_{1k} = 0$ and display it in the cell $(1,k)$ with the assumption, that the cost of $(1,k)$ cell is the next minimum cost. Now cross off both the first row and jth column and proceed similarly until all rim requirements are satisfied.

[1] By "cross off" a row or column we mean that we shall ignore that row or column for further consideration.

► **Example 11.3.2** *Find out an initial B.F.S. of the following balanced T.P. using row minima method.*

	D_1	D_2	D_3	D_4	a_i	
O_1	4	2	5	3	6	
O_2	5	4	3	2	13	Capacity
O_3	1	4	6	5	9	
b_j	7	8	5	8	28	
	Demand					

Solution: It is displayed in the transportation table given below:

Table 11.5

	D_1	D_2	D_3	D_4	a_i			
O_1	4	6; 2	5	3	6			
O_2	5	0; 4	5; 3	8; 2	13	13	5	
O_3	7; 1	2; 4	6	5	9	9	9	9
b_j	7	8	5	8	28 / 28			
	7	2	5	8				
	7	2	5					
	7	2						

Explanation. B.F.S. is given in the Table 11.5.

Lowest cost in first row is $c_{12} = 2$. $\min(a_1, b_2) = \min(6, 8) = 6$. Set $x_{12} = 6$ and allocate it in cell $(1, 2), a_1 < b_2$; therefore the capacity of O_1 is exhausted completely and hence cross off the first row and diminish b_2 by a_1 which is shown in Table 11.5. Give a shade on the first row and ignore it for future computation.

Lowest cost in the second row is $c_{24} = 2$. $\min(a_2, b_4) = \min(13, 8) = 8$. Set $x_{24} = 8$ and allocate it in the cell $(2, 4)$; $b_4 < a_2$. Therefore the capacity of O_2 will not be exhausted; give a shade on the fourth column and ignore the column for future computation since the demand of D_4 is satisfied completely. Cross off the fourth column and diminish a_2 by b_4. Reconsider the second row. The next lowest cost in the row is $c_{23} = 3$. $\min(a_2 - b_4, b_2) = \min(13 - 8, 5) = 5$. Set $x_{23} = 5$ and allocate it in cell the $(2, 3)$. As $a_2 - b_4 = 5 = b_3$, the capacity of O_2 is exhausted as well as the demand of D_3 is satisfied completely. Therefore the solution will be degenerate. The next lowest cost of the second row is $c_{22} = 4$. Set $x_{22} = 0$ and allocate it in the cell $(2, 2)$. Cross off the second row and third column simultaneously. Now complete the Table 11.5 and all the rim requirements are satisfied now. Solution obtained is a B.F.S., because the number of variables are $3 + 4 - 1 = 6$ and the set of corresponding cells do not contain a loop.

Column Minima Method: The technique used in the column minima method is same as that in the row minima method.

11.3.3 Matrix Minima Method or (Cost Minima Method)

Step 1. Select the smallest cost in the cost matrix. Let it be c_{ij}. Set $x_{ij} = \min(a_i, b_j)$ and allocate it in the cell (i, j). This is the maximum feasible amount that can be allocated in the cell (i, j).

Step 2. If $a_i < b_j$, the capacity of the origin O_i will be exhausted completely. Cross off the ith row and diminish b_j by a_i.

If $b_j < a_i$, the demand of the destination D_j will be satisfied completely. Cross off the jth column and diminish a_i by b_j.

Step 3. If $a_i = b_j$, the capacity of the origin O_i will be exhausted and the demand of D_j will be satisfied simultaneously. Set $x_{ij} = a_i = b_j$; allocate it in the cell (i, j). Cross off either the ith row or jth column but not the both. Of course, we may drop both the ith row and jth column by inserting a basic variable 0 (zero) at a cell corresponding to the lowest cost of those row and column.

Step 4. Apply the same technique in the reduced transportation table until all rim requirements are satisfied. At any stage, if the minimum cost is not unique, make any arbitrary choice among the minima.

▶ **Example 11.3.3** *Determine an initial B.F.S. to the following balanced T.P. using matrix minima method:*

	D_1	D_2	D_3	D_4	a_i	
O_1	5	3	6	2	19	
O_2	4	7	9	1	37	Capacity
O_3	3	4	7	5	34	
b_j	16	18	31	25	90	
		Demand				

where O_i and D_j denote the ith origin and jth destination respectively.

Explanation: B.F.S. is given in the Table 11.6.

Table 11.6

	D_1	D_2	D_3	D_4	a_i			
O_1	(5)	18 (3)	1 (6)	(2)	19	19	1	1
O_2	(4)	(7)	12 (9)	25 (1)	37	12	12	12
O_3	16 (3)	(4)	18 (7)	(5)	34	34	34	18
b_j	16	18	31	25	90 / 90			
	16	18	31					
	16		31					
			31					

The smallest cost is $c_{24} = 1$. Set $x_{24} = \min(a_2, b_4) = \min(37, 25) = 25$ and allocate it in the cell $(2, 4)$. $b_4 < a_2$, therefore cross off the fourth column and diminish a_2 by b_4 and give a shade on the fourth column and ignore it for future computation.

The smallest cost in the reduced table is c_{12} or c_{31}. Let us select $c_{12} = 3$ as the smallest cost and allocate $x_{12} = \min(a_1, b_2) = \min(19, 18) = 18$ in the cell $(1,2), b_2 < a_1$, therefore cross off the second column; give shade on the 2nd column and diminish a_1 by b_2. Proceed similarly until all rim requirements are satisfied and Table 11.6 gives the B.F.S. It is a B.F.S. because the number of variables are $4 + 3 - 1 = 6$ and the cells corresponding to the feasible solution do not contain a loop. Here the solution is not unique.

11.3.4 Vogel's Approximation Method: (Unit Penalty Method)

Step 1. Select the lowest and next to lowest cost for each row and determine the difference between them for each row and display them within the first bracket against the respective rows. If there are two or more with same lowest costs, difference may be taken to be zero (in some cases it may produce good result). Compute, similarly the difference for each column and display them within the bracket against the respective columns.

Step 2. Find the largest value of the differences and find out the row or column for which the difference is maximum. Let the maximum difference corresponds to ith row. Select the lowest cost in the ith row. Let it be c_{ij}. Allocate $x_{ij} = \min(a_i, b_j)$ in the cell (i, j) which is the maximum feasible amount that can be allocated in the cell (i, j). If the maximum difference is not unique, select any one of them.

Step 3. If $a_i < b_j$, cross off the ith row and diminish b_j by a_i. If $b_j < a_i$, cross off the jth column and diminish a_i by b_j. If $a_i = b_j$ allocate $x_{ij} = a_i = b_j$ in cell (i, j) and cross off either ith row or jth column but not the both. Of course, we can omit both the ith row and jth column simultaneously by inserting a basic variable 0 (zero) to one of the cells of the corresponding row or column possessing the next minimum cost and the solution will be degenerate then.

Step 4. Recompute the row and column differences for the reduced transportation table. Repeat the procedure discussed above until all rim requirements are satisfied.

▶ **Example 11.3.4** *Obtain an initial B.F.S. to the balanced T.P. given below using Vogel's approximation method.*

		Warehouses				
		D_1	D_2	D_3	D_4	a_i
	O_1	19	30	50	10	7
Factory	O_2	70	30	40	60	9
	O_3	40	8	70	20	18
	b_j	5	8	7	14	34

Factory capacity (column a_i); Demands (row b_j)

Initial B.F.S.: Here the initial basic feasible solution using by VAM is being shown in a single table in a very compact manner which will save some time and labour.

The initial B.F.S. obtained by using the method is non-degenerate and unique and the initial solution is $x_{11} = 5$, $x_{14} = 2$, $x_{23} = 7$, $x_{24} = 2$, $x_{32} = 8$ and $x_{34} = 10$.

Explanation: Step 1: Select the lowest and next to lowest cost for each row and each column and determine the difference between them for each row and column and display them within the first bracket against the respective rows and columns. Here all the differences have been shown within the first compartment. Maximum difference is 22 which occurs at the second column and the minimum cost of that column is $c_{32} = 8$. Allocate $\min(18, 8) = 8$ in the cell $(3, 2)$. The demand of D_2 has been satisfied and shade the second column as shown in the Table 11.7. The resulting cost matrix will be obtained after deleting the cost components of the second column.

Table 11.7

	D_1	D_2	D_3	D_4	a_i			
O_1	5 / 19	30	50	2 / 10	7(9)	7(9)	2(40)	2(40)
O_2	70	30	7 / 40	2 / 60	9(10)	9(20)	9(20)	9(20)
O_3	40	8 / 8	70	10 / 20	18(12)	10(20)	10(50)	
b_j	5(21)	8(22)	7(10)	14(10)				
	5(21)		7(10)	14(10)				
			7(10)	14(10)				
			7(10)	4(50)				

Step 2: Applying the same technique in the resulting matrix, the capacities, demands and the differences of the cost components have been shown in the second compartment. Maximum difference is 21 which occurs in the first column and the lowest cost of that column is $c_{11} = 19$. Allocate $\min(7, 5) = 5$ in the cell $(1, 1)$. The demand of D_1 has been satisfied and shade the first column as shown in the table. The resulting cost matrix will be obtained deleting the cost components of the first column.

Step 3: Proceeding in the same way, we get the maximum difference 50 which occurs in the third row and minimum cost is $c_{34} = 20$. Allocate $\min(10, 14) = 10$ in the cell $(3, 4)$ and the capacity of O_3 will be exhausted and the resulting matrix will be obtained deleting the cost components of the third row; shade the third row as shown in the table. Using the same technique, ultimately we obtain the initial B.F.S. where all capacities have been exhausted and all the demands will be met [as the problem is a balanced transportation problem].

► **Example 11.3.5** *Find the initial basic feasible solution of the following balanced T.P. by Vogel's approximation method and calculate the cost for that solution.*

	A	B	C	a_i
F_1	10	9	8	8
F_2	10	7	10	7
F_3	11	9	7	9
F_4	12	14	10	4
b_j	10	10	8	

[Kalyani (H)'82]

Solution: I.B.F.S. by VAM.

Table 11.8

	A	B	C	a_i				
F_1	[6] 10	[2] 9	8	8(1)	8(1)	8(1)	8(1)	6
F_2	10	[7] 7	10	7(3)				
F_3	11	[1] 9	[8] 7	9(2)	9(2)	1(2)		
F_4	[4] 12	14	10	4(2)	4(2)	4(2)	4(2)	4
b_j	10(0)	10(2)	8(1)	28 / 28				
	10(1)	3(0)	8(1)					
	10(1)	3(0)						
	10(2)	2(5)						
	10							

Here the solution is not unique since in the second compartment maximum difference occurs in two rows. I.B.F.S. is $x_{11} = 6$, $x_{12} = 2$, $x_{22} = 7$, $x_{32} = 1$, $x_{33} = 8$, $x_{41} = 4$. There are six components of B.F.S. and $6 = 4 + 3 - 1$ and all the components are positive. Hence the solution is a non-degenerate B.F.S. and the cost $= 6\times10+2\times9+7\times7+1\times9+8\times7+4\times12 = 60+18+49+9+56+48 = 240$.

► **Example 11.3.6** *Find the initial basic feasible solution of the following balanced T.P. with the help of matrix minima method and VAM method and compare their corresponding costs.*

	D_1	D_2	D_3	D_4	a_i
O_1	15	28	13	21	18
O_2	22	15	19	14	14
O_3	16	12	14	31	13
O_4	24	23	15	30	20
b_j	16	15	10	24	

Solution: (i) Matrix Minima Method:

Table 11.9

	D_1	D_2	D_3	D_4	a_i				
O_1	[8] 15	28	[10] 13	21	18	18	8	8	
O_2	22	15	19	[14] 14	14	14	14		
O_3	16	[13] 12	14	31	13				
O_4	[8] 24	[2] 23	15	[10] 30	20	20	20	20	20
b_j	16	15	10	24	65 / 65				
	16	2	10	24					
	16	2		24					
	16	2		10					
	8	2		10					

Here $m+n=4+4=8$, and the number of basic cells $7=4+4-1$, and the set of cells containing positive components do not form a loop. Hence the solution is a non-degenerate basic feasible solution. I.B.F.S. by matrix minima method which is unique and the solution is $x_{11}=8$, $x_{13}=10$, $x_{24}=14$, $x_{32}=13$, $x_{41}=8$, $x_{42}=2$, $x_{44}=10$ and the cost

$$= 8\times 15+10\times 13+14\times 14+13\times 12+8\times 24+2\times 23+10\times 30$$

$$= 120+130+196+156+196+46+300=1144.$$

(ii) I.B.F.S. by VAM

Table 11.10

	D_1	D_2	D_3	D_4	a_i				
O_1	(16) 15	28	13	(2) 21	18(2)	18(6)	18(6)	18(6)	2(7)
O_2	22	15	19	(14) 14	14(1)	14(1)			
O_3	16	(13) 12	14	31	13(2)	13(4)	13(4)		
O_4	24	(2) 23	(10) 15	(8) 30	20(8)	10(1)	10(1)	10(1)	10(7)
b_j	16(1)	15(3)	10(1)	24(7)	65 / 65				
	16(1)	15(3)		24(7)					
	16(1)	15(11)		10(9)					
	16(9)	2(5)		10(9)					
		2(5)		10(9)					

Here also the solution is a non-degenerate basic feasible solution. The solution is $x_{11}=16$, $x_{14}=2$, $x_{24}=14$, $x_{32}=13$, $x_{42}=2$, $x_{43}=10$, $x_{44}=8$ and here the solution is not unique. Cost is

$$= 16\times 15+2\times 21+14\times 14+13\times 12+2\times 23+10\times 15+8\times 30$$

$$= 240+42+196+156+46+150+240=1070 \text{ units.}$$

VAM gives a better solution then the matrix minima method.

▶ **Example 11.3.7** *Find the initial basic feasible solution of the following balanced T.P. by row minima method and VAM and compare their corresponding costs.*

	D_1	D_2	D_3	D_4	D_5	a_i	
O_1	4	7	0	3	6	14	
O_2	1	2	−3	3	8	9	Supply
O_3	3	−1	4	0	5	17	
b_j	8	3	8	13	8		

Demand

Solution: (i) Row minima method:

Table 11.11

	D_1	D_2	D_3	D_4	D_5	a_i				
O_1	4	7	8 0	6 3	6	14	6			
O_2	8 1	1 2	−3	3	8	9	9	9	1	
O_3	3	2 −1	4	7 0	8 5	17	17	17	17	17
b_j	8	3	8	13	8	40 / 40				
	8	3		13	8					
	8	3		7	8					
		3		7	8					
		2		7	8					

The number of rows and columns are five and three; $5 + 3 - 1 = 7$. Here the number of basic cells are seven and the set of cells do not form a loop. And all components of the basic variables are positive and hence the solution is a non-degenerate B.F.S. and unique. I.B.F.S. $= x_{13} = 8$, $x_{14} = 6$, $x_{21} = 8$, $x_{22} = 1$, $x_{32} = 2$, $x_{34} = 7$, $x_{35} = 8$ and the cost is

$$= 8 \times 0 + 6 \times 3 + 8 \times 1 + 1 \times 2 + 2 \times (-1) + 7 \times 0 + 8 \times 5$$
$$= 0 + 18 + 8 + 2 - 2 + 0 + 40 = 66 \text{ units.}$$

(ii) By VAM:

Table 11.12

	D_1	D_2	D_3	D_4	D_5	a_i				
O_1	7 4	7	0	3	7 6	14(3)	14(1)	14(1)	14(2)	14(2)
O_2	1 1	2	8 −3	3	8	9(4)	1(1)	1(2)	1(7)	
O_3	3	3 −1	4	13 0	1 5	17(1)	17(1)	14(3)	1(2)	1(2)
b_j	8(2)	3(3)	8(3)	13(3)	8(1)	40 / 40				
	8(2)	3(3)		13(3)	8(1)					
	8(2)			13(3)	8(1)					
	8(2)				8(1)					
	7(1)				8(1)					

Here the solution is not unique but here also the solution is non-degenerate and the B.F.S. $x_{11} = 7$, $x_{15} = 7$, $x_{21} = 1$, $x_{23} = 8$, $x_{32} = 3$, $x_{34} = 13$, $x_{35} = 1$ and the cost

$$= 7 \times 4 + 7 \times 6 + 1 \times 1 + 8 \times (-3) + 3 \times (-1) + 13 \times 0 + 1 \times 5$$

$$= 28 + 42 + 1 - 24 - 3 + 0 + 5 = 49 \text{ units.}$$

VAM gives better result then the row minima method.

▶ **Example 11.3.8** *Obtain the initial B.F.S. to the following transportation problem by Vogel's approximation method and prove that the solution is degenerate.*

	D_1	D_2	D_3	D_4	a_i
O_1	4	6	5	2	6
O_2	6	4	1	4	10
O_3	5	2	3	1	12
O_4	4	6	7	8	14
b_j	9	16	10	7	42

Table 11.13

	D_1	D_2	D_3	D_4	a_i			
O_1	4	6	5	(6) 2	6(2)	6(2)	6(2)	
O_2	6	4	(10) 1	4	10(3)			
O_3	5	(12) 2	3	1	12(1)	12(1)		
O_4	(9) 4	(4) 6	(0) 7	(1) 8	14(2)	14(2)	14(2)	14(2)
b_j	9(0)	16(2)	10(2)	7(1)	42 / 42			
	9(0)	16(4)	0(2)	7(1)				
	9(0)	4(0)	0(2)	7(6)				
	9	4	0	1				

Here the solution is a degenerate B.F.S. The solution is not unique because the basic variables can be put in other cells so that the set of cells do not contain a loop. I.B.F.S. $x_{14} = 6$, $x_{23} = 10$, $x_{32} = 12$, $x_{41} = 9$, $x_{42} = 4$, $x_{43} = 0$, $x_{44} = 1$ and the cost

$$= 6 \times 2 + 10 \times 1 + 12 \times 2 + 9 \times 4 + 4 \times 6 + 0 \times 7 + 1 \times 8$$

$$= 12 + 10 + 24 + 36 + 24 + 0 + 8 = 114 \text{ units.}$$

Explanation. Taking the difference of two lowest costs of the first column as 0 (zero) all the differences are placed in the first compartment. Maximum difference is 3 which occurs in the second row and the lowest cost of the second row is $c_{23} = 1$. Allocate $\min(10, 10) = 10$ in the cell $(2, 3)$ and the supply of O_2 will be exhausted and the demand of D_3 will be met simultaneously. Then we can cross off either the second row and put '0' in the third column or cross off the third column and put '0' in the second row. Proceeding similarly as in the above problem ultimately we get a degenerate B.F.S. which is not unique one; $m + n = 4 + 4 = 8$; therefore $m + n - 1 = 7$, there is one of which is zero and the set of cells do not form a loop. Hence the B.F.S. is degenerate one.

Note. In some cases, if the difference of two lowest costs be taken as zero, we may get a better reult which we can notice in some problems. But Prof. Hadley had taken the difference between the lowest and next to lowest cost which had been shown in the problem solved in the table (9 – 14) of page 309 of 'Linear Programming' written by him.

Remarks. We have discussed five methods of finding an initial B.F.S. to a problem. From the discussions it is clear that the solutions are not unique. The basic feasible solution obtained by using north-west corner rule may be far from optimal as the costs are totally ignored during computation. The solutions obtained by matrix minima method and Vogel's approximation method are good enough because generally these require lesser number of iterations to reach at the optimal stage. Therefore in order to solve a problems it is better to find an initial B.F.S. by using either matrix minima method or VAM method. Out of these, generally VAM requires lesser number of iterations. Hence if there is no instruction in finding the initial B.F.S. of a problem by using a particular method, we should always find the initial B.F.S. by VAM.

Exercise 11A

1. Obtain an initial B.F.S. to the T.P. using north-west corner rule.

(*a*)

	D_1	D_2	D_3	D_4	a_i
O_1	4	3	2	5	6
O_2	6	1	4	3	9
O_3	7	2	4	6	7
b_j	4	6	6	6	

(*b*)

	D_1	D_2	D_3	D_4	a_i
O_1	9	8	5	7	12
O_2	4	6	8	7	14
O_3	5	8	9	5	16
b_j	8	18	13	3	

Prove that initial B.F.S. to the T.P. (b) is degenerate. Find an initial B.F.S. to the problem (b) using matrix minima method and prove that the B.F.S. will not be degenerate then.

2. Determine an initial B.F.S. to the following problem using (a) row minima method, (b) matrix minima method and (c) Vogel's approximation method.

	D_1	D_2	D_3	D_4	a_i	
O_1	21	16	25	13	11	
O_2	17	18	14	23	13	Capacity
O_3	32	27	18	41	19	
b_j	6	10	12	15	43	
	Demand					

[C.U.M.Com.'86; C.U.(P)'89]

3. Obtain an initial B.F.S. to the transportation problem given in Ex. (11.3.8) by north-west corner rule and matrix minima method and prove that the solution is degenerate in the second case.

4. For the following problem obtain the different starting solutions by adopting the north-west corner rule and Vogel's approximation method respectively and find out which solution is better?

				a_i
	5	1	8	12
	2	4	0	14
	3	6	7	4
b_j	9	10	11	

[C.U.(P)'81]

[*Hints.* In north-west corner rule solution is $x_{11} = 9$, $x_{12} = 3$, $x_{22} = 7$, $x_{23} = 7$, $x_{33} = 4$ and the corresponding cost is $5\times9+1\times3+7\times4+0\times7+7\times4 = 104$ units and the solution due to VAM is $x_{11} = 2$, $x_{12} = 10$, $x_{21} = 3$, $x_{23} = 11$, $x_{31} = 4$ and the corresponding cost is $5\times2+1\times10+2\times3+0\times11+3\times4 = 38$ units.]

5. Find the initial basic feasible solution to the following transportation problem by adopting (a) row minima method, (b) matrix minima method, (c) VAM and prove that row minima method and VAM will give better result.

	D_1	D_2	D_3	D_4	a_i
O_1	9	5	4	7	12
O_2	2	5	8	3	15
O_3	6	2	3	4	13
b_j	6	11	13	10	40

6. For a problem in which $a_1 = 3$, $a_2 = 4$, $a_3 = 7$ and $b_1 = 1$, $b_2 = 3$, $b_3 = 3$, $b_4 = 2$, $b_5 = 5$, find the initial B.F.S. and show that it is a degenerate B.F.S.

[C.U.M.Com.'88,'91]

Answers

1. (a) $x_{11} = 4$, $x_{12} = 2$, $x_{22} = 4$, $x_{23} = 5$, $x_{33} = 1$, $x_{34} = 6$.
   (b) $x_{11} = 8$, $x_{12} = 4$, $x_{22} = 14$, $x_{23} = 0$, $x_{33} = 13$, $x_{34} = 3$.
   (c) $x_{13} = 12$, $x_{21} = 8$, $x_{22} = 5$, $x_{23} = 1$, $x_{32} = 13$, $x_{34} = 3$.
2. (a) $x_{14} = 11$, $x_{21} = 1$, $x_{23} = 12$, $x_{31} = 5$, $x_{32} = 10$, $x_{34} = 4$.
   (b) Same as (a).
   (c) $x_{14} = 11$, $x_{21} = 6$, $x_{22} = 3$, $x_{24} = 4$, $x_{32} = 7$, $x_{33} = 12$.

3. (a) $x_{11} = 6, x_{21} = 3, x_{22} = 7, x_{32} = 9, x_{33} = 3, x_{43} = 7, x_{44} = 7$.
   (b) $x_{11} = 6, x_{22} = 0, x_{23} = 10, x_{32} = 5, x_{34} = 7, x_{41} = 3, x_{42} = 11$, solution is not unique.

5. Solution is $x_{13} = 12, x_{21} = 6, x_{24} = 9, x_{32} = 11, x_{33} = 1, x_{34} = 1$ in row minima method and VAM. Cost = 116 units. $x_{13} = 11, x_{14} = 1, x_{21} = 6, x_{24} = 9, x_{32} = 11, x_{33} = 2$, cost 118 units.

6. As the cost components are not given, we shall have to find the initial B.F.S. using only North-West Corner Rule. I.B.F.S. is $x_{11} = 1, x_{12} = 2, x_{22} = 1, x_{23} = 3$, either x_{24} or $x_{33} = 0, x_{34} = 2, x_{35} = 3$. Solution is degenerate.

11.4 Optimality Conditions

So long we have discussed the methods of determining the initial B.F.S. The next problem before us is to find whether the solution obtained, is optimal or not. A transportation problem is a maximization problem. Hence at the optimal stage, $c_{ij} - z_{ij} \geq 0$ for all cells corresponding to non-basic variables, i.e., $z_{ij} - c_{ij} \leq 0$ for all cells corresponding to non-basic variables. Therefore to test the optimality of the problem, we shall have to determine the values of $z_{ij} - c_{ij}$ for all cells corresponding to non-basic variables. If $z_{ij} - c_{ij} \leq 0$ for all cells corresponding to non-basic varibles, the solution is optimal. We know from theorem (A.2.14) that for all basic cells $z_{ij} - c_{ij} = 0$. This property is very useful in determining the values of $z_{ij} - c_{ij}$ for all cells corresponding to non-basic variables. If the conditions $z_{ij} - c_{ij} \leq 0$ are not satisfied for all non-basic cells, we shall have to proceed further to get an optimal solution which we shall discuss in future.

11.4.1 Determination of net evaluation ($\mathbf{z_{ij} - c_{ij}}$) (u v method)

In determining the net evaluation we shall make use of the property of the duality theory.

The original T.P. is

$$\text{minimize,} \quad z = \sum_{i=1}^{m} \sum_{j=1}^{n} c_{ij} x_{ij}$$

subject to

$$\sum_{j=1}^{n} x_{ij} = a_i, i = 1, 2, \cdots, m$$

$$\sum_{i=1}^{m} x_{ij} = b_j, j = 1, 2, \cdots, n,$$

where there are $(m+n)$ constraints all of which are equations and out of them only $(m+n-1)$ equations are independent. Hence there are $(m+n)$ dual variables to the primal (original) problem of which *one variable can be selected arbitrarily and all variables are unrestricted* in sign [as all the primal constraints are equations]. If the dual variables are

$$\mathbf{w} = (u_1, u_2, \cdots, u_m, v_1, \cdots, v_n) = (\mathbf{u}, \mathbf{v}) \tag{11.4.1}$$

the dual constraints are given by

$$u_i + v_j \leq c_{ij} \tag{11.4.2}$$

$u_i[i = 1, 2, \cdots, m]$ and $v_j[j = 1, 2, \cdots, n]$ are unrestricted in sign.

Now if B^{-1} is the basis inverse of the primal problem at the optimal stage and $\mathbf{c}_B$ is the associated cost vector, the dual optimal solution is given by

$$\mathbf{w} = \mathbf{c}_B B^{-1} \tag{11.4.3}$$

[from (theorem 10.7.1)] $\mathbf{w}$ is expressed here as a row vector with $(m+n)$ components. If $\mathbf{a}_j$ is a column vector corresponding to a non-basic variable then

$$z_j - c_j = \mathbf{c}_B B^{-1}\mathbf{a}_j - c_j. \tag{11.4.4}$$

Therefore the net evaluation $z_{ij} - c_{ij}$, corresponding to non-basic cells in a transportation problem are given by

$$\begin{aligned} z_{ij} - c_{ij} &= \mathbf{c}_B B^{-1}\mathbf{a}_{ij} - c_{ij} \\ &= \mathbf{w}\mathbf{a}_{ij} - c_{ij} \\ &= (\mathbf{u}, \mathbf{v})\mathbf{a}_{ij} - c_{ij}, \end{aligned} \tag{11.4.5}$$

where $\mathbf{a}_{ij}$ is the column vector of the co-efficient matrix associated with x_{ij}.

$\mathbf{a}_{ij} = \mathbf{e}_i + \mathbf{e}_{m+j}$. Hence from (11.4.5) we get

$$\begin{aligned} z_{ij} - c_{ij} &= (u_1, u_2, \cdots, u_m, v_1, v_2, \cdots, v_n)(\mathbf{e}_i + \mathbf{e}_{m+j}) - c_{ij} \\ &= u_i + v_j - c_{ij}[i = 1, 2, \cdots, m; j = 1, 2, \cdots, n.] \end{aligned} \tag{11.4.6}$$

For the basic cells, the net evaluations $z_{ij} - c_{ij} = 0$, i.e., if x_{rs} be a basic variable then

$$z_{rs} - c_{rs} = 0 \tag{11.4.7}$$

$$\text{or,} \quad u_r + v_s - c_{rs} = 0$$

$$\text{or,} \quad u_r + v_s = c_{rs} \text{ for all basic cells.} \tag{11.4.8}$$

If we select arbitrarily one of the values of the dual variables $(u_1, u_2, \cdots, u_m, v_1, v_2, \cdots, v_n)$ as zero, all other values can be determined using the relation (11.4.8). And once all quantities are known, the net evaluations $z_{ij} - c_{ij}$, given by the formula in (11.4.6) can be calculated. All these calculations can be done easily from the transportation table.

11.4.2 Numerical Calculation of the Net Evaluations corresponding to the Non-basic Cells

Below given is a transportation table involving 4 origins and 4 destinations in which the cost components are displayed in their proper places and all basic cells are marked by circular black spots. The solution (not given in the table) is a *basic feasible solution.* The problem before us is to calculate numerically all $z_{ij} - c_{ij}$ corresponding to the non-basic cells.

Table 11.14

				u_i
• 6	• 9	-1 8	3 7	3
-1 4	• 6	• 4	• 7	0
-6 9	1 5	• 4	5 2	0
1 3	6 1	-1 6	• 8	1
v_j: 3	6	4	7	

From (11.4.8) we get for a basic cell (r, s)

$$u_r + v_s = c_{rs}$$

and for non-basic cells (i, j)

$$z_{ij} - c_{ij} = u_i + v_j - c_{ij} \quad \text{[From (11.4.6)]}.$$

We know that we can select arbitrarily one of the values of $u_i[i = 1, 2, \cdots, m]$ and $v_j[j = 1, 2, \cdots, n]$.

For simple computation we may take the value of one variable equal to zero.

In the above table, cells $(1, 1)$, $(1, 2)$, $(2, 2)$, $(2, 3)$, $(2, 4)$, $(3, 3)$ and $(4, 4)$ are marked with circular black spots. The cells are basic cells and the solution will be a B.F.S. because the set of cells do not contain a loop and the number of cells $(4 + 4 - 1) = 7$.

In the second row there are three basic cells. Let us take $u_2 = 0$ [in general a variable u_i or v_j is taken to be zero, the corresponing row or column of which contains the maximum basic cells].

For basic cell $(2, 2)$

$$u_2 + v_2 = c_{22}$$

$$\text{or,} \quad 0 + v_2 = 6 \text{ which implies that } v_2 = 6.$$

Similarly for basic cell $(2, 3)$

$$u_2 + v_3 = c_{23}$$

$$\text{or,} \quad 0 + v_3 = 4 \text{ which implies that } v_3 = 4.$$

and for basic cell $(2, 4)$

$$u_2 + v_4 = c_{24}$$

$$\text{or,} \quad 0 + v_4 = 7 \text{ which implies that } v_4 = 7.$$

Now for basic cell (3, 3)

$$u_3 + v_3 = c_{33}$$
$$\text{or,} \quad u_3 + 4 = 4 \text{ which implies that } u_3 = 0.$$

For basic cell (4, 4)

$$u_4 + v_4 = c_{44}$$
$$\text{or,} \quad u_4 + 7 = 8 \text{ which implies that } u_4 = 1.$$

Again for basic cell (1, 2),

$$u_1 + v_2 = c_{12}$$
$$\text{or,} \quad u_1 + 6 = 9 \text{ which implies that } u_1 = 3.$$

and last of all for basic cell (1, 1),

$$u_1 + v_1 = c_{11}$$
$$\text{or,} \quad 3 + v_1 = 6 \text{ which implies that } v_1 = 3.$$

Thus we have calculated all values of variables $u_1, u_2, u_3, u_4; v_1, v_2, v_3, v_4$. Then using the relation

$$z_{ij} - c_{ij} = u_i + v_j - c_{ij}$$

we can calculate all net evaluations corresponding to the non-basic cells.

For example,

$$z_{13} - c_{13} = u_1 + v_3 - c_{13} = 3 + 4 - 8 = -1$$
$$z_{14} - c_{14} = u_1 + v_4 - c_{14} = 3 + 7 - 7 = 3$$

and so on.

In the transportation table, the values of all u_i and v_j are listed outside of the block as given in the table and all net evaluations corresponding to non-basic cells are displayed on the north-east corner of the respective cells. Following this rule, all net evaluations corresponding to the non-basic cells can be calculated.

Note. There is another method of computing $z_{ij} - c_{ij}$ which is known as *stepping stone algorithm.* But in digital computer *u v method* is widely used.

11.4.3 Determination of the Entering Cell and the Entering Vector

To test the optimality of a B.F.S. at any stage, we require to calculate the values of all net evaluations corresponding to the non-basic cells. If all net evaluations are non-positive quantities, the solution is optimal. But if at least one of them is positive, the solution is not optimal. As in the simplex method, now we shall have to proceed further to get an optimal solution. The first problem before us is to select a vector which will enter in the basis and will move the solution towards optimality. The vector which will enter in the basis is the entering vector and the corresponding cell in the transportation table is the new *basic cell.* If $\mathbf{a}_{pk}$ be the entering vector, the cell (p, k) will be the new basic cell which will enter in the set of basic cells. And due to this, a cell will leave the set of basic cells. This cell is known as the *leaving cell or departing cell* and the vector corresponding to the departing cell is known as *departing vector.* The entering vector is selected in the following manner.

If $\max_{ij}(z_{ij} - c_{ij}, z_{ij} - c_{ij} > 0) = z_{pk} - c_{pk} > 0$, i.e., if the positive maximum of the net evaluations occurs at cell (p, k), $\mathbf{a}_{pk}$ is the entering vector and cell (p, k) is the entering cell. If the maximum is not unique we may select any one cell corresponding to the maximum value of the net evaluations.

In the table (11.14) $\max_{ij}(z_{ij} - c_{ij}, z_{ij} - c_{ij} > 0) = 6 > 0$ which occurs at the cell $(4, 2)$. Therefore, in the next iteration, the cell $(4, 2)$ will be the new basic cell, and$\mathbf{a}_{42}$ will be the vector to enter in the next basis.

11.4.4 Determination of the Departing Cell and the value of the Basic Variable in the Entering Cell

After the selection of entering cell, we can identify the cell which will leave the set of basic cells, geometrically from the transportation table. Let the cell (p, k) be the entering cell. The number of (basic) cells including the cell (p, k) is $m + n - 1 + 1 = m + n$. Evidently the set of column vectors corresponding to these $(m + n)$ cells are linearly dependent. Therefore it is always possible to construct a loop connecting the cell (p, k) and the set or any subset of the basic cells. Construct the loop by trial and error method and the loop is unique.

Now allocate a value $\theta > 0$ [a variable] in the cell (p, k) and readjust the basic variables in the ordered set of cells containing the simple loop by adding and subtracting the value θ alternately from the corresponding quantities such that all rim requirements are satisfied. If the ordered set of cells containing the simple loop are

$$(p, k), (p, t), (t, r), (r, s) \text{ etc.}$$

the new variables corresponding to the cells are θ, $x_{pt} - \theta$, $x_{tr} + \theta$, $x_{rs} - \theta$ etc. where x_{pt} etc. are the basic variables in the current iteration.

Now select the maximum value of θ in such a way that readjusted values of the variables vanish at least in one cell {excluding the cell (p, k)} of the ordered set *and all other variables remain non-negative.* Let us assume that the variable vanishes in the cell (r, s) satisfying all conditions stated above, i.e., $x_{rs} - \theta = 0$ which gives the value of $\theta = x_{rs}$. This is the value of new basic variable to be allocated in the new basic cell (p, k). The cell (r, s) is the departing cell and it will be a non-basic cell during next iteration. The vector $\mathbf{a}_{rs}$ will leave the basis and the new basic variables will be $x_{rs}, x_{pt} - x_{rs}, x_{tr} + x_{rs}$, i.e., $\theta, x_{pt} - \theta, x_{tr} + \theta$ etc. corresponding to the cells $(p, k), (p, t), (t, r)$ respectively. All basic variables in the cells not in the loop remain unchanged. It may so happen that for maximum value of θ, the readjusted variables may vanish for more than one cell. In that case it is not possible to select uniquely the cell which will leave the set of basic cells. Select arbitrarily one of the cells as a departing cell and write down the value 0 (zero) as new basic variable in all other such cells and the solution in the next iteration will be degenerate. The method of solving a degenerate problem will be discussed in future.

11.4.5 Computational Procedure

To solve a transportation problem, proceed step by step as mentioned below:

Step 1. Determine an initial B.F.S. of the given problem using any one of methods discussed previously. But to get the optimal solution quickly, in general use either matrix minima method or Vogel's approximation method.

Step 2. Calculate all net evaluations corresponding to non-basic cells and display them on the north-east corner of the corresponding non-basic cells. If all net evaluations are non-positive quantities at any iteration, the solution is optimal. Then calculate the corresponding minimum cost using the relation $\min z = \hat{z} = \Sigma \hat{c}_{ij}\hat{x}_{ij}$ where $\hat{x}_{ij}$ are the components of optimal solution and $\hat{c}_{ij}$ are the corresponding transportation cost per unit commodity. If at least one net evaluation is positive, the solution is not optimal.

Step 3. If the solution is not optimal, determine the entering cell and the value of θ which will be allocated in the entering cell and the cell which will leave the basis. Construct a new transportation table with readjusted basic variables. Calculate again, all net evaluations corresponding to the non-basic cells and display them on the north-east corner of the corresponding non-basic cells. Test the optimality of the solution. If the solution is optimal, calculate the minimum cost of transportation. If the solution is not optimal, proceed similarly until the optimality conditions are satisfied.

This method of solving a transportation problem is known as MODI METHOD.

Solved Problems

▶ **Example 11.4.1** *Obtain the optimum basic feasible solution to the transportation problem given in* Ex. 11.3.3 *and find out the corresponding cost of transportation.* [C.U.,App.Math.'77; C.U.(P)'83,'92,'93]

Solution: In table (11.6) an initial B.F.S. is given which is obtained using matrix minima method.

We have to test first whether the solution is optimal or not. For this we shall have to calculate the net evaluations for all non-basic cells. The calculated values are displayed on north-east corner of respective non-basic cells in the table 11.15(A).

Table 11.15(A)

				u_i
−3 / 5	[18] / 3	[1] / 6	−4 / 2	0
θ, 1 / 4	−1 / 7	[12] − θ / 9	[25] / 1	3
[16] − θ / 3	0 / 4	[18] + θ / 7	−6 / 5	1
v_j: 2	3	6	−2	

Table 11.15(B)

				u_i
−3 / 5	[18] / 3	[1] / 6	−3 / 2	0
[12] / 4	−2 / 7	−1 / 9	[25] / 1	2
[4] / 3	0 / 4	[30] / 7	−5 / 5	1
v_j: 2	3	6	−1	

During calculation of the net evaluations we have taken arbitrarily $u_1 = 0$ and have calculated all net evaluations. All net evaluations corresponding to the non-basic cells are not less than or equal to zero. Net evaluation corresponding to non-basic cell $(2, 1)$ is 1. Therefore the solution is not optimal.

11.4.6 Selection of Entering Vector and Entering Cell

Since there is only one cell (non-basic) $(2, 1)$ for which net evaluation is positive therefore the vector $\mathbf{a}_{21}$ corresponding to the cell $(2, 1)$ is the entering vector and cell $(2, 1)$ will be the entering cell (new basic cell in the next iteration).

11.4.7 Value of the Variable which is to be allocated in the new basic cell and the Vector which will leave the basis

Construct a loop connecting the cell $(2, 1)$ and the set of basic cells or any subset of the basic cells of the origin solution. In this problem the ordered set of cells $(2, 1), (2, 4), (3, 3), (3, 1)$ are said to form a loop. Now allocate a value $\theta > 0$ [a variable] in the cell $(2, 1)$ and readjust the basic variables in the ordered set of cells forming the loop, by adding and subtracting θ alternately as given in the table 11.15(A), such that all rim requirements are satisfied properly. Now select the maximum value of θ in such a way that the values of the readjusted variables vanish at least in one cell containing the loop {except the cell $(2, 1)$} and *variables remain non-negative in other cells.*

From the table, it is clear that value of $\theta = 12$ and for that the value of variable in the cell $(2, 3)$ is zero. Therefore the cell $(2, 3)$ will leave the set of basic cells and the vector $\mathbf{a}_{23}$ will leave the basis. With this known value of $\theta = 12$ construct the new transportation table 11.15(B). All basic variables in the cells not in the loop remain unchanged and again calculate the net evaluations corresponding to the non-basic cells. Calculated values are displayed in table 11.15(B). All calculations are performed with the assumption that $u_1 = 0$. Here all net evaluations are non-positive quantities. Therefore the solution obtained is optimal. The optimal solution (non-degenerate) is given by $x_{12} = 18$, $x_{13} = 1$, $x_{21} = 12$, $x_{24} = 25$, $x_{31} = 4$ and $x_{33} = 30$ and the corresponding cost of transportation is given by

$$\hat{z} = 3 \times 18 + 6 \times 1 + 4 \times 12 + 1 \times 25 + 3 \times 4 + 7 \times 30 = 355 \text{ units.}$$

Note 1. As at the optimal stage, net evaluation, corresponding to the non-basic cell $(3, 2)$ is zero, then alternative optimal solution exists.

2. If we calculate $c_{ij} - z_{ij}$, the vector to enter in the next basis is $\mathbf{a}_{pk}$, where $\min_{ij}(c_{ij} - z_{ij}, c_{ij} - z_{ij} < 0) = c_{pk} - z_{pk}$ and at the optimal stage, all $c_{ij} - z_{ij} \geq 0$ for all non-basic cells.

▶ **Example 11.4.2** *Determine the optimal solution to the problem given in* Ex. 11.3.4 *and find the minimum cost of transportation.* [C.U.(P)'99]

Table 11.16(A)

				u_i
[5] 19	−32 30	−60 50	[2] 10	0
−1 θ 70	18 [7] 30	[2] 40	−θ 60	50
−11 [8] 40	−θ 8	−70 70	[10] +θ 20	10
v_j 19	−2	−10	10	

Table 11.16(B)

				u_i
[5] 19	−32 30	−42 50	[2] 10	0
−19 70	[2] 30	[7] 40	−18 60	32
−11 40	[6] 8	−52 70	[12] 20	10
v_j 19	−2	8	10	

First we have to test whether the initial B.F.S. is optimal or not. For this we shall have to calculate the net evaluations corresponding to all non-basic cells. The calculated values are displayed in the table 11.16(A). During computation of net evaluations we have taken arbitrarily $u_1 = 0$. All net evaluations corresponding to non-basic cells are not less than or equal to zero. Net evaluation corresponding to the non-basic cell $(2,2)$ is 18. Therefore the solution is not optimal. We shall have to construct the next table to get an optimal solution.

11.4.8 Selection of Entering Vector and Entering Cell

Since there is only one cell $(2,2)$ for which net evaluation is positive, therefore the vector $\mathbf{a}_{22}$, corresponding to the cell $(2,2)$ is the entering cell. This cell $(2,2)$ will be a basic cell in the next iteration.

11.4.9 Value of the Variable which is to be allocated in the new basic cell and the vector which will leave the basis

Construct a loop connecting the cell $(2,2)$ and the set of basic cells or any subset of basic cells. In this problem the ordered set of cells $(2,2)$, $(2,4)$, $(3,4)$ and $(3,2)$ are said to form a simple look [Ignore the intermediate basic cell (2, 3)]. Now allocate a value $\theta > 0$, a variable in the cell $(2,2)$ and readjust the basic variables in the ordered set of cells forming a simple loop by adding and subtracting θ alternately as given in the table 11.16(A) such that all rim requirements are satisfied properly. Now select the maximum value of θ in such a way that the values of the readjusted variables vanish at least in one cell containing the loop [except in the cell $(2,2)$] and *variables remain non-negative in other cells.* From the table it is clear that $\theta = 2$ and for that the value of the variable in the cell $(2,4)$ is zero. Therefore the cell $(2,4)$ will leave the set of basic cells and the vector $\mathbf{a}_{24}$ will leave the basis in the next iteration. With the known value $\theta = 2$ construct the new transportation table and again calculate the net evaluations corresponding to non-basic cells. Calculated values are displayed in the table 11.16(B) and all calculations are non-positive quantities. Hence the solution obtained is optimal. The optimal solution is given by $x_{11} = 5$, $x_{14} = 2$, $x_{22} = 2$, $x_{23} = 7$, $x_{32} = 6$ and $x_{34} = 12$ and the corresponding cost of transportation is given by

$$\hat{z} = 19 \times 5 + 10 \times 2 + 30 \times 2 + 40 \times 7 + 8 \times 6 + 20 \times 2 = 743 \text{ units.}$$

► **Example 11.4.3** *Obtain the initial B.F.S. to the following transportation problem by matrix minima method and then find out an optimal solution and the corresponding cost of transportation.*

	D_1	D_2	D_3	D_4	a_i
O_1	5	4	6	14	15
O_2	2	9	8	6	4
O_3	6	11	7	13	8
b_j	9	7	5	6	27

Solution: Initial B.F.S. calculated with the help of matrix minima method is given in table 11.17(A). Now calculate all net evaluations corresponding to the non-basic cells with the assumption $u_1 = 0$. All net evaluations are not non-positive. Hence the solution is not optimal.

Cell $(2, 4)$ has the positive net evaluation 3. Thus in the next iteration, cell $(2, 4)$ will be the new basic cell. Construct the loop as shown in the table 11.17(A). Loop is simple and unique. Insert the value $\theta > 0$ in the cell $(2, 4)$ and readjust the basic variables in the cells containing the loop accordingly as given in the table 11.17(A).

Table 11.17(A)

					u_i
	5 +θ 5	7 4	3 −θ 6	−2 14	0
	4 −θ 2	−8 9	−5 8	θ 3 6	−3
	0 6	−6 11	2 +θ 7	6 −θ 13	1
v_j	5	4	6	12	

Table 11.17(B)

					u_i
	8 5	7 4	−3 6	−5 14	0
	1 −θ 2	−8 9	−8 8	3 +θ 6	−3
	θ 3 6	−3 11	5 7	3 −θ 13	4
v_j	5	4	3	9	

Now the maximum value of θ will be 3 and cell $(1, 3)$ will leave the basic cell and all other variables remain non-negative. Construct the table 11.17(B) with the value of $\theta = 3$. Calculate all net evaluations corresponding to the non-basic cells with the assumption $u_1 = 0$. Cell $(3, 1)$ will be the new basic cell. Construct a simple loop with cell $(3, 1)$ as shown in the table 11.17(B). Insert the value $\theta > 0$ in the cell $(3, 1)$ and readjust the basic variables as shown in the table 11.17(B). Maximum value of $\theta = 1$, cell $(2, 1)$ will leave the set of basic cells and all other variables remain non-negative. With $\theta = 1$ construct the table 11.17(C).

Table 11.17(C)

					u_i
	8 5	7 4	0 6	−2 14	−1
	−3 2	−11 9	−8 8	4 6	−7
	1 6	−6 11	5 7	3 13	0
v_j	6	5	7	13	

Calculate all net evaluations corresponding to the non-basic cells in the table 11.17(C) with the assumption $u_2 = 0$. All net evaluations are non-positive. Hence the solution is optimal and the optimal solution is $x_{11} = 8$, $x_{12} = 7$, $x_{24} = 4$, $x_{31} = 1$, $x_{33} = 5$, $x_{34} = 2$ and the minimum cost of transportation is $\hat{z} = 5 \times 8 + 4 \times 7 + 6 \times 4 + 6 \times 1 + 7 \times 5 + 13 \times 2 = 159$ units.

Note. As the net evaluation corresponding to the non-basic cell $(1, 3)$ is zero, an alternative optimal solution exists.

▶ **Example 11.4.4** *Solve the following balanced T.P. by using VAM to determine the initial B.F.S.*

	D_1	D_2	D_3	D_4	a_i
O_1	4	2	7	–1	27
O_2	3	0	2	4	33
O_3	5	3	4	5	23
O_4	3	5	4	–2	17
b_j	31	24	25	20	100 \ 100

Solution: Determination of I.B.F.S. by VAM.

Table 11.18

	D_1	D_2	D_3	D_4	a_i				
O_1	[24] 4	2	7	[3] –1	27(3)	27(3)	24(2)	24(3)	
O_2	[7] 3	[24] 0	[2] 2	4	33(2)	33(2)	33(2)	9(1)	9(1)
O_3	5	3	[23] 4	5	23(1)	23(1)	23(1)	23(1)	23(1)
O_4	3	5	4	[17] –2	17(5)				
b_j	31(0)	24(2)	25(2)	20(1)	100 \ 100				
	31(1)	24(2)	25(2)	3(5)					
	31(1)	24(2)	25(2)						
	31(1)		25(2)						
	7(2)		25(2)						

Here the solution is non-degenerate but not unique.

Optimality Test:

Table 11.19

	D_1	D_2	D_3	D_4	u_i
O_1	[24] 4	−1 / 2	−4 / 7	[3] −1	1
O_2	[7] 3	[24] 0	[2] 2	−6 / 4	0
O_3	0 / 5	−1 / 3	[23] 4	−5 / 5	2
O_4	0 / 3	−5 / 5	−2 / 4	[17] −2	0
v_j	3	0	2	−2	

Since there are three basic variables in the second row then let us take $u_2 = 0$; with u_2 zero we calculate all u_i and v_j which are shown in the table 11.19. Now we calculate all $z_{ij} - c_{ij}$ for all non-basic cells. All $z_{ij} - c_{ij} \leq 0$; and for non-basic cell $(4, 1)$ $z_{41} - c_{41} = 0$. Then the alternative optimal solution exists. One optimal solution is $x_{11} = 24$, $x_{14} = 3$, $x_{21} = 7$, $x_{22} = 24$, $x_{23} = 2$, $x_{33} = 23$, $x_{44} = 17$ and the min cost $= 24 \times 4 + 3 \times (-1) + 7 \times 3 + 24 \times 0 + 2 \times 2 + 23 \times 4 + 17 \times -2 = 96 - 3 + 21 + 0 + 4 + 92 - 34 = 196$ units.

► **Example 11.4.5** *Solve the following balanced T.P. by using VAM to determine the initial B.F.S.*

	D_1	D_2	D_3	D_4	a_i
O_1	9	8	5	7	12
O_2	4	6	8	7	14
O_3	5	8	9	5	16
b_j	8	18	13	3	42 / 42

Solution: Determination of I.B.F.S. by VAM.

Table 11.20

	D_1	D_2	D_3	D_4	a_i		
O_1	9	8	[12] 5	7	12(2)		
O_2	4	[14] 6	8	7	14(2)	14(2)	
O_3	[8] 5	[4] 8	[1] 9	[3] 5	16(0)	16(0)	16
b_j	8(1)	18(2)	13(3)	3(2)	42 / 42		
	8(1)	18(2)	1(1)	3(2)			
	8.	4	1	3			

I.B.F.S. by VAM is not unique but it is a non-degenerate B.F.S.

Optimality Test: In the third row there are four basic variables then we take $u_3 = 0$; with u_3 zero we calculate all u_i and v_j which are shown in the table (11.21). Now we calculate all $z_{ij} - C_{ij}$ for non-basic cells which are shown in the table (11.21).

Table 11.21

	D_1	D_2	D_3	D_4	u_i
O_1	−8 / 9	−4 / 8	[12] 5	−6 / 7	−4
O_2	−1 / 4	[14] 6	−1 / 8	−4 / 7	−2
O_3	[8] 5	[4] 8	[1] 9	[3] 5	0
v_j	5	8	9	5	

All $z_{ij} - c_{ij} < 0$ at all non-basic cells. Then we reach at the optimal stage. Optimal solution is $x_{13} = 12$, $x_{22} = 14$, $x_{31} = 8$, $x_{32} = 4$, $x_{33} = 1$, $x_{33} = 3$ and the optimal solution is unique. The min cost $= 12\times5+14\times6+8\times5+4\times8+1\times9+3\times5 = 60 + 84 + 40 + 32 + 9 + 15 = 240$ units.

► **Example 11.4.6** *Find the I.B.F.S. to the following transportation problem using north-west corner rule and prove that the optimal solution is non-degenerate though the initial solution is degenerate.*

	D_1	D_2	D_3	D_4	a_i
O_1	9	8	5	7	12
O_2	4	6	8	7	14
O_3	5	8	9	5	16
b_j	8	18	13	3	42

Solution: Initial B.F.S. is given in the table 11.21(A). Solution is degenerate with $x_{23} = 0$. Calculate the net evaluations corresponding to the non-basic cells with the assumption $u_1 = 0$. Net evaluations are 5, 3 and 3 corresponding to the non-basic cells $(1,3)$, $(2,1)$ and $(3,1)$ respectively. Hence the solution given is not optimal. Put $\varepsilon > 0$ [very small +ve quantity] in the cell $(2,3)$ instead of zero. As the net evaluation 5 in the cell $(1,3)$ is the positive maximum, then the cell $(1,3)$ will be the new basic cell. Construct a loop as given in the table 11.21(A). Put $\theta > 0$ in the cell $(1,3)$ and readjust the basic variables in the cells containing the loop. Now the maximum value of θ will be ϵ such that basic variable in the cell $(2,3)$ will be zero and all other variables remain positive. Thus the cell $(2,3)$ will

Table 11.21(A)

(Boxed allocations in [ ]; net evaluations of non-basic cells shown before the "/"; cost after the "/".)

				u_i
[8] / 9	[4] −θ / 8	θ 5 / 5	−1 / 7	0
3 / 4	[14] +θ / 6	[ε] −θ / 8	−3 / 7	−2
3 / 5	−1 / 8	[13] / 9	[3] / 5	−1
v_j: 9	8	10	6	

Table 11.21(B)

				u_i
[8] −θ / 9	[4] −ε / 8	[ε] +θ / 5	−6 / 7	0
3 / 4	[14] +ε / 6	−5 / 8	−8 / 7	−2
θ 8 / 5	4 / 8	[13] −θ / 9	[3] / 5	4
v_j: 9	8	5	1	

leave the set of basic cells. With the value of $\theta = \epsilon$, construct the table 11.21(B) and again calculate the net evaluations corresponding to the non-basic cells. All net evaluations are not non-positive. Thus the solution is not optimal. Net evaluation corresponding to the cell $(3,1)$ is 8 which is a positive maximum. Thus the cell $(3,1)$ will be the new basic cell. Proceed accordingly as given in the table 11.21(B). Maximum value of θ will be 8 and cell $(1,1)$ will leave the set of basic cells. With the value of $\theta = 8$ construct the table 11.21(C). Now it is interesting to note that the solution will remain non-degenerate if we put $\epsilon = 0$ at this stage. Thus we put $\epsilon = 0$ and get the basic variables in the table 11.21(C).

Table 11.21(C)

				u_i
−8 / 9	[4] −θ / 8	[8] +θ / 5	−6 / 7	−4
−5 / 4	[14] / 6	−5 / 8	−8 / 7	−6
[8] / 5	θ 4 / 8	[5] −θ / 9	[3] / 5	0
v_j: 5	12	9	5	

Table 11.21(D)

				u_i
−8 / 9	−4 / 8	[12] / 5	−6 / 7	−4
−1 / 4	[14] / 6	−1 / 8	−4 / 7	−2
[8] / 5	[4] / 8	[1] / 9	[3] / 5	0
v_j: 5	8	9	5	

Now proceed step by step and in the table 11.21(D) we get the optimal solution because all net evaluations corresponding to the non-basic cells are non-positive. The optimal solution is

$x_{13} = 12$, $x_{22} = 14$, $x_{31} = 8$, $x_{32} = 4$, $x_{33} = 1$, $x_{34} = 3$ and $\min z = 5 \times 12 + 6 \times 14 + 5 \times 8 + 8 \times 4 + 9 \times 1 + 5 \times 3 = 240$.

The solution is non-degenerate as the number of positive basic variables is $(6 = 3 + 4 - 1)$ and the set of cells do not contain a loop.

11.4.10 Unbalanced Transportation Problems

There are two types of unbalanced T.P.:

1. When $\sum_{i=1}^{m} a_i > \sum_{j=1}^{n} b_j$, i.e., the total available capacities of m-origins are greater than the total demands of n-destinations. But this problem can be converted into a balanced transportation problem using the following device:

   (a) Imagine a fictitious or fake $(n+1)$th destination D_{n+1}.

   (b) Assume that the demand of the destination D_{n+1} is

   $$b_{n+1} = \sum_{i=1}^{m} a_i - \sum_{j=1}^{n} b_j.$$

   (c) Assume that the cost components $c_{i,n+1} = 0$ for all i $[i = 1, 2, \cdots, m]$ i.e., $c_{1,n+1} = c_{2,n+1} = \cdots = c_{m,n+1} = 0$.

With this assumption, the T.P. will be a balanced problem having m-origins and $(n+1)$ destinations. Due to the assumption $c_{i,n+1} = 0$, $[i = 1, 2, \cdots, m]$ the minimum cost of transportation remains unaffected and the total demands of the destinations will be satisfied completely though the capacities of the origins will not be exhausted completely. Below an unbalanced T.P. is given:

	D_1	D_2	D_3	a_i
O_1	2	3	4	6
O_2	4	3	1	8
O_3	2	2	5	6
b_j	5	4	7	

In the problem,

$$\sum_{i=1}^{3} a_i = a_1 + a_2 + a_3 = 6 + 8 + 6 = 20$$

$$\text{and} \quad \sum_{j=1}^{3} b_j = b_1 + b_2 + b_3 = 5 + 4 + 7 = 16.$$

$$\sum_{i=1}^{3} a_i = 20 > 16 = \sum_{j=1}^{3} b_j.$$

This problem can be converted into a balanced T.P. in the following manner:

	D_1	D_2	D_3	D_4	a_i
O_1	2	3	4	0	6
O_2	4	3	1	0	8
O_3	2	2	5	0	6
b_j	5	4	7	4	20

$$b_4 = \sum_{i=1}^{3} a_i - \sum_{j=1}^{3} b_j = 20 - 16 = 4 \text{ and } c_{14} = c_{24} = c_{34} = 0.$$

Now this transportation problem can be solved as in the previous methods.

In this problem, an initial B.F.S. obtained, using the matrix minima method, is $x_{11} = 2$, $x_{14} = 4$, $x_{22} = 1$, $x_{23} = 7$, $x_{31} = 3$ and $x_{32} = 3$, and the minimum cost of transportation such that the total demands of the destinations will be satisfied is 25. Multiple optimal solutions exist in the problem and one of the optimal solutions is $x_{11} = 3$, $x_{14} = 3$, $x_{23} = 7$, $x_{24} = 1$, $x_{31} = 2$ and $x_{32} = 4$, i.e., $x_{11} = 3$, $x_{23} = 7$, $x_{31} = 2$ and $x_{32} = 4$ as actually no quantity is transported to the destination D_4.

2. When $\sum_{j=1}^{n} b_j > \sum_{i=1}^{m} a_i$, i.e., the total demands of n destinations are greater than the total available capacities of m-origins. Here also the problem can be converted into an ordinary balanced transportation problem but the total demands of n-destinations will not be satisfied completely. Though we cannot satisfy all demands, we can still allocate the materials available at the origins to the destinations in such a way that minimizes the cost of transportation.

   To convert it into a balanced transportation problem:

   (a) Imagine a fake $(m + 1)$th origin O_{m+1}.

   (b) Assume that the capacity of the origin O_{m+1} is

   $$a_{m+1} = \sum_{j=1}^{n} b_j - \sum_{i=1}^{m} a_i.$$

   (c) Assume that the components $c_{m+1,j} = 0$ for all j.

The problem will be ultimately a transportation problem having $(m+1)$ origins and n-destinations. Now the problem can be solved as in the previous case.

11.4.11 Solution of Unbalanced Transportation Problems

▶ **Example 11.4.7** *Determination of the initial B.F.S of the following unbalanced transportation problem by VAM.*

	D_1	D_2	D_3	D_4	a_i	
O_1	14	19	11	20	10	
O_2	19	12	14	17	15	Supply
O_3	14	16	11	18	12	
b_j	8	12	16	14		
			Demand			

Here $\Sigma b_j = 50 > \Sigma a_i = 37$ then the problem is an unbalanced problem. $(50 - 37) = 13$; we assume a fictitious or fake origin O_4 having supply capacity 13 and the cost components of transporting these 13 units is 0, and we can rewrite the problem in the manner.

	D_1	D_2	D_3	D_4	a_i
O_1	14	19	11	20	10
O_2	19	12	14	17	15
O_3	14	16	11	18	12
O_4	0	0	0	0	13
b_j	8	12	16	14	50 / 50

Initial basic feasible by VAM.

Table 11.22

	D_1	D_2	D_3	D_4	a_i				
O_1	14	19	[10] 11	20	10(3)	10(3)	10(3)		
O_2	19	[12] 12	[2] 14	[1] 17	15(2)	15(2)	3(3)	3(3)	3(3)
O_3	[8] 14	16	[4] 11	18	12(3)	12(3)	42(3)	12(3)	4(7)
O_4	0	0	0	[13] 0	13(0)				
b_j	8(14)	12(12)	16(11)	14(17)	50 / 50				
	8(0)	12(4)	16(0)	1(1)					
	8(0)		16(0)	1(1)					
	8(5)		6(3)	1(1)					
			6(3)	1(1)					

Here the solution is a non-degenerate basic feasible solution.

Solve the above problem by taking the initial basic feasible solution obtain by VAM. Find the amount which is not supplied and find the destination to which the fake amount has been supplied.

Optimality test: This is a minimization problem.

Table 11.23

	D_1	D_2	D_3	D_4	u_i
O_1	0 14	–10 19	[10] 11	–6 20	–3
O_2	–2 19	[12] 12	[2] 14	[4] 17	0
O_3	[8] 14	–7 16	[4] 11	–4 18	–3
O_4	0 0	–5 0	–3 0	[13] 0	–17
v_j	17	12	14	17	

In the second row there are three basic variables then we shall take $u_2 = 0$, with $u_2 = 0$ we have calculated all $u_i (i = 1, 2, 3, 4)$ and $v_j (j = 1, \cdots, 4)$ as shown in the table. Now we calculate all $z_{ij} - c_{ij}$ for non-basic cells.

All $z_{ij} - c_{ij}$ are ≤ 0 and $z_{11} - c_{11} = 0$. Then we reach at the optimal stage and alternative optimal solution exist.

One optimal solution is $x_{13} = 10$, $x_{22} = 12$, $x_{23} = 2$, $x_{24} = 1$, $x_{31} = 8$, $x_{33} = 4$ and the fake amount 13 units is not supplied to the destination D_4, and

$$\text{Min cost} = 10 \times 11 + 12 \times 12 + 2 \times 14 + 1 \times 17 + 8 \times 14 + 4 \times 11$$
$$= 110 + 144 + 28 + 17 + 112 + 44 = 455 \text{ units.}$$

Remark: Destination D_4 will be want of 13 units.

▶ **Example 11.4.8** *Identical products are produced in three factories and sent to four warehouses for delivery to the customers. The costs of transportation and capacities are given by the cost matrix as*

Factories	W_1	W_2	W_3	W_4	a_i (Capacities)
F_1	3	8	7	4	30
F_2	5	2	9	5	50
F_3	4	3	6	2	80
b_j (Demands)	20	60	55	40	

(a) *Find an optimal schedule of delivery for minimization of cost of transportation.*

(b) *Find the idle capacity of the warehouses.* [C.U.(Pass)'95]

(c) *Do you anticipate any alternative optimum solution for the problem? How can the same be identified?* [C.U.M.Com.'89]

Solution: Total supply $= 30 + 50 + 80 = 160 \neq 175 = 20 + 60 + 55 + 40 =$ Total demands. Thus, it is an unbalanced transportation problem with

$$\sum_{i=1} a_i < \sum_{j=1} b_j.$$

Now $175 - 160 = 15$. Now assuming a fictitious factory F_4 with production capacity 15 with the cost of transportation per unit zero we can adjust it as a balanced T.P. with the cost matrix with supply and demand as given below:

	W_1	W_2	W_3	W_4	a_i
F_1	3	8	7	4	30
F_2	5	2	9	5	50
F_3	4	3	6	2	80
F_4	0	0	0	0	15
b_j	20	60	55	40	175 / 175

Initial B.F.S. with VAM: Initial B.F.S. using VAM is given in the following table in a compact form which will be able to save both labour and time with B.F.S. $x_{11} = 20$, $x_{13} = 10$, $x_{22} = 50$, $x_{32} = 10$, $x_{33} = 30$, $x_{34} = 40$ and $x_{43} = 15$ which is non-degenerate one.

Table 11.24

	W_1	W_2	W_3	W_4	a_i				
F_1	[20] 3	8	[10] 7	4	30(1)	30(1)	30(1)	30(1)	30(4)
F_2	5	[50] 2	9	5	50(3)	50(3)			
F_3	4	[10] 3	[30] 6	[40] 2	80(1)	80(1)	80(1)	70(2)	30(2)
F_4	0	0	[15] 0	0	15(0)				
b_j	20(3)	60(2)	55(6)	40(2)					
	20(1)	60(1)	40(1)	40(2)					
	20(1)	10(5)	40(1)	40(2)					
	20(1)		40(1)	40(2)					
	20(1)		40(1)						

Optimality Test: Initial B.F.S. is the optimal solution which has been shown in the following table where all $z_{ij} - c_{ij} \leq 0$ for all non-basic cells which indicates that the solution is optimal.

Table 11.25

					u_i
	[20] c 3	−4 c 8	[10] c 7	−1 c 4	1
	−4 c 5	[50] c 2	−4 c 9	−4 c 5	−1
	−2 c 4	[10] c 3	[30] c 6	[40] c 2	0
	−4 c 0	−3 c 0	[15] c 0	−4 c 0	−6
v_j	2	3	6	2	

Then (a) the optimal schedule is $x_{11} = 20$, $x_{13} = 10$, $x_{22} = 50$, $x_{32} = 10$, $x_{33} = 30$, $x_{34} = 40$ and $x_{43} = 15$ and Min cost $= 3 \times 20 + 7 \times 10 + 2 \times 50 + 3 \times 10 + 6 \times 30 + 2 \times 40 + 0 \times 15 = 520$ units.

(b) Actually there is no factory F_4 and there is 15 units demand short and the optimal variable $x_{43} = 15$ is nothing but the idle capacity of the warehouse W_3, i.e., W_3 will be 15 units short of demand to make it full.

(c) The optimal solution is unique, i.e., there is no alternative optimum solution which we can anticipate from the fact that there exists no $z_{ij} - c_{ij} = 0$ corresponding to a non-basic cell at the optimal stage.

▶ **Example 11.4.9** *Solve the following balanced T.P. by finding I.B.F.S by matrix minima method and by VAM.*

Problem is shown in the tabe which the readers can easily realise.

Solution: Determination of I.B.F.S. by matrix minima method.

Table 11.26

	D_1	D_2	D_3	D_4	a_i				
O_1	c 2	c 5	c 0	[12] c −3	12				
O_2	[7] c 4	c 6	c 8	[8] c 1	15	15	15	7	7
O_3	[3] c 4	[8] c 0	[3] c 4	c 5	14	14	6	6	6
O_4	c 2	c 6	[9] c 1	c 4	9	9	9	9	
b_j	10	8	12	20	50 \ 50				
	10	8	12	8					
	10		12	8					
	10		12						
	10		3						

Optimality Test: Initial basic feasible solution is not unique.

Table 11.27

	D_1	D_2	D_3	D_4	u_i
O_1	−2 / 2	−9 / 5	[θ] 0 / 0	[12] −θ / −3	−4
O_2	[7] −θ / 4	−6 / 6	−4 / 8	[8] +θ / 1	0
O_3	[3] +θ / 4	[8] / 0	[3] −θ / 4	−6 / 5	0
O_4	−1 / 2	−9 / 6	[9] / 1	−8 / 4	−3
v_j	4	0	4	1	

All $z_{ij} - c_{ij} \leq 0$ in the table 11.27; $x_{13} - c_{13} = 0$. Thus the solution $x_{14} = 12$, $x_{21} = 7$, $x_{24} = 8$, $x_{31} = 3$, $x_{32} = 8$, $x_{33} = 3$, $x_{43} = 9$ is optimal and min cost $= 33$ units and alternative optimal solution exists. We now find an alternative optimal solutiòn. We put θ in this cell (1, 3) and draw a simple loop as shown in the table 11.27 and readjusted the basic variables as shown in the table 11.27. Now θ to be taken in such a way that all basic variables remain positive and vanishes in a cell. If we take $\theta = 3$ then the conditions will be satisfied, variable in the cell (3, 3) vanishes, with the value of $\theta = 3$, the next solution is

Table 11.28

	D_1	D_2	D_3	D_4	u_i
O_1	−2 / 2	−9 / 5	[3] / 0	[9] / −3	0
O_2	[4] / 4	−6 / 6	−4 / 8	[11] / 1	4
O_3	[6] / 4	[8] / 0	0 / 4	−4 / 5	4
O_4	−3 / 2	−11 / 6	[9] / 1	−8 / 4	−1
v_j	0	−4	0	−3	

Here in the table 11.28 all $z_{ij} - c_{ij} \leq 0$ for the non-basic cells and $z_{33} - c_{33} = 0$. Thus alternative optimal solution exists. One alternative optimal solution is $x_{13} = 3$, $x_{14} = 9$, $x_{21} = 4$, $x_{24} = 11$, $x_{31} = 6$, $x_{32} = 8$, $x_{43} = 9$ and

$$\begin{aligned}\text{min cost} &= 3 \times 0 + 9 \times (-3) + 4 \times 4 + 11 \times 1 + 6 \times 4 + 8 \times 0 + 9 \times 1 \\ &= 0 - 27 + 16 + 11 + 24 + 0 + 9 = 33 \text{ units.}\end{aligned}$$

I.B.F.S. by VAM.

Table 11.29

	D_1	D_2	D_3	D_4	a_i				
O_1	2	5	0	[12] −3	12(3)	12(3)			
O_2	[7] 4	6	8	[8] 1	15(3)	15(3)	15(3)	7(4)	
O_3	[3] 4	[8] 0	[3] 4	5	14(4)	6(0)	6(0)	6(0)	6(0)
O_4	2	6	[9] 1	4	9(1)	9(1)	9(1)	9(1)	9(1)
b_j	10(0)	8(5)	12(1)	20(4)	50 / 50				
	10(0)		12(1)	20(4)					
	10(2)		12(3)	8(3)					
	10(2)		12(3)						
	3(2)		12(3)						

The initial basic feasible solution is not unique.

Optimality Test: Same as in the case of matrix minima method.

Objective and Short Answer Type Questions with Answers
Transportation Problems

1. (a) Formulate mathematically a transportation problem (balanced) as a L.P.P. having m origins and n destinations $(m, n \geq 2)$.
   (b) What are the number of constraints and variables in a balanced T.P.?
   (c) What is the number of independent constraints in a balanced T.P.?
   (d) Is a constraint of a T.P. (balanced) an equation or an inequation?
   [All problems have m origins and n destinations] [C.U.(H)'94]

   [*Ans.* (b) $m + n$, mn; (c) $m + n - 1$; (d) Equations.]

2. What is an unbalanced T.P.? How can you convert it into a balanced T.P.?

3. Which one of the statements is true?
   (a) A T.P. is strictly a maximization problem.
   (b) A T.P. is strictly a minimization problem.
   (c) A T.P. may be a maximization or a minimization problem.

   [*Ans.* (c) is true. T.P. is generally a minimization problem. But a maximization problem can also be solved using T.P. technique]

4. Prove that there exists at least one feasible solution in a balanced T.P.

   [*Ans.* Theorem (11.1.1).]

5. Prove that there exists a finite optimal solution in each balanced T.P.

   [*Ans.* Theorem (11.1.2).]

6. Define a loop in a transportation table. What is the nature of a loop in a transportation table?

[*Ans.* All loops are simple.]

7. (a) What is the number of basic variables in a balanced T.P. with m origins and n destinations? [C.U.(H)'96]
   (b) What is the maximum number of positive components in a B.F.S. of a T.P. (balanced) with m origins and n destinations?

[*Ans.* (a) $m+n-1$; (b) $m+n-1$.]

8. How can you establish geometrically that a set of vectors $\mathbf{a}_{ij}$ associated with the variables x_{ij} in a T.P. are linearly independent?

[*Ans.* If the set of cells c_{ij} or any subset of the cells corresponding to the vectors $\mathbf{a}_{ij}$ of the transportation table cannot be ordered so as to form a loop, the set of vectors are said to be linearly independent.]

9. How can you detect that in a T.P. the solution is optimal? What is the criterion for the existence of multiple optimal solutions?

[*Ans.* If at any table all $z_{ij}-c_{ij} \leq 0$ corresponding to non-basic cells, we can say that the solution is optimal.

If at the optimal stage, at least one $z_{ij}-c_{ij}=0$ corresponding to a non-basic variable, the problem is said to have multiple optimal solutions.]

10. In solving a T.P. what is the utility of constructing loops in a transportation table?

[*Ans.* First, it is possible to determine whether a given set of solution is a B.F.S. Secondly, it is possible to determine which vector will leave the basis and the maximum possible amount that can be allocated in the new basic cell.]

11. Which one of the statements is true?
    (a) In a T.P. with m origins and n destinations (balanced) the number of independent constraints is (a) $m+n$, (b) mn, (c) $m+n-1$.
    (b) In a transportation table $(m, n \geq 2)$ (a) it is always possible to construct a loop containing $(m+n)$ cells, (b) it is not possible.

[*Ans.* (a) $m+n-1$, (b) It is always possible.]

12. In a transportation problem with 3 origins and 4 destinations, may the variables $x_{11}, x_{22}, x_{33}, x_{34}, x_{24}, x_{23}$ be considered as basic variables? Give reasons.

[*Ans.* No; The set of corresponding cell (1,1), (2,2), (3,3), (3,4), (2,4) and (2,3) are said to form a loop. Thus the vectors of the variables are not linearly independent and thus the set of variables cannot be considered as basic variables.]

13. In a transportation problem with 4 origins and 3 destinations, may the variables $x_{11}, x_{21}, x_{22}, x_{23}, x_{32}$ and x_{43} be considered as a set of basic variables? Give reasons.

[*Ans.* Yes; The number of variables is $6 = 4+3-1$ and the set of corresponding cells (1,1), (2,1), (2,2), (2,3), (3,2) and (4,3) cannot be ordered so as to form a loop. Hence the variables may be considered as a set of basic variables.]

14. In a T.P. with 4 origins and 4 destinations, may the variables x_{11}, x_{22}, x_{23}, x_{24}, x_{31}, x_{34}, x_{42}, x_{44} be considered as basic variables? Give reasons.

[*Ans.* No; The number of variables is greater than $(4+4-1=7)$. Hence the variables cannot be considered as a set of basic variables.]

15. Find the initial B.F.S. of the following T.P. using North-West corner rule and prove that (ii) is degenerate.

(*a*)

	D_1	D_2	D_3	D_4	a_i
O_1	4	6	9	5	16
O_2	4	2	7	1	14
O_3	2	5	2	8	10
b_j	12	7	6	15	

(*b*)

	D_1	D_2	D_3	D_4	a_i
O_1	9	7	4	2	20
O_2	2	9	8	6	15
O_3	6	7	5	8	15
b_j	14	21	6	9	

[*Ans.* (a) $x_{11}=12$, $x_{12}=4$, $x_{22}=3$, $x_{23}=6$, $x_{24}=5$, $x_{34}=10$.

(b) $x_{11}=14$, $x_{12}=6$, $x_{22}=15$, $x_{33}=6$, $x_{34}=9$ and either $x_{23}=0$ or $x_{32}=0$.]

16. Find the initial B.F.S. of the following T.P. using (a) Row minima method (b) Matrix-minima method and (c) VAM.

(*a*)

			a_i
30	20	10	50
5	15	25	50
30	30	40	b_j

(*b*)

			a_i
6	7	8	28
9	3	5	12
13	17	10	b_j

(a) $x_{13}=40$, $x_{12}=10$, $x_{21}=30$, $x_{22}=20$ [solution is same in matrix minima method and VAM].

(b) i. $x_{11}=13$, $x_{12}=15$, $x_{22}=2$, $x_{23}=10$.

ii. $x_{11}=13$, $x_{12}=5$, $x_{13}=10$, $x_{22}=12$.

iii. Same as (ii).

Exercise 11B

1. (a) Determine the initial B.F.S. of the problem using row minima method and hence obtain the optimal feasible solution and the corresponding minimum cost of transportation.

   (b) Determine the initial B.F.S. of the problem using matrix minima method and then find out the optimal feasible solution and the corresponding minimum cost of transportation.

(a)

	D_1	D_2	D_3	a_i
O_1	5	4	3	5
O_2	2	1	5	4
O_3	6	2	1	5
b_j	2	6	6	14

(b)

	D_1	D_2	D_3	a_i
O_1	7	5	4	12
O_2	5	4	3	10
O_3	4	1	5	8
b_j	10	9	11	30

2. Find the minimum cost of transportation in the following problem:

[C.U.(P)'84,'88] [C.U.M.Com.'88 costs are multiplied by 4]

	D_1	D_2	D_3	D_4	a_i
O_1	10	7	3	6	3
O_2	1	6	8	3	5
O_3	7	4	5	3	7
b_j	3	2	6	4	15

Use row minima method in determining an initial B.F.S. Solve the problem also using north-west corner rule of determination of an initial B.F.S.

3. A company has three plants A, B, C and three warehouses X, Y, Z. The number of units available at the plants are 60, 70, 80 respectively. The demands at X, Y, Z are 50, 80, 80 respectively. The unit cost of transportation are as follows: [C.U.'80]

	X	Y	Z
A	8	7	3
B	3	8	9
C	11	3	5

Find the allocation so that the total transportation cost is minimum.

4. Find the optimal solution and the corresponding cost of transportation in the following T.P.

(a)

	D_1	D_2	D_3	a_i
O_1	2	4	3	2
O_2	1	5	4	4
O_3	2	6	3	6
O_4	4	2	5	8
b_j	7	6	7	20

(b)

	1	2	3	4	
1	23	27	16	18	30
2	12	17	20	51	40
3	22	28	12	32	53
b_j	22	35	25	41	

[J.U.'84; C.U.(P)'87]

(c)

	1	2	3	4	
1	5	3	6	4	30
2	3	4	7	8	15
3	9	6	5	8	15
b_j	10	25	18	7	

[C.U.(P)'84,'86,'90; J.U.M.Sc.'83]

5. Solve the following balanced transportation problem:

	D_1	D_2	D_3	D_4	a_i	
O_1	6	4	2	7	8	
O_2	5	1	4	6	14	
O_3	6	5	2	5	9	Supply
O_4	4	3	2	1	11	
b_j	7	13	12	10	42	
		Demand				

6. Solve the following unbalanced transportation problem:

	D_1	D_2	D_3	a_i
O_1	4	5	6	12
O_2	3	1	5	11
O_3	2	4	4	7
b_j	6	5	8	

7. A firm manufacturing a single product has three plants I, II, III. The three plants have produced 60, 35 and 40 units respectively during this month. The firm had made a commitment to sell 22 units to customer A, 45 units to customer B, 20 units to customer C, 18 units to customer D and 30 units to customer E. Find the minimum cost of shifting the manufactured product to the five customers. The cost matrix is given below:

		Customer				
		A	B	C	D	E
	I	4	1	3	4	4
Plant	II	2	3	2	2	3
	III	3	5	2	4	4

8. Solve the transportation problem and prove that optimal solutions degenerate and multiple optimal solutions exist.

	D_1	D_2	D_3	D_4	a_i
O_1	5	2	4	6	9
O_2	2	4	1	5	11
O_3	4	2	3	1	8
b_j	6	7	7	8	28

9. A company has factories at A, B and C which supply warehouses at D, E, F and G. Monthly factory capacities are 160, 150 and 190 units respectively. Monthly warehouse requirements are 80, 90, 110 and 160 units respectively. Unit transportation cost (in rupees) are as follows:

	D	E	F	G
A	42	48	38	37
B	40	49	52	51
C	39	38	40	43

Determine the optimum distribution for this company to minimize the transportation cost and determine the idle factory capacity or capacities.

10. Origins O_1, O_2, O_3 and O_4 have surpluses of 30, 50, 75 and 20 empty freight cars respectively and the destinations D_1, D_2, D_3, D_4, D_5 and D_6 are in need of 20, 40, 30, 10, 50 and 25 empties respectively. The cost per unit of moving

an empty from the origins to the destinations is given below:

Destinations

	D_1	D_2	D_3	D_4	D_5	D_6
O_1	1	2	1	4	5	2
O_2	2	3	2	1	4	3
Origins O_3	4	2	5	9	6	2
O_4	3	1	7	3	4	6

Determine the minimum possible cost of transporting empty cars from excess origins to "deficit" destinations.

11. Solve the following two balanced T.P. problems by using VAM determine the initial B.F.S.

(a)

	D_1	D_2	D_3	D_4	a_i
O_1	1	2	3	4	6
O_2	4	3	2	0	8
O_3	0	2	2	1	10
b_j	4	6	8	6	24

(b)

	D_1	D_2	D_3	D_4	a_i
O_1	5	8	3	6	30
O_2	4	5	7	4	50
O_3	6	2	4	5	40
b_j	30	20	40	30	120

[C.U.(H)'80]

12. (a) An oil corporation has got three refineries P, Q and R and it has to send petrol to four different depots A, B, C and D. The cost of shipping 1 gal. of petrol from each refinery to each depot is given below. The requirement of the depots and the available petrol at the refineries are also given. Find the minimum cost of shipping after obtaining an initial solution by VAM.

Depot

		A	B	C	D	a_i
	P	10	12	15	8	130
Refinery	Q	14	11	9	10	150
	R	20	5	7	18	170
	b_j	90	100	140	120	

(b) In a transportation problem the cost matrix is given:

	D_1	D_2	D_3
O_1	7	3	4
O_2	2	2	3
O_3	3	4	6

A basic F.S. is given, $x_{13} = 2$, $x_{22} = 1$, $x_{23} = 2$, $x_{31} = 4$, $x_{33} = 1$. Is the solution optimal? Find the minimum cost of the problem.

13. There are three sources which store a given product. The sources supply these products to four dealers. The capacities of the sources, and the demands of the dealers are given. Capacities $S_1 = 150$, $S_2 = 40$, $S_3 = 80$. Demands

$D_1 = 90$, $D_2 = 70$, $D_3 = 50$, $D_4 = 60$. The cost matrix is given. Find the minimum cost of transportation.

	D_1	D_2	D_3	D_4
S_1	27	23	31	69
S_2	10	45	40	32
S_3	30	54	35	57

[C.U.(H)'84,'87]

14. Solve the following balanced T.P.

(a)

	D_1	D_2	D_3	D_4	a_i
F_1	15	20	13	21	15
F_2	22	15	19	14	12
F_3	16	12	14	31	13
F_4	24	23	15	30	20
b_j	16	10	10	24	

(b)

	D_1	D_2	D_3	D_4	a_i
O_1	21	13	17	24	11
O_2	19	15	13	17	15
O_3	23	11	22	20	18
O_4	16	18	21	14	11
b_j	13	17	11	14	

(c)

	1	2	3	4	a_i
1	2	3	11	7	6
2	1	0	6	1	1
3	5	8	15	9	10
b_j	7	5	3	2	

[Kalyani(H)'87]

(d)

	D_1	D_2	D_3	D_4	D_5	a_i
O_1	3	5	6	9	7	9
O_2	5	2	4	3	6	6
O_3	4	1	3	2	5	10
b_j	2	5	6	7	5	

[*Ans.* (a) Optimal solution $x_{11} = 3$, $x_{14} = 12$, $x_{24} = 12$, $x_{31} = 3$, $x_{32} = 10$, $x_{41} = 10$, $x_{43} = 10$. Alternative optimal solutions exist. Min cost = 1023 units.

(b) Optimal solution $x_{11} = 11$, $x_{23} = 11$, $x_{24} = 4$, $x_{32} = 17$, $x_{34} = 1$, $x_{41} = 2$, $x_{44} = 9$. Min cost = 807 units. Alternative optimal solution exists.

(c) Optimal solution $x_{12} = 5$, $x_{13} = 1$, $x_{23} = 1$, $x_{31} = 7$, $x_{33} = 1$, $x_{34} = 2$. Min cost = 100.

(d) Optimal solution $x_{11} = 2$, $x_{13} = 2$, $x_{15} = 5$, $x_{24} = 6$, $x_{32} = 5$, $x_{33} = 4$, $x_{34} = 1$. Min cost = 90 units. Alternative optimal solutions exist.]

15. A company has three factories A, B, C which supply five wholesale dealers with small car fans. The production capacities of the factories, the demands of the customers assumed constant and the distribution costs are given below. The object is to supply the wholesalers with their demands in the cheapest way. Find the optimal solution.

	a	b	c	d	e	a_i
A_1	5	7	10	5	3	5
A_2	8	6	9	12	14	10
A_3	10	9	8	10	15	10
b_j	3	3	10	5	4	25

16. Solve the following unbalanced T.P. problem:

	D_1	D_2	D_3	D_4	a_i
O_1	4	6	2	3	12
O_2	8	2	3	4	10
O_3	5	7	6	5	8
b_j	9	8	6	5	

17. Solve the following balanced transportation problem:

(a)

	D_1	D_2	D_3	D_4	a_i
O_1	18	22	14	19	15
O_2	23	16	15	20	22
O_3	18	24	15	24	13
b_j	10	10	18	12	50 / 50

(b) Prove that the initial solution obtained by N.W.C rule is the optimal solution but initial solution will not be an optimal one in VAM or Matrix minima method.

	D_1	D_2	D_3	
O_1	6.1	5	6.1	40
O_2	3	1	2	50
O_3	2	1	1	10
	20	50	30	

(c) Obtain the optimal solution of the following transportation problem:

	D_1	D_2	D_3	D_4	D_5	a_i
O_1	2	1	7	4	3	5
O_2	2	1	4	5	3	9
O_3	5	2	9	9	1	9
b_j	5	2	6	7	3	23

[C.U.(H)'89,'96]

[*Hints.* Initial solution by VAM (taking the difference between two lowest costs zero) $x_{14} = 5$, $x_{21} = 1$, $x_{23} = 6$, $x_{24} = 2$, $x_{31} = 4$, $x_{32} = 2$, $x_{35} = 3$ is the optimal solution and min cost = 83 units.]

18. A manufacturer has distribution centres located at Agra, Allahabad and Calcutta. These centres have available 40, 20 and 40 units of his product. His retail outlets require the following number of units; $A = 25$, $B = 10$, $C = 20$, $D = 30$ and $E = 25$. The cost per unit in rupees between each centre and outlet is given in the following table:

Distribution centres	Retail outlets				
	A	B	C	D	E
Agra	55	30	40	50	40
Allahabad	35	30	100	45	60
Calcutta	40	60	95	35	30

(a) Determine the optimal shipping cost.

(b) Do you anticipate any alternative solution? Give reason for your answer.

(c) Find the idle capacity of the retail outlets, if any. [C.U.,M.Com.'89]

19. Initial B.F.S. are given using VAM in the problems. Find the optimal solution and minimum cost in each case. See Solved Ex. 11.4.2.

20. Messrs Hindustan Construction Company Limited require 3, 3, 4 and 5 million cubic feet of fill at four earthen dam sites D_1, D_2, D_3 and D_4 in the district of Birbhum, West Bengal. The company can transfer the fill from three mounds M_1, M_2 and M_3 where 2, 6 and 7 million cubic feet of fill are available respectively. The cost of transporting one million cubic feet of fill from the mounds to the dam sites (expressed in Rs. Lakhs) are shown in the cost matrix given below:

		Dam sites			
		D_1	D_2	D_3	D_4
	M_1	15	10	17	18
Mounds	M_2	16	13	12	13
	M_3	12	17	20	11

(a) Formulate the above problem as a linear programming problem.

(b) Find the optimal solution of the problem. [C.U.,M.Com.'85]

21. The cost matrix of a transportation problem is given below. There is a pre-condition that if all demands of a particular destination be not met one unit of penalty is to be charged for each unit of goods not supplied. Find the minimum cost of transportation and mention in which destination or destinations, demand will not be met.

	D_1	D_2	D_3	D_4	Capacity
O_1	12	9	13	7	14
O_2	9	7	11	7	10
O_3	12	10	8	9	11
Demands	9	12	13	6	

[*Hints*. This is actually an unbalanced problem which can be made balanced in the following manner:

	D_1	D_2	D_3	D_4	
O_1	12	9	13	7	14
O_2	9	7	11	7	10
O_3	12	10	8	9	11
O_4	1	1	1	1	5
	9	12	13	6	

where O_4 is a fake origin but the cost components are 1 (one) each instead of 0 (zero).]

Answers

1. (a) min $z = 30$ for $x_{13} = 5, x_{21} = 2, x_{22} = 2, x_{32} = 4, x_{33} = 1$ (multiple solutions exist).
   (b) min $z = 107$ for $x_{12} = 1, x_{13} = 11, x_{21} = 10, x_{23} = 0, x_{32} = 8$.
2. min $z = 47$ for $x_{13} = 3, x_{21} = 3, x_{24} = 2, x_{32} = 2, x_{33} = 3, x_{34} = 2$.
3. min $z = 750$ for $x_{13} = 60, x_{21} = 50, x_{22} = 0, x_{23} = 20, x_{32} = 80$.
4. (a) min $z = 47$ for $x_{11} = 2, x_{21} = 4, x_{31} = 1, x_{33} = 5, x_{42} = 6, x_{43} = 2$ (multiple solutions exist).
   (b) Min cost $= 2221$ at $x_{14} = 30, x_{21} = 5, x_{22} = 35, x_{31} = 17, x_{33} = 25, x_{34} = 11$.
   (c) Min cost $= 231$ at $x_{12} = 23, x_{14} = 7, x_{21} = 10, x_{22} = 2, x_{23} = 3, x_{33} = 15$ (Multiple solution exist).
5. min $z = 86$ for $x_{11} = 5, x_{13} = 3, x_{21} = 1, x_{22} = 13, x_{33} = 9, x_{41} = 1, x_{44} = 10$ (alternative solutions exist).
6. min $z = 57$ for $x_{13} = 1, x_{22} = 5, x_{23} = 6, x_{31} = 6, x_{33} = 1$.
7. min $z = 290$ for $x_{12} = 45, x_{15} = 15, x_{21} = 2, x_{24} = 18, x_{25} = 15, x_{31} = 20, x_{33} = 20$ (alternative solution exists).
8. min $z = 47$ for $x_{11} = 2, x_{12} = 7, x_{14} = 0, x_{21} = 4, x_{23} = 7, x_{34} = 8$.
9. min $z = 17050$ for $x_{14} = 160, x_{21} = 80, x_{22} = 10, x_{32} = 80$ and $x_{33} = 110$; 60 units remain idle in the factory B.
10. min $z =$ Rs. 430 for $x_{11} = 20, x_{13} = 10, x_{23} = 20, x_{24} = 10, x_{25} = 20, x_{32} = 40, x_{35} = 10, x_{36} = 25, x_{45} = 20$ (alternative solutions exists).
11. (a) min $z = 28$ for $x_{12} = 6, x_{23} = 2, x_{24} = 6, x_{31} = 4, x_{32} = 0, x_{33} = 6$.
    (b) min $z = 420$ for $x_{13} = 30, x_{21} = 30, x_{24} = 20, x_{32} = 20, x_{33} = 10, x_{34} = 10$.
12. (a) min cost $= 3640$ for $x_{11} = 90, x_{14} = 40, x_{23} = 70, x_{24} = 80, x_{32} = 100, x_{33} = 70$.
    (b) Solution is not optimal. Min cost $= 33$ for $x_{13} = 2, x_{23} = 3, x_{31} = 4, x_{32} = 1, x_{33} = 0$.
13. min $z = 8190$ for $x_{11} = 30, x_{12} = 70, x_{13} = 50, x_{24} = 40, x_{31} = 60, x_{34} = 20$.
15. min cost $= 183$ units for $x_{14} = 1, x_{15} = 4, x_{21} = 3, x_{22} = 3, x_{23} = 4, x_{33} = 6, x_{34} = 4$.
16. min cost $= 87$ for $x_{11} = 1, x_{13} = 6, x_{14} = 5, x_{22} = 8, x_{31} = 8$ (alternative solution exists).
17. (a) min cost $= 835$ units at $x_{13} = 3, x_{14} = 12, x_{22} = 10, x_{23} = 12, x_{31} = 10, x_{33} = 3$ (alternative solution exists).
    (b) $x_{11} = 20, x_{12} = 20, x_{22} = 30, x_{23} = 20, x_{33} = 10$. Min cost $= 302$.
18. (a) Optimal shipping cost $= 3525$ for $x_{12} = 10, x_{13} = 20, x_{15} = 10, x_{21} = 20, x_{34} = 25$ and $x_{35} = 15$.
    (b) No alternative optimal solution as no $z_{ij} - c_{ij} = 0$ for non-basic cells.
    (c) 5 units at A and 5 units at D.
20. Minimize, $z = 15x_{11} + 10x_{12} + \cdots + 18x_{14} + 16x_{21} + \cdots + 13x_{24} + 12x_{31} + \cdots + 11x_{34}$ subject to

$$\sum_{j=1}^{4} x_{ij} = a_i, a_1 = 2, a_2 = 6, a_3 = 7[i = 1, 2, 3]$$

$$\sum_{i=1}^{3} x_{ij} = b_j, b_1 = 3, b_2 = 3, b_3 = 4, b_4 = 5[j = 1, 2, 3, 4]$$

and $\sum a_i = \sum b_j = 15$.

Min cost $= 174$ for $x_{12} = 2, x_{22} = 1, x_{23} = 4, x_{24} = 1, x_{31} = 3, x_{34} = 4$.

21. Optimal solution $x_{12} = 8, x_{14} = 6, x_{21} = 6, x_{22} = 4, x_{33} = 11$; 3 and 2 units of demands will not be met for destinations D_1 and D_3 respectively. Min cost $= 289$ units.

11.5 Assignment Problems

Assignment problem is a particular type of a transportation problem where n origins are to be assigned to an equal number of destinations *in one to one basis* such that the assignment cost (or profit) is minimum (or maximum). Consider the following examples. In a factory there are n jobs and n workers. c_{ij}, $[i, j = 1, 2, \cdots, n]$ is the assignment cost if the ith job is assigned to jth man. The problem is to find the minimum assignment cost such that all n jobs are assigned to n workers in one to one basis, i.e., only one job is assigned to a particular worker. In a medical firm there are n medical representatives and they are to be sent in n different localities for business purpose. c_{ij}, is the profit component if the ith person is sent to the jth destination. Determine the assignment such that the total profit is maximum.

11.5.1 Mathematical Formulation of an Assignment Problem

Let x_{ij} be a variable defined by:

$x_{ij} = 1$ if ith origin is assigned to jth destination.

$x_{ij} = 0$ if ith origin is not assigned to jth destination.

Now the assignment problem is

Optimize (minimize or maximize)

$$z = \sum_{i=1}^{n}\sum_{j=1}^{n} c_{ij}x_{ij}$$

subject to

$$\sum_{j=1}^{n} x_{ij} = a_i = 1, \quad i = 1, 2, \cdots, n$$

$$\sum_{i=1}^{n} x_{ij} = b_j = 1, \quad j = 1, 2, \cdots, n$$

$$\text{and } \sum_{i=1}^{n} a_i = \sum_{j=1}^{n} b_j = n.$$

From the above discussions it is clear that the problem before us is to select n cells in a $(n \times n)$ transportation table, only one cell in each row and each column, such that the sum of the corresponding costs (or profits) be minimum (or maximum). Obviously, the solution obtained is a degenerate solution. To get the solution, we shall state and prove an important theorem.

Thus an assignment problem is a particular type of L.P.P. with B.F.S. always degenerate.

Theorem 11.5.1 *Given a cost or profit matrix $C = [c_{ij}]_{n\times n}$, if we form another matrix $C^* = [c^*_{ij}] = c_{ij} - u_i - v_j$, u_i and v_j are arbitrary chosen constants the solution of C will be identical with that of C^*.*

Let

$$z = \sum_{i=1}^{n}\sum_{j=1}^{n} c_{ij}x_{ij} \quad \text{and} \quad z^* = \sum_{i=1}^{n}\sum_{j=1}^{n} c_{ij}^* x_{ij}$$

$$z^* = \sum_{i=1}^{n}\sum_{j=1}^{n} (c_{ij} - u_i - v_j)x_{ij}$$

$$= \sum_{i=1}^{n}\sum_{j=1}^{n} c_{ij}x_{ij} - \sum_{i=1}^{n}\sum_{j=1}^{n} u_i x_{ij} - \sum_{i=1}^{n}\sum_{j=1}^{n} v_j x_{ij}$$

$$= z - \sum_{i=1}^{n} u_i - \sum_{j=1}^{n} v_j \quad \left[\because \sum_{j=1}^{n} x_{ij} = 1 = \sum_{i=1}^{n} x_{ij}\right].$$

As $\sum_{i=1}^{n} u_i$ and $\sum_{j=1}^{n} v_j$ are constants then z^* and z differ only by a constant. From the above theorem we conclude that if in an assignment problem we subtract a number from every element of a row or column of the cost (or profit) matrix, then an assignment which is optimum for the transformed matrix will be optimum for the original matrix also.

11.5.2 Computational Procedure

Without going details into the theoretical developments of the problems, we shall chalk out the computational procedure for a minimization problem, step by step based mainly on the property discussed above.

(1) If the problem is a maximization problem, then convert it into a minimization problem by changing the cost matrix C to $-C$. Subtract the minimum cost element of each row from all other elements of the respective row. Then subtract the minimum element of each column from all other elements of the respective column of the resulting cost matrix. [These two operations can be interchanged]. A set of k zeros ($k \geq n$) will be obtained in the new cost matrix. Connect all the zeros by a *minimum* number of horizontal and vertical lines. Let the number of lines be N. There are two cases to discuss.

(i) If $N = n$, then the optimality conditions are satisfied, i.e., it is possible to select a single zero in each row and each column of the transportation table and hence $c_{ij}^* x_{ij} = 0$ which is the optimal value of z^* (the minimization problem). Now select only n zeros in such a way that there is one and only one zero in each row and each column. To do this, initially cross off all zeros which lie on the points of intersection of the lines. Next select a row or a column containing only one zero and cross off all zeros of the corresponding column or row. Proceed in the way to get n zeros. The sum of the cost components of the original cost matrix corresponding to n zeros of the final matrix gives the minimum cost. If the selection of n zeros is unique, the solution is unique. Otherwise the solution is not unique.

(ii) If $N < n$, then the optimality conditions will not be satisfied. Imagine a new cost matrix, *whose elements are the elements of the remaining rows and columns of the penaltimate table which are not covered by the lines.* Subtract the smallest element of the new matrix from all the elements of this matrix and add that smallest element of the matrix to all the elements which lie on the points of intersection of the lines. Now connect all the zeros of the new matrix by a *minimum* number of horizontal and vertical lines. If the number of lines be $N = n$, then the optimality conditions are satisfied and then proceed as in the above case (i) to get the minimum cost. If $N < n$, then proceed similarly as in the case (ii) until the number of lines is n. Then find out the minimum cost etc. and max $(z) = -\min(-z)$.

N.B. This method of solving an assignment problem is called Hungarian Method.

Worked out Examples

► **Example 11.5.1** *Find out the optimal (minimum) assignment cost from the following cost matrix.*

	I	*II*	*III*	*IV*
A	9	6	6	5
B	8	7	5	6
C	8	6	5	7
D	9	9	8	8

This is a problem of minimization.

Step 1. Subtract the lowest element 5 of the first row from all the elements of the first row. Similarly subtract the lowest element of each row from all the elements of we corresponding row to get the matrix in table 11.30(A).

Step 2. Subtract the lowest element of each column from all elements of the corresponding column of matrix in table 11.30(A) to get the matrix in table 11.30(B)

Table 11.30(A)

$$\begin{bmatrix} 4 & 1 & 1 & 0 \\ 3 & 2 & 0 & 1 \\ 3 & 1 & 0 & 2 \\ 1 & 1 & 0 & 0 \end{bmatrix}$$

Table 11.30(B)

$$\begin{bmatrix} 3 & 0 & 1 & 0 \\ 2 & 1 & 0 & 1 \\ 2 & 0 & 0 & 2 \\ 0 & 0 & 0 & 0 \end{bmatrix}$$

Step 3. Connect all the zeros of matrix in table 11.30(B) by a minimum number of horizontal and vertical lines. Here the number of lines are 4. Hence the optimality conditions are satisfied.

Step 4. Write all zeros in a 4×4 transportation table 11.31(B) in their respective positions and cross off (omit) the zeros at the points of intersection of the lines.

Table 11.31(A)

3	0	1	0
2	1	0	1
2	0	0	2
0	0	0	0

Table 11.31(B)

			✓0
		✓0	
	✓0	0	
✓0			0

Step 5. Select 4 zeros in such a way that there is only one zero in each row and each column. Select the first column containing only a single zero at cell $(4, 1)$ and then cross off all the zeros of the fourth row. Now select the single zero at cell $(3, 2)$ in the 2nd column and cross off all zeros of the third row. Then select the zeros at $(2, 3)$ and cross off all zeros of the third row. Then select the zeros at $(2, 3)$ and $(1, 4)$. Each of the zeros is the single zero in each row and each column.

Step 6. Hence the optimal assignment is unique and assignment is $D \to I$, $C \to II$, $B \to III$ and $A \to IV$ and the optimal assignment cost = The sum of the cost components of the original cost matrix corresponding to 4 zeros in the final table $= 9 + 6 + 5 + 5 = 25$ units. Here the solution is unique.

▶ **Example 11.5.2** *Find the optimal assignment and the corresponding assignment cost from the following cost matrix.* *[C.U.(P)'94]*

	A	*B*	*C*	*D*	*E*
1	9	8	7	6	4
2	5	7	5	6	8
3	8	7	6	3	5
4	8	5	4	9	3
5	6	7	6	8	5

This is a minimization problem.

Step 1. Subtract the lowest element 4 from all the elements of the first row. Similarly subtract the lowest element of each row from all the elements of the corresponding row to get the matrix in table 11.32

Table 11.32

5	4	3	2	0
0	2	0	1	3
5	4	3	0	2
5	2	1	6	0
1	2	1	3	0

Step 2. Now subtract the lowest element of each column from all the elements of the corresponding column in table 11.32 to get the matrix in table 11.32.

Table 11.33(A)

5	2	3	2	0
0	0	0	1	3
5	2	3	0	2
5	0	1	6	0
1	0	1	3	0

Table 11.33(B)

4	2	2	1	0
0	1	0	1	4
5	3	3	0	3
4	0	0	5	0
0	0	0	2	0

Table 11.33(C)

				✓0
~~0~~		✓0		
			✓0	
	✓0	~~0~~		
✓0	~~0~~	~~0~~		

Step 3. Connect all the zeros of the matrix of the table 11.33(A) by a minimum number of horizontal and vertical lines. Here the number of lines are 4 which is less than the number of rows or columns of the matrix. Hence the optimality conditions are not satisfied. Now subtract the smallest element 1 (one) from the elements of the remaining rows and columns of the table 11.33(A) which are *not covered by the lines* and add that smallest element to all the elements which lie on the points of intersection of the lines to get the matrix in table 11.33(B).

Step 4. Now connect again all zeros of the matrix in table 11.33(B) by a *minimum* number of horizontal and vertical lines. The number of lines is five now. Hence the optimality conditions are satisfied.

Step 5. Write down all the zeros in a 5×5 transportation table in their respective positions after removing all zeros at the points of intersection of the lines from the table 11.33(C).

Step 6. Now select five zeros in such a way that there is only one zero in each row and each column. Select the first row containing only a single zero at cell $(1,5)$. Now select the zero at cell $(3,4)$. There are no other zeros in the third row and fourth column. Now further selection of zeros is not unique. Now select arbitrarily the zero (0) at cell $(2,3)$ and cross off the zeros of the second row and third column. Then select the zero at cell $(5,1)$ and cross off the remaining zero in the fifth row. Last of all select the zero at cell $(4,2)$ and the selection is complete. The optimal assignment is $1 \to E$, $2 \to C$, $3 \to D$, $4 \to B$, $5 \to A$ and the optimal assignment cost is $4+5+3+5+6 = 23$ units.

Note. Here the solution is not unique. Other optimal assignments are

$$1 \to E, 2 \to A, 3 \to D, 4 \to C, 5 \to B$$
$$\text{and} \quad 1 \to E, 2 \to A, 3 \to D, 4 \to B, 5 \to C$$

and the minimum cost remains same in each case.

▶ **Example 11.5.3** *Find the optimal assignment and the optimal assignment cost from the following cost matrix.*

	A	B	C	D	E
1	−14	−6	−22	−11	−6
2	−18	−22	−14	−15	−9
3	−18	−12	−9	−12	−12
4	−10	−22	−16	−22	−8
5	−16	−16	−14	−10	−10

Step 1. Select the lowest cost (−22) from the first row and subtract this element from all other elements of the first row. Now the same procedure will be followed for all other rows and the resulting matrix is.

Table 11.34

	A	B	C	D	E
1	8	16	0	11	16
2	4	0	8	7	13
3	0	6	9	6	6
4	12	0	6	0	14
5	0	0	2	6	6

Step 2. Now the same procedure will be followed in each column and the resulting matrix is

Table 11.35

	A	B	C	D	E
1	8	16	0	11	10
2	4	0	8	7	7
3	0	6	9	6	0
4	12	0	6	0	8
5	0	0	2	6	0

Step 3. Now try to connect all zeros by minimum numbers of horizontal and vertical lines. Here the minimum number of lines are five. Thus we reach at the optimal stage.

Table 11.36

	A	B	C	D	E
1			✓0		
2		✓0			
3	0̸				✓0
4				✓0	
5	✓0				0̸

Now select all the zeros (except the zeros at the points of intersection) and put them at their respective positions.

Step 4. There is one zero in cell $(1, 3)$, there are no other zeros in the first row and the third column. Similarly select the zero in cell $(2, 2)$, now we select the zero in $(4, 4)$. Henceforth the selection of zeros are not unique; one selection of zeros is $(5, 1)$ and $(3, 5)$. Therefore one optimal assignment is $1 \to C$, $2 \to B$, $3 \to E$, $4 \to D$, and $5 \to A$ and the optimal assignment cost is $= -[22 + 22 + 12 + 22 + 16] = -94$ units.

Optimal assignment is not unique. Another optimal assignment is $1 \to C$, $2 \to B$, $3 \to A$, $4 \to D$ and $5 \to E$ and the optimal assignment cost is $= -[22 + 22 + 18 + 22 + 10] = -94$ units.

▶ **Example 11.5.4** *Find the optimal assignment profit from the following profit matrix:*

	D_1	D_2	D_3	D_4	D_5
O_1	2	4	3	5	4
O_2	7	4	6	8	4
O_3	2	9	8	10	4
O_4	8	6	12	7	4
O_5	2	8	5	8	8

This is a maximization problem. We know that $\max z = -\min(-z) = -\min z^*$ where $z^* = -z$. Hence first of all, change the sign of each component of the profit matrix and the new matrix is

	D_1	D_2	D_3	D_4	D_5
O_1	−2	−4	−3	−5	−4
O_2	−7	−4	−6	−8	−4
O_3	−2	−9	−8	−10	−4
O_4	−8	−6	−12	−7	−4
O_5	−2	−8	−5	−8	−8

Step 1. Subtract the lowest element (-5) from all other elements of the first row. Similarly subtract the lowest element of each row from all other elements of that row to get the matrix in table 11.37.

Table 11.37

	D_1	D_2	D_3	D_4	D_5
O_1	3	1	2	0	1
O_2	1	4	2	0	4
O_3	8	1	2	0	6
O_4	4	6	0	5	8
O_5	6	0	3	0	0

Step 2. Now subtract the lowest elements of each column from all other elements of that column and the resulting matrix is

Table 11.38

	D_1	D_2	D_3	D_4	D_5
O_1	2	1	2	0	1
O_2	0	4	2	0	4
O_3	7	1	2	0	6
O_4	3	6	0	5	8
O_5	5	0	3	0	0

Step 3. Connect all the zeros of the matrix of the table 11.38 by a *minimum* number of horizontal and vertical lines. Here the number of lines are 4 which is less than the number of columns or rows of the matrix. Hence the optimality conditions are not satisfied. Now subtract the smallest element 1 (one) from the elements of the remaining rows and columns of penaltimate table 11.39(A) which are not covered by the lines and add that smallest element (one) to all the elements which lie on the points of intersection of the lines to get the new matrix in table 11.39(B).

Table 11.39(A)

2	1	2	0	1
0	4	2	0	4
7	1	2	0	6
3	6	0	5	8
5	0	3	0	0

Table 11.39(B)

2	0	2	0	0
0	3	2	0	3
7	0	2	0	5
3	5	0	5	7
6	0	4	1	0

Step 4. Now connect all the zeros of the matrix in table 11.39(B) by a minimum number of horizontal and vertical lines. The number of lines are 5. Hence the optimality conditions are satisfied.

Table 11.40

	0 ✗		0 ✗	0 ✓
0 ✓			0 ✗	
	0 ✗		0 ✓	
		0 ✓		
	0 ✓			0 ✗

Step 5. Write down all the zeros in a 5×5 transportation table in their respective positions in table 11.40.

Step 6. Select 5 zeros in such a way that there is only one zero in each row and each column. Select first column containing only a single zero at cell $(2,1)$ and cross off all zeros remaining in the second row. Now select the zero at cell $(4,3)$. There are no other zeros in the fourth row and 3rd column. Now select arbitrarily the zero at cell $(3,4)$ and cross off all the remaining zeros in 3rd row and fourth column. In the same way select the zeros at cells $(1,5)$ and $(5,2)$. Here the selection of zeros is not unique. Hence multiple optimal solutions exist.

Step 7. One of the optimal assignments is

$$O_1 \to D_5, O_2 \to D_1, O_3 \to D_4, O_4 \to D_3 \text{ and } O_5 \to D_2$$

and the maximum profit $\max(z) = -\min z^* = -\min(-z)$

$$= -(-4-7-10-12-8) = 41 \text{ units.}$$

Other optimal assignments are

$$O_1 \to D_2, O_2 \to D_1, O_3 \to D_4, O_4 \to D_3, O_5 \to D_5$$

$$\text{and } O_1 \to D_4, O_2 \to D_1, O_3 \to D_2, O_4 \to D_3, O_5 \to D_5.$$

▶ **Example 11.5.5** *There are five pumps available for developing five wells. The efficiency of each pump in producing the maximum yield at each well is shown in the following table. In what way the pumps be assigned to the wells so as to maximize the over all efficiency?* *[C.U.(H)'96]*

	I	*II*	*III*	*IV*	*V*
A	45	40	65	25	55
B	50	30	25	60	30
C	25	20	10	20	40
D	35	25	30	25	20
E	80	60	50	70	50

This is a profit maximization problem. We know that $\max z = -\min(-z)$. Therefore, initially we shall have to change the sign of the all components of the profit matrix and the resulting matrix is

	I	*II*	*III*	*IV*	*V*
A	−45	−40	−65	−25	−55
B	−50	−30	−25	−60	−30
C	−25	−20	−10	−20	−40
D	−35	−25	−30	−25	−20
E	−80	−60	−50	−70	−50

Step 1. Now subtract the lowest element of each row from all other elements of that row and the resulting matrix is:

	I	II	III	IV	V
A	20	25	0	40	10
B	10	30	35	0	30
C	15	20	30	20	0
D	0	10	5	10	15
E	0	20	30	10	30

Step 2. Now subtract the lowest element of each column from all other elements of that column and the resulting matrix is

Table 11.41

	I	II	III	IV	V
A	20	15	0	40	10
B	10	20	35	0	30
C	15	10	30	20	0
D	0	0	5	10	15
E	0	10	30	10	30

Step 3. Now try to connect all the zeros by minimum number horizontal and vertical lines. Here the minimum number of lines is five, thus we reach at the optimal stage.

Table 11.42

	I	II	III	IV	V
A			0✓		
B				0✓	
C					0✓
D	✗0	0✓			
E	0✓				

Step 4. Now put all the zeros at their corresponding positions after omitting the zeros at the points of intersection. Now select the zero at the cell $(1, 3)$ and cross off all the zeros of the first row and third column. Here there is no such zero. In the same way we shall select the zeros in cell $(2, 4)$, $(3, 5)$ and $(5, 1)$ and $(4, 2)$ such that there is only one zero in each row and each column and the selection is unique. The optimal assignment is $A \to III$, $B \to IV$, $C \to V$, $D \to II$ and $E \to I$. $\min(-z) = -65 - 60 - 40 - 25 - 80 = -270$, $\max z = -\min(-z) = -(-270) = 270$ units with the optimal assignment given above.

11.6 Restricted Assignment

During assignment, some restrictions are being adopted considering from the economical, commercial or physical points of view. Such types of assignment problems are called restricted assignment problems. The following solved problem will give some idea about it.

▶ **Example 11.6.1** *In a factory there are five operators A, B, C, D, E and the five machines I, II, III, IV, V the operating cost is given if ith operator operates the jth machine $[i, j = 1, 2, \cdots, 5]$. But there is a restriction that C cannot be allowed to operate the third machine and similarly B cannot be allowed to operate the fifth machine. The cost matrix is given below. Find the optimal assignment and the optimal assignment cost.*

Machines

		I	II	III	IV	V
	A	24	29	18	32	19
	B	17	26	34	22	21
Operators	C	27	16	28	17	25
	D	22	18	28	30	24
	E	28	16	31	24	27

Solution: **Step 1.** Since C cannot operate the third machine and B cannot operate the fifth machine we simply ignore the the cells $(3, 3)$ and $(2, 5)$ and ignoring the cells we simply solve in the usual manner and the matrix is there fore

Table 11.43

	I	II	III	IV	V
A	24	29	18	32	19
B	17	26	34	22	—
C	27	16	—	17	25
D	22	18	28	30	24
E	28	16	31	24	27

Step 2. After the row and column operations the resulting matrix is:

Table 11.44

	I	II	III	IV	V
A	6	11	0	13	0
B	0	9	17	4	
C	11	0		0	8
D	4	0	10	11	5
E	12	0	15	7	10

Now try to connect all zeros by minimum number of horizontal and vertical lines. Here the minimum number of lines is four, less then five. Thus we do not reach at the optimal stage. Now select the lowest element from the uncovered matrix and subtract it from all other uncovered elements and add it with the elements of the points of intersection of the lines. And the resulting matrix is:

Table 11.45

	I	II	III	IV	V
A	6	15	0	13	0
B	0	13	17	4	
C	11	4		0	8
D	4	0	6	7	1
E	8	0	11	3	6

Here the minimum number of lines are four; thus we donot reach at the optimal stage. The same operation will be followed and the resulting matrix is:

Table 11.46

	I	II	III	IV	V
A	7	16	0	13	0
B	0	13	16	3	
C	12	5		0	8
D	4	0	5	6	0
E	8	0	10	2	5

Here the number of lines is five; thus we reach at the optimal stage. Now we shall put the zeros at their corresponding position after omitting the zeros at the points of intersection.

Table 11.47

	I	II	III	IV	V
A			0✓		Ø
B	0✓				
C				0✓	
D					0✓
E		0✓			

The selection of zeros is given in the table such that there is one zero in each row and each column and the selection is unique. The optimal assignment is $A \to III$, $B \to I$, $C \to IV$, $D \to V$ and $E \to II$ and the optimal assignment cost $= 18 + 17 + 17 + 24 + 16 = 92$.

11.7 Unbalanced Assignment Problem

An assignment problem is said to be unbalanced if the number of *Assignee* and the number of *Assignment* be not same and the cost or profit matrix be not a square matrix. Here either the number of assignee is greater than the number of assignment and vice-versa. But the problem can also be solved by converting the matrix (cost or profit) into a square matrix by introducing either fake assignee or assignment whichever is necessary and assuming all costs components zero corresponding to that assignee or assignment. The following solved problem will give a clear idea how to solve such type of problems.

► **Example 11.7.1** *In a factory, there are six machines (of same type) and five workers. The handling costs for the ith worker* ($i = 1, \cdots, 5$) *to handle the jth machine* ($j = 1, 2, \cdots, 6$) *are given below in the form of a matrix. Find the optimal assignment and the minimum cost of handling the machines and find which machine will remain unused.*

	M_1	M_2	M_3	M_4	M_5	M_6
W_1	12	8	10	14	11	18
W_2	14	14	8	15	17	12
W_3	9	11	13	15	6	12
W_4	11	9	9	11	8	14
W_5	10	12	15	13	10	12

The problem can be made a balanced problem by introducing a fake worker W_6 with all cost component $c_{6j} = 0[j = 1, 2, \cdots, 6]$ and the square matrix is given below.

	M_1	M_2	M_3	M_4	M_5	M_6
W_1	12	8	10	14	11	18
W_2	14	14	8	15	17	12
W_3	9	11	13	15	6	12
W_4	11	9	9	11	8	14
W_5	10	12	15	13	10	12
W_6	0	0	0	0	0	0

Proceeding in the similar manner as in the case of the previous solved problems, we obtain the optimal assignment table (as the minimum number of lines required to connect all zeros is six) and the optimal assignments which are given in the following tables:

Table 11.48(A)

	M_1	M_2	M_3	M_4	M_5	M_6
W_1	4	0	2	4	4	8
W_2	6	6	0	3	10	2
W_3	2	4	6	6	0	3
W_4	2	0	0	0	0	3
W_5	0	2	5	1	1	0
W_6	2	2	2	0	3	0

Table 11.48(B)

	M_1	M_2	M_3	M_4	M_5	M_6
W_1		✓0				
W_2			✓0			
W_3					✓0	
W_4		×0		✓0		
W_5	✓0					×0
W_6				×0		✓0

The optimal unique assignment is $W_1 \to M_2$, $W_2 \to M_3$, $W_3 \to M_5$, $W_4 \to M_4$, $W_5 \to M_1$, M_6 remains unused.

The min cost $= 8 + 8 + 6 + 11 + 10 = 43$ units.

Exercises 11C

1. Find the optimal assignment cost from the following cost matrices:

(a)

	A	B	C	D
I	4	5	4	3
II	3	2	2	6
III	4	5	3	5
IV	2	4	2	6

(b)

	A	B	C	D
I	9	12	10	10
II	8	12	7	9
III	4	6	7	8
IV	8	4	5	5

2. Find the optimal assignment profit from the following (a), (b) profit matrices and optimal cost for (c), (d) cost matrices.

(a)

	I	II	III	IV
A	6	4	3	12
B	5	7	6	2
C	4	5	8	7
D	6	7	8	12

(b)

	I	II	III	IV
A	7	5	4	3
B	8	2	6	4
C	5	3	2	1
D	5	4	1	8

(c)

	I	II	III	IV
A	5	3	1	8
B	7	9	2	6
C	6	4	5	7
D	5	7	7	6

[C.U.(P)'87,'89]

(d)

	I	II	III	IV
A	10	12	19	11
B	5	10	7	8
C	12	14	13	11
D	8	15	11	9

[V.U.(P)'91] [V.U.(P)'90]

3. A department has four sub-ordinates and four tasks are to be performed. The sub-ordinates differ in efficiency and the tasks differ in their intrinsic difficulties. The estimate of time (in man hours) each man would take to perform each task is given by:

		Task			
		I	*II*	*III*	*IV*
	1	18	26	17	11
	2	13	28	14	26
Sub-ordinate	3	38	19	18	15
	4	19	26	24	10

[C.U.(H)'87;C.U.(P)'90]

How should the tasks be allotted to men so as to optimize the total man hours?

4. A marketing manager has five salesmen and 5 sales districts. Considering the capabilities of the salesman and the nature of the districts, the marketing manager estimates that sales per month (in hundred rupees) for each salesman in each district would be as follows:

		District				
		A	*B*	*C*	*D*	*E*
	1	32	38	40	28	40
	2	40	24	28	21	36
Salesman	3	41	27	33	30	37
	4	22	38	41	36	36
	5	29	33	40	35	39

Find the assignment of the salesmen to districts that will produce maximum sales.

5. A car hire company has one car in each of the five depots a, b, c, d and e. A customer in each of the five towns A, B, C, D and E requires a car. The distance (in km) between the depots (origins) and the towns (destinations) where the customers are, given by the following distance matrix:

	a	b	c	d	e
A	20	40	30	50	40
B	70	40	60	80	40
C	20	90	80	100	40
D	80	60	120	70	40
E	20	80	50	80	80

How should the cars be assigned to the customers so as to minimize the distance travelled? [Cal. M.Sc. (App. Math.)'77]

6. A company has five jobs to be done on five machines; any job can be done on any machine. The time in hours taken by the machines for the different jobs are as given below. Assign the machine to a job so as to minimize the total machine hours.

		Job				
		1	2	3	4	5
Machine	A	11	6	14	16	17
	B	7	13	22	7	10
	C	10	7	2	2	2
	D	4	10	8	6	11
	E	13	15	16	10	18

7. A company has 5 jobs to be done. The following matrix shows the return in Rs. of assigning ith machine $(i = 1, 2, \cdots, 5)$ to the jth job $(j = 1, 2, \cdots, 5)$. Assign the five jobs to five machines so as to maximize the expected profit.

		Job				
		1	2	3	4	5
Machine	A	5	11	10	12	4
	B	2	4	6	3	5
	C	3	1	5	14	6
	D	6	14	4	11	7
	E	7	9	8	12	5

8. Find the optimal assignment for the assignment problem with the given cost matrix. [C.U.(P)'80]

	M_1	M_2	M_3	M_4	M_5
J_1	8	4	2	6	1
J_2	0	9	5	5	4
J_3	3	8	9	2	6
J_4	4	3	1	0	3
J_5	9	5	8	9	5

9. Find the optimal assignment for the assignment problem with the given cost matrix. [C.U.(H)'80]

	J_1	J_2	J_3	J_4
M_1	10	9	7	8
M_2	5	8	7	7
M_3	5	4	6	5
M_4	2	3	4	3

10. The heads of the department has five jobs V, W, X, Y and Z. The number of hours each man would take to perform each job is given in the matrix. How would the jobs be allocated to minimize the total time?

	V	W	X	Y	Z
A	3	5	10	15	8
B	4	7	15	18	8
C	8	12	20	20	12
D	5	5	8	10	6
E	10	10	15	25	10

11. Find the optimal assignments for the assignment problem with the following cost matrix: [C.U.(P)'81,'99]

	M_1	M_2	M_3	M_4	M_5
J_1	3	8	2	10	3
J_2	8	7	2	9	7
J_3	6	4	2	7	5
J_4	8	4	2	3	5
J_5	9	10	6	9	10

12. Solve the following unbalanced assignment problem for the assignment with the following cost matrix:

	J_1	J_2	J_3	J_4
A	10	12	8	6
B	6	9	12	14
C	3	8	7	12

	J_1	J_2	J_3	J_4
A	10	12	8	6
B	6	9	12	14
C	3	8	7	12
D	0	0	0	0

[*Hints.* Convert the cost matrix as given in the second matrix as a balanced (4×4) assignment problem with a fictitious row with all costs zero and then solve it in the usual method.]

13. Consider the problem of assigning four operators to four machines. The assignment costs in rupees are given below. Operator 1 cannot be assigned to machine 3 and operator 3 cannot be assigned to machine 4. Find the optimal cost of assignment. [C.U.(H)'83,'93,'95]

Operator \ Machine	1	2	3	4
1	5	5	–	2
2	7	4	2	3
3	9	3	5	–
4	7	2	6	7

[*Hints.* As operator 1 cannot be assigned to machine 3 and operator 3 cannot be assigned to machine 4 then during computation no zero will be considered in the cells $(1, 3)$ and $(3, 4)$ and thus solve in the usual method.]

14. Find the optimal assignment for the assignment problem with the following cost matrix: [C.U.(P)'83,'90]

	1	2	3	4
1	1	4	6	3
2	9	7	10	9
3	4	5	11	7
4	8	7	8	5

15. In a factory there are four workers and five machines and the operating cost for each worker to handle each machine are given below. Find the optimal assignment and the assignment cost and find which machine will remain idle:

	I	*II*	*III*	*IV*	*V*
A	12	15	14	11	16
B	15	12	14	13	15
C	16	18	12	15	15
D	12	14	13	13	14

16. Five workers perform five jobs and the operating cost are given below, but there is restriction that the worker *A* cannot perform the third job and the worker *C* cannot perform the second job. Find the optimal assignment and the optimal assignment cost.

	I	*II*	*III*	*IV*	*V*
A	14	18	–	16	16
B	12	17	15	15	16
C	14	–	13	16	20
D	12	15	16	14	19
E	15	22	13	14	16

17. Solve the following unbalanced assignment problems where the cost matrix is given below:

	I	*II*	*III*	*IV*
A	13	16	14	21
B	19	16	17	23
C	15	24	13	17
D	11	14	21	23
E	15	17	19	17

18. Solve the two following assignment problems where in the first problem matrix is a cost matrix and the second problem the matrix is a profit matrix:

(a)

	1	2	3	4	5
A	13	15	19	17	20
B	23	26	21	19	15
C	13	17	21	24	15
D	17	13	19	26	17
E	26	19	24	29	11

(b)

	1	2	3	4	5
A	10	2	−7	8	6
B	0	8	−5	5	4
C	8	3	2	1	5
D	1	1	4	2	3
E	8	2	1	−1	4

19. Solve the following restricted assignment problems where the assignment cost matrix is given below:

	I	II	III	IV	V
A	24	29	18	32	19
B	17	26	34	22	–
C	27	16	–	17	25
D	22	18	28	30	24
E	28	16	31	24	27

Solve the above problem taking the matrix to be a profit matrix.

Answers

1. (a) $I \to D$, $II \to B$, $III \to C$, $IV \to A$ and Min. cost = 10 units.
   (b) $I \to D$, $II \to C$, $III \to A$, $IV \to B$ and Min. cost = 25 units.
2. (a) $A \to IV$, $B \to II$, $C \to III$, $D \to I$ and Max. profit = 33 units. Multiple solutions exist.
   (b) $A \to II$, $B \to III$, $C \to I$, $D \to IV$ and Max. profit = 24 units. Multiple solutions exist.
   (c) $A \to III$, $B \to IV$, $C \to II$, $D \to I$ and Min. cost = 16 units.
   (d) $A \to II$, $B \to III$, $C \to IV$, $D \to I$ and Min. cost = 38 units.
3. $1 \to III$, $2 \to I$, $3 \to II$, $4 \to IV$ and Min. time 59 hours.
4. $1 \to B$, $2 \to A$, $3 \to E$, $4 \to C$, $5 \to D$ and Max. profit = Rs.19,100. Multiple solutions exist.
5. $a \to E$, $b \to B$, $c \to A$, $d \to D$, $e \to C$ and Min. distance 200 km. Multiple solutions exist.
6. $A \to 2$, $B \to 5$, $C \to 3$, $D \to 1$, $E \to 4$ and Min. time is 32 hours.
7. $A \to 3$, $B \to 5$, $C \to 4$, $D \to 2$, $E \to 1$ and Max. profit is Rs.50.
8. $J_1 \to M_5$, $J_2 \to M_1$, $J_3 \to M_4$, $J_4 \to M_3$, $J_5 \to M_2$ and Min. cost = 9 units.
9. $M_1 \to J_3$, $M_2 \to J_1$, $M_3 \to J_2$, $M_4 \to J_4$ and Min. cost = 19 units.
10. $A \to X$, $B \to W$, $C \to V$, $D \to Y$, $E \to Z$ and Min. time = 45 hours.

11. $J_1 \to M_5$, $J_2 \to M_3$, $J_3 \to M_2$, $J_4 \to M_4$, $J_5 \to M_1$ Min cost 21 units.

12. $A \to J_4$, $B \to J_2$, $C \to J_1$ and Min. cost = 18 units.

13. $1 \to 4$, $2 \to 3$, $3 \to 2$, $4 \to 1$ and Min. cost = Rs.14.

14. $1 \to 1$, $2 \to 3$, $3 \to 2$, $4 \to 4$ and Min. cost = 21 units.

15. $A \to IV$, $B \to II$, $C \to III$, $D \to I$ and the fifth machine will remain idle.

16. $A \to V$, $B \to I$, $C \to III$, $D \to II$ and $E \to IV$, Min cost = 70 units.

17. $A \to I, C \to III, D \to II, E \to IV$ and Min. cost = 57 units. (There will be no-assignment for B). Multiple solutions exist.

18. (a) $A \to 4$, $B \to 3$, $C \to 1$, $D \to 2$, $E \to 5$ and Min. cost = 75 units.

    (b) $A \to 4$, $B \to 2$, $C \to 5$, $D \to 3$, $E \to 1$ and Max. profit = 33 units.

19. $A \to III$, $B \to I$, $C \to IV$, $D \to V$, $E \to II$ and Min. cost = 92 units.

    $A \to II$, $B \to III$, $C \to I$, $D \to IV$, $E \to V$ and Max. profit = 147 units.

Appendix A

Development of the Simplex Theory

A.1 Introduction

In the simplex theory, the initial basic problem is to find a basis (which always exists there) for which the solution is feasible. Let the constraints be

$$\mathbf{a}_1 x_1 + \mathbf{a}_2 x_2 + \cdots + \mathbf{a}_n x_n = \mathbf{b} \tag{A.1.1}$$

and the objective function $z = \mathbf{c}\mathbf{x}$, where $\mathbf{c} = (c_1, c_2 \cdots, c_n)$

Let $\mathbf{x}_B = B^{-1}\mathbf{b} \geq \mathbf{0}$ and

$$\begin{aligned} \mathbf{x}_B &= [x_{\beta 1}, x_{\beta 2}, \cdots, x_{\beta m}], \\ \mathbf{c}_\beta &= (c_{B1}, c_{B2}, \cdots, c_{Bm}) \end{aligned}$$

as $B = (\beta_1, \beta_2 \cdots, \beta_m)$, where $\beta_i (i = 1, 2, \cdots, m)$ are some of the vectors $\in A$ $[j = 1, 2, \cdots, n]$. Now after finding such basis, we shall calculate z_B corresponding to such basis and next problem is to test whether this is an optimal solution and if not, we shall have to develop the theories such that the optimal condition be satisfied or the problem has said to have an unbounded solution etc. The developments of the theories are given below.

A.2 Determination of the improved value of the objective function

A.2.1 Problem of Maximization

Theorem A.2.1 *Let $\mathbf{x}_B = B^{-1}\mathbf{b}$ be a B.F.S to a L.P.P. and the corresponding value of the objective function $z_B = \mathbf{c}_B\mathbf{x}_B$. If for some jth columns corresponding to non-basic variables, the condition $z_j - c_j < 0$ holds with at least one $y_{ij} > 0$, $[i = 1, 2, \cdots, m]$ then it is possible to obtain a new B.F.S. which improves the value of the objective function by replacing one vector of B by $\mathbf{a}_j$.*

Proof. Let

$$\begin{aligned} A &= (\mathbf{a}_1, \mathbf{a}_2, \cdots, \mathbf{a}_n) \\ \text{and} \quad B &= (\beta_1, \beta_2, \cdots, \beta_r, \cdots, \beta_m) \text{ be a basis matrix} \end{aligned}$$

$$\mathbf{x}_B = B^{-1}\mathbf{b} \text{ is the B.F.S.}$$

$$\text{and} \quad z_B = \mathbf{c}_B\mathbf{x}_B = \sum_{i=1}^{m} \mathbf{c}_{Bi}x_{Bi},$$

the value of the objective function corresponding to the B.F.S. $\mathbf{x}_B = B^{-1}\mathbf{b}$. We have

$$A\mathbf{x} = \mathbf{b} = B\mathbf{x}_B. \tag{A.2.1}$$

As $\mathbf{a}_j$ is a column vector corresponding to a non-basic variable x_j therefore

$$\mathbf{a}_j = B\mathbf{y}_j = \beta_1 y_{1j} + \beta_2 y_{2j} + \cdots + \beta_r y_{rj} + \cdots + \beta_m y_{mj}.$$

Assuming $y_{rj} \neq 0$

$$\beta_r = \frac{1}{y_{rj}}[\mathbf{a}_j - \beta_1 y_{1j} - \cdots - \beta_{r-1}, j y_{r-1}, j - \beta_{r+1}, j y_{r+1}, j - \cdots - \beta_m y_{mj}] \tag{A.2.2}$$

From (A.2.1)

$$\beta_1 x_{B1} + \beta_2 x_{B2} + \cdots + \beta_r x_{Br} + \cdots + \beta_m x_{Bm} = \mathbf{b}$$

$$\text{or,} \quad \sum_{\substack{i=1 \\ i\neq r}}^{m} \beta_i x_{Bi} + \beta_r x_{Br} = \mathbf{b}. \tag{A.2.3}$$

Putting the value of β_r from (A.2.2) in (A.2.3) we get

$$\sum_{\substack{i=1 \\ i\neq r}}^{m} \beta_i x_{Bi} + x_{Br}\left(\frac{\mathbf{a}_j}{y_{rj}} - \frac{\beta_1 y_{ij}}{y_{rj}} \cdots \frac{y_{ij}}{y_{rj}}\beta_i - \cdots - \frac{\beta_m y_{mj}}{y_{rj}}\right) = \mathbf{b}$$

$$\text{or} \quad \sum_{\substack{i=1 \\ i\neq r}}^{m} \left(x_{Bi} - \frac{y_{ij}}{y_{rj}}x_{Br}\right)\beta_i + \frac{x_{Br}}{y_{rj}}\mathbf{a}_j = \mathbf{b}. \tag{A.2.4}$$

As $\{\beta_1, \beta_2, \beta_r, \cdots, \beta_m\}$ is a basis set of vectors $\in A$ and

$$\mathbf{a}_j = \beta_i y_{ij} + \beta_2 y_{2j} + \cdots + \beta_r y_{rj} + \cdots + \beta_m y_{mj}$$

with $y_{rj} \neq 0$ then $\mathbf{a}_j$ can replace β_r to form a new basis and hence from (A.2.3) and (A.2.4) a new B.S.

$\mathbf{x}'_B = [x'_{B1}, x'_{B2}, \cdots, x'_{Bm}]$ is obtained which is given by

$$\left.\begin{aligned} x'_{Bi} &= x_{Bi} - \frac{y_{ij}}{y_{rj}}x_{Br}, \quad i = 1, 2, \cdots, m, \quad i \neq r \\ x'_{Br} &= \frac{x_{Br}}{y_{rj}}, \quad i = r \end{aligned}\right\} \tag{A.2.5}$$

Now we put some restrictions to make (A.2.5) a feasible solution. As $x_{Br} \geq 0$, therefore $x'_{Br} \geq 0$ if and only if $y_{rj} > 0$. Again for $y_{rj} > 0$ and $y_{ij} \leq 0$, $x_{Bi} \geq 0$ for $[i = 1, 2, \cdots, m$ and $i \neq r]$.

But if $y_{rj} > 0$ and $y_{ij} > 0$ then $x'_{Bi} \geq 0$ provided

$$x_{Bi} - \frac{y_{ij}}{y_{rj}} x_{Br} \geq 0, \quad \text{or,} \quad \frac{x_{Br}}{y_{rj}} \leq \frac{x_{Bi}}{y_{ij}}. \tag{A.2.6}$$

If we select

$$\frac{x_{Br}}{y_{rj}} = \min_i \left(\frac{x_{Bi}}{y_{ij}}, y_{ij} > 0 \right) \tag{A.2.7}$$

then the condition (A.2.6) will be satisfied and solution given in (A.2.5) will be a new B.F.S. Now we are to prove that for this B.F.S. the value of the objective function will be improved.

Let the value of the objective function be z' for the B.F.S. given in (A.2.5). Then

$$z' = \mathbf{c}'_B \mathbf{x}'_B = \sum_{i=1}^{m} c'_{Bi} x'_{Bi},$$

where c'_{Bi} is the co-efficient of the corresponding basic variable x'_{Bi}.

Obviously $c'_{Bi} = c_{Bi}$ for $i = 1, 2, \cdots, m, i \neq r$ and $c'_{Br} = c_j$ for $i = r$. [As β_r is replaced by $\mathbf{a}_j$]

$$\begin{aligned}
z' &= \sum_{i=1}^{m} c'_{Bi} x'_{Bi} \\
&= \sum_{\substack{i=1 \\ i \neq r}}^{m} c_{Bi} \left(x_{Bi} - \frac{y_{ij}}{y_{rj}} x_{Br} \right) + c_j \frac{x_{Br}}{y_{rj}} \quad \text{[putting the values from (A.2.5)]} \\
&= \sum_{i=1}^{m} c_{Bi} \left(x_{Bi} - \frac{y_{ij}}{y_{rj}} x_{Br} \right) + c_j \frac{x_{Br}}{y_{rj}} \quad \left[\text{as } x_{Bi} - \frac{y_{ij}}{y_{rj}} x_{Br} = 0 \text{ for } i = r \right] \\
&= \sum_{i=1}^{m} c_{Bi} x_{Bi} + \frac{x_{Br}}{y_{rj}} \left(c_j - \sum_{i=1}^{m} c_{Bi} y_{ij} \right) \\
&= z_B - \frac{x_{\beta r}}{y_{rj}} (z_j - c_j) \quad \left[\text{from (8.2.5) since } \sum_{i=1}^{m} c_{Bi} y_{ij} = z_j \right].
\end{aligned} \tag{A.2.8}$$

Now as $(x_{\beta r}/y_{rj}) \geq 0$, and $z_j - c_j < 0$ therefore $z' \geq z_B$ and if $(x_{\beta r}/y_{rj}) = 0$ then $z' = z_B$.

Thus under the conditions stated above the value of the objective function is being improved or at least remain same.

[In a problem of minimization, the value of the objective function will be diminished, provided $z_j - c_j > 0$ for columns corresponding to the non-basic variables with at least one $y_{ij} > 0$.]

The quantity y_{rj} is known as the *key element* or *pivot element*. The value of y_{rj} is determined from the condition that in the jth column,

$$\min_i \left(\frac{x_{Bi}}{y_{ij}}, y_{ij} > 0 \right) = \frac{x_{Br}}{y_{rj}}. \tag{A.2.9}$$

The jth column vector $\mathbf{a}_j$ is the *vector to enter in the next basis.* The rule for determination of the key element is known as *Minimum ratio rule.* The rule for determination of $\mathbf{a}_j$ will be discussed later.

Now if in a current B.F.S. $x_{Br} = 0$ for which $y_{rj} > 0$ then from mini-ratio rule

$$\min_i \left(\frac{x_{Bi}}{y_{ij}}, y_{ij} > 0 \right) = \frac{x_{Br}}{y_{rj}} = 0$$

and then $z' = z_B$, i.e., the value of the objective function will not improve due to transformation. As $x_{Br} = 0$, the current B.F.S. is degenerate and the next B.F.S. will be also degenerate because $x'_{Br} = \frac{x_{Br}}{y_{rj}} = 0$.

But if $x_{Br} = 0$, and $y_{rj} \leq 0$ then from the mini-ratio rule it is evident that this variable will not enter in the calculation of $\frac{x_{Bi}}{y_{ij}}$ and the new B.F.S. may not be degenerate, though the current B.F.S. is degenerate.

If y_{rj} be the key element then β_r, the vector in the rth position of the basis matrix B is being replaced by $\mathbf{a}_j$ and the vector β_r is the *vector* to leave the current basis.

The new basis matrix $\hat{B} = (\beta_1, \beta_2, \cdots, \beta_{r-1}, \mathbf{a}_j, \beta_{r+1}, \cdots, \beta_m)$.

A.2.2 Conditions for a solution to be optimal

Theorem A.2.2 *Given a basic feasible solution* $\mathbf{x}_B = B^{-1}\mathbf{b}$ *with* $z_B = \mathbf{c}_B\mathbf{x}_B$ *to the L.P.P. Maximize* $\mathbf{z} = \mathbf{cx}$ *subject to* $A\mathbf{x} = \mathbf{b}$, $\mathbf{x} \geq \mathbf{0}$ *such that* $z_j - c_j \geq 0$ *for every column of* $\mathbf{a}_j$ *in* A. *Then* z_B *is the maximum value and* $\mathbf{x}_B$ *is the optimal B.F.S.*

Proof.

$$A = (\mathbf{a}_1, \mathbf{a}_2, \cdots, \mathbf{a}_n)$$
$$B = (\beta_1, \beta_2, \cdots, \beta_m)$$

where $\beta_1, \beta_2, \cdots, \beta_m$ are some of

$$\mathbf{a}_1, \mathbf{a}_2, \cdots, \mathbf{a}_n \text{ and } A\mathbf{x} = \mathbf{b}. \tag{A.2.10}$$

$\mathbf{x}_B = B^{-1}\mathbf{b}$ is the B.F.S. and the value of objective function is $z_B = \mathbf{c}_B\mathbf{x}_B$, $z_j - c_j \geq 0$ for all column vectors $\mathbf{a}_j$. It is required to prove that z_B is the optimal value of the objective function corresponding to the B.F.S. $\mathbf{x}_B$.

Let $\mathbf{x}' = [x'_1, x'_2, \cdots, x'_n]$ be any F.S. (may be a B.F.S.) of the problem which makes the objective function z, to

$$z' = \sum_{j=1}^{n} c_j x'_j.$$

We are to prove that $z' \leq z_B$.

We have

$$A\mathbf{x}' = \mathbf{b} = B\mathbf{x}_B \quad [\because \mathbf{x}' \text{ is a solution set of } A\mathbf{x} = \mathbf{b}]$$

$$\text{or,} \quad \mathbf{x}_B = B^{-1}(A\mathbf{x}') = (B^{-1}A)\mathbf{x}'$$

$$= \mathbf{Y}\mathbf{x}' \quad [\mathbf{Y} = \{\mathbf{y}_1, \mathbf{y}_2, \cdots, \mathbf{y}_n\}] \quad [\text{as } B^{-1}\mathbf{a}_j = \mathbf{y}_j]$$

$$= (\mathbf{y}_1, \mathbf{y}_2, \cdots, \mathbf{y}_n)[x'_1, x'_2 \cdots, x'_n] \tag{A.2.11}$$

Therefore

$$[x_{B1}, x_{B2}, \cdots, x_{Bm}] = \begin{bmatrix} y_{11} & y_{12} & \cdots & y_{1j} & y_{1n} \\ y_{21} & y_{22} & \cdots & y_{2j} & y_{2n} \\ \cdots & \cdots & \cdots & \cdots & \cdots \\ y_{i1} & y_{i2} & \cdots & y_{ij} & y_{in} \\ y_{m1} & y_{m2} & \cdots & y_{mj} & y_{mn} \end{bmatrix} \begin{bmatrix} x'_1 \\ x'_2 \\ \vdots \\ x'_j \\ x'_n \end{bmatrix}.$$

Equating we get

$$x_{Bi} = \sum_{j=1}^{n} y_{ij} x'_j. \tag{A.2.12}$$

Now $z_j - c_j \geq 0$ for all columns $\mathbf{a}_j$ of A. Therefore $c_j \leq z_j$ for all j

$$\text{or,} \quad c_j x'_j \leq z_j x'_j \quad \text{as } x'_j \geq 0$$

$$\text{or,} \quad \sum_{j=1}^{n} c_j x'_j \leq \sum_{j=1}^{n} x'_j \left(\sum_{i=1}^{m} c_{Bi} y_{ij} \right) \quad \left[\because z_j = \sum_{i=1}^{m} c_{Bi} y_{ij} \right]$$

$$\text{or,} \quad z' \leq \sum_{i=1}^{m} c_{Bi} \sum_{j=1}^{n} x'_j y_{ij}$$

$$\text{or,} \quad z' \leq \sum_{i=1}^{m} c_{Bi} x_{Bi} \quad [\text{from (A.2.12)}]$$

$$\text{or,} \quad z' \leq \mathbf{c}_B \mathbf{x}_B$$

$$\text{or,} \quad z' \leq z_B \tag{A.2.13}$$

Hence z_B is the optimal value of the objective function corresponding to B.F.S. $\mathbf{x}_B$. Hence the theorem is proved.

[For a problem of minimization, the necessary and sufficient condition is $z_j - c_j \leq 0$ for all columns $\mathbf{a}_j$ in A.]

A.2.3 Another Proof

We know

$$z' = z_B - \frac{x_{Br}}{y_{rj}}(z_j - c_j) \text{ from (A.2.8).}$$

Since $\frac{x_{Br}}{y_{rj}} \geq 0$ and all $z_j - c_j \geq 0$ then $z' \leq z_B$ and will be equal only when $\frac{x_{Br}}{y_{rj}} = 0$ or $z_j - c_j = 0$ for all values of j. Hence for all $z_j - c_j \geq 0$, z_B is the optimal value of the L.P.P.

Note: Consider the value of $z_j - c_j$ for a column vector $\beta_r (r = 1, 2, \cdots, m)$ of the basic variables. We know

$$\beta_i = B\mathbf{y}_j \quad [\because \mathbf{a}_j = B\mathbf{y}_j]$$

$$\text{or,} \quad \beta_i = \beta_1 y_{1j} + \beta_2 y_{2j} + \cdots + \beta_i y_{ij} + \cdots + \beta_m y_{mj}$$

from which we get $\mathbf{y}_j = (0, 0, \cdots, 1, 0, \cdots, 0) = \mathbf{e}_i$ (an unit vector with i-th component unity)

$$c_{Bi} = c_j.$$

Then

$$\begin{aligned} z_j - c_j &= \mathbf{c}_B \mathbf{y}_j - \mathbf{c}_{Bi} \\ &= \mathbf{c}_B \mathbf{e}_i - \mathbf{c}_{Bi} \\ &= \mathbf{c}_{Bi} - \mathbf{c}_{Bi} = 0. \end{aligned} \tag{A.2.14}$$

Thus $z_j - c_j = 0$ for all columns corresponding to basic variables [a very important property to remember].

A.2.4 Condition for the non-existence of finite optimal value of the objective function

Theorem A.2.3 *In a linear programming problem,*

$$\textit{Maximize,} \quad z = \mathbf{c}\mathbf{x}, \qquad \textit{subject to } A\mathbf{x} = \mathbf{b}, \mathbf{x} \geq \mathbf{0}$$

there exists no finite optimal value of objective function, if for at least one column corresponding to a non-basic variable $z_j - c_j < 0$ *and* $y_{ij} \leq 0$ *for all* i.

Proof.

$$\begin{aligned} A &= (\mathbf{a}_1, \mathbf{a}_2, \cdots, \mathbf{a}_n), \text{ co-efficient matrix} \\ B &= (\beta_1, \beta_2, \cdots, \beta_m), \text{ a basis matrix} \end{aligned}$$

and let $\mathbf{x}_B = B^{-1}\mathbf{b} = [x_{B1}, x_{B2}, \cdots, x_{Bm}]$ is a B.F.S.

Therefore

$$B\mathbf{x}_B = \mathbf{b}, \quad \text{or,} \quad \sum_{i=1}^{m} \beta_i x_{Bi} = \mathbf{b}. \tag{A.2.15}$$

The value of the objective function corresponding to the B.F.S. is $z_B = \mathbf{c}_B \mathbf{x}_B$.

Let $\mathbf{a}_j$ be a column vector of a non-basic variable with $z_j - c_j < 0$ and $y_{ij} \leq 0$ for all i.

Then for any $\lambda > 0$

$$\sum_{i=1}^{m} \beta_i x_{Bi} - \lambda \mathbf{a}_j + \lambda \mathbf{a}_j = \mathbf{b}. \tag{A.2.16}$$

As $\mathbf{a}_j$ is a column vector of a non-basic variable therefore

$$\mathbf{a}_j = \sum_{i=1}^{m} \beta_i y_{ij}. \tag{A.2.17}$$

Putting the value of $\mathbf{a}_j$ from (A.2.17) in (A.2.16), it may be expressed as

$$\sum_{i=1}^{m} \beta_i x_{Bi} - \lambda \sum_{i=1}^{m} \beta_i y_{ij} + \lambda \mathbf{a}_j = \mathbf{b}$$

$$\text{or} \quad \sum_{i=1}^{m} \beta_i (x_{Bi} - \lambda y_{ij}) + \lambda \mathbf{a}_j = \mathbf{b} \tag{A.2.18}$$

which indicates that

$$\mathbf{x}' = [x_{B1} - \lambda y_{1j}, x_{B2} - \lambda y_{2j}, \cdots, x_{Bm} - \lambda y_{mj}, \lambda]$$

is a solution set of the equation $A\mathbf{x} = \mathbf{b}$ [with all other variables zero].

As $\lambda > 0$ and $y_{ij} \leq 0$ for all i, then $x_{Bi} - \lambda y_{ij} \geq 0$ for all i and hence $\mathbf{x}'$ is a feasible solution of $A\mathbf{x} = \mathbf{b}$. But the solution set $\mathbf{x}'$ is a non-basic solution with $(m+1)$ variables non-negative.

Let z' be the value of objective function corresponding to the solution set $\mathbf{x}'$.

Then

$$\begin{aligned} z' = \mathbf{c}'_B\mathbf{x}' &= \sum_{i=1}^{m} c_{Bi}x'_{Bi} + c_j\lambda \quad [\text{where } x'_{Bi} = x_{Bi} - \lambda y_{ij}] \\ &= \sum_{i=1}^{m} c_{Bi}(x_{Bi} - \lambda y_{ij}) + c_j\lambda = \sum_{i=1}^{m} c_{Bi}x_{Bi} + \lambda\left(c_j - \sum_{i=1}^{m} c_{Bi}y_{ij}\right) \\ &= z_B - \lambda(z_j - c_j) \end{aligned} \tag{A.2.19}$$

As $\lambda > 0, z_j - c_j < 0$ therefore $z' \to \infty$ if $\lambda \to \infty$.

Hence no finite value of the objective function exists under the conditions stated above. The solution set of this type is known as an *unbounded solution* to the problem.

[**Note:** Some of the components of $\mathbf{x}'$ may be zero. For example if $x_{Br} = 0$ and $y_{rj} = 0$, then $x'_{Br} = 0$. But still the solution is non-basic. Because the column vectors, $\beta_1, \beta_2, \cdots, \beta_{r-1}, \mathbf{a}_j, \beta_{r+1}, \cdots, \beta_m$ do not form a new basis as y_{rj}, the co-efficient of β_r in the expression of $\mathbf{a}_j = \sum_{i=1}^{m} \beta_i y_{ij}$, is zero.]

A.2.5 Condition for the existence of multiple optimal solutions

1. In a simplex table, at the optimal stage, if there exist some columns corresponding to non-basic variables, where $z_j - c_j = 0$, $y_{ij} > 0$ for at least one i then there are multiple optimal B.F.S. of the problem.

   Let

   $$\mathbf{x}_B = [x_{B1}, x_{B2}, \cdots, x_{Bm}]$$

   be an optimal B.F.S. and $\hat{z}$ be the optimal value of the objective function. The value of B.F.S. in the next iteration is

   $$\mathbf{x}'_B = [x'_{B1}, x'_{B2}, \cdots, x'_{Bm}]$$

   where

   $$x'_{Br} = \frac{x_{Br}}{y_{rj}} \quad \text{and} \quad x'_{Bi} = x_{Bi} - \frac{y_{ij}}{y_{rj}}x_{Bi}, \quad i = 1, 2, \cdots, m, \; i \neq r.$$

   $$\text{and} \quad z' = z_B - \frac{x_{Br}}{y_{rj}}(z_j - c_j),$$

where z' is the value of the objective function in the next iteration. As $\mathbf{x}_B$ is an optimal B.F.S. then there is no column vector $\mathbf{a}_j$ for which $z_j - c_j < 0$. Now $z_j - c_j = 0$ for some columns of the non-basic variables.

Then

$$z' = z_B - \frac{x_{Br}}{y_{rj}} \times 0 = z_B.$$

Then z' is also the optimal value of objective function corresponding to the B.F.S. $\mathbf{x}'_B$.

Then an alternative optimal B.F.S. exists. Hence there exist multiple optimal solutions of the problem under the conditions stated above.

Note: The above discussion is related to a problem of maximization.

2. At the optimal stage, if at least for one column corresponding to non-basic variables, the condition $z_j - c_j = 0$ with $y_{ij} \le 0$ for all i, holds, then there exists an alternative non-basic optimal solution of the problem.

Let

$$\mathbf{x}_B = [x_{B1}, x_{B2}, \cdots, x_{Bm}]$$

be an optimal B.F.S. of the problem with the optimal value of the objective function z_B. As $y_{ij} \le 0$, for all i, in a j-th column vector of the non-basic variable, then there exists a non-basic feasible solution,

$$\mathbf{x}' = [x_{B1} - \lambda y_{1j}, x_{B2} - \lambda y_{2j}, \cdots, x_{Bm} - \lambda y_{mj}, \lambda], \quad \lambda > 0$$

and the value of the objective function is

$$z' = z_B - \lambda(z_j - c_j)$$

corresponding to the solution set $\mathbf{x}'$.

As $z_j - c_j = 0$, then $z' = z_B$. Therefore z' is also the optimal value of the objective function for a non-basic F.S. $\mathbf{x}'$ with $(m+1)$ variables. Hence there exists an alternative non-basic optimal solutuion to the problem.

A.2.6 Selection of the vector to enter in the next basis, the key element and the vector to leave the current basis

We have

$$\begin{aligned} z' &= z_B - \frac{x_{Br}}{y_{rj}}(z_j - c_j), \quad y_{rj} > 0 \\ &= z_B - \frac{x_{Br}}{y_{rj}}(z_j - c_j). \end{aligned}$$

From the above deduction, it is evident that for $z_j - c_j < 0$, the value of the objective function will be improved provided $\frac{x_{Br}}{y_{rj}} > 0$. If $z_j - c_j < 0$, only in one column, say kth column, the kth vector $\mathbf{a}_k$ is taken as the incoming vector. Now if in that column,

$$\min_i \left(\frac{x_{Bi}}{y_{ik}}, y_{ik} > 0 \right) = \frac{x_{Br}}{y_{rk}} \qquad \text{(A.2.20)}$$

which occurs at $i = r$ then y_{rk} is the key element and β_r, the rth vector in the basis, is the departing vector. But if the minimum ratio is not unique, then the selection of the key element and the departing vectors are not unique. An arbitrary choice of key row corresponding to minimum ratio may be done and almost all problems can be solved using this technique. The rth row of the simplex table is known as *key* or *pivot row* and the kth column is known as *key or pivot column* and the intersection element of the key row and key column which is y_{rk} is the key on pivot element.

If $z_j - c_j < 0$ for more than one column and

$$z_k - c_k = \min_j (z_j - c_j, z_j - c_j < 0) \tag{A.2.21}$$

then $\mathbf{a}_k$ will be the entering vector. *If the minimum is not unique, select arbitrarily any vector as incoming vector corresponding to the minimum.* Then the selection of key element and departing vector are to be done by the previous method. This method is widely applied. No difficulty arises here in selecting the entering vector and as it is easier to select the entering vector, this method is effectively used by the electronic computers to solve the problems. *But in this method once a vector enters in the basis may leave the basis during subsequent iterations.*

In the case of degenerate problems, the selection of entering vector can be done by other methods which is not discussed here.

A.2.7 Important Observations

To calculate a new B.F.S., i.e., to construct a new basis matrix, a vector β_r of B is to be replaced by a vector $\mathbf{a}_k$ of A, which is one of the column vectors associated with the non-basic variables. The vector β_r which is being removed, is the *vector to leave the current basis* and the vector $\mathbf{a}_k$ which is being inserted in the new basis matrix, is the *vector to enter in the next basis.* The method of selection of $\mathbf{a}_k$ and β_r has been discussed previously. The improved B.F.S. is given by (A.2.5) of theorem (A.2.1) [Putting $j = k$]

$$\left.\begin{aligned} x'_{Bi} &= x_{Bi} - \frac{y_{ik}}{y_{rk}} x_{Br}, \quad i = 1, 2, \cdots, m,\ i \neq r \\ x'_{Br} &= \frac{x_{Br}}{y_{rk}}, \quad y_{rk} > 0, i = r. \end{aligned}\right\} \tag{A.2.22}$$

The element y_{rk} is the key element.

Due to the transformation all the elements y_{ij} change to y'_{ij} $[i = 1, 2, \cdots, m,$ $j = 1, 2, \cdots, n]$ and y'_{ij} can be expressed by the relations

$$\left.\begin{aligned} &\text{(i)}\quad y'_{ij} = y_{ij} - \frac{y_{ik}}{y_{rk}} y_{rj} \quad [i = 1, 2, \cdots, m; i \neq r] \\ &\text{(ii)}\quad y'_{rj} = \frac{y_{rj}}{y_{rk}}, \quad y_{rk} > 0, i = r. \end{aligned}\right\} \tag{A.2.23}$$

In particular when $i = k$

$$\text{(iii)}\quad y'_{ik} = y_{ik} - \frac{y_{ik}}{y_{rk}} y_{rk} = 0 \quad [i = 1, 2, \cdots, m; i \neq r]$$

$$\text{and (iv)}\quad y'_{rk} = \frac{y_{rk}}{y_{rk}} = 1, i = r.$$

Hence due to the transformation, the kth column vector y'_k in the next table will be a unit vector $\mathbf{e}_r$.

The proofs for the transformed formulae are given below.

Proof. Let the basis $B = (\beta_1, \beta_2, \cdots, \beta_r, \cdots, \beta_m)$ at any stage. Then for any non-basis vector $\mathbf{a}_j$,

$$\mathbf{a}_j = y_{1j}\beta_1 + y_{2j}\beta_2 + \cdots + y_{rj}\beta_r + \cdots + y_{mj}\beta_m = \sum_{i=1}^{m} y_{ij}\beta_i \tag{A.2.24}$$

and for $\mathbf{a}_k$ (the vector to enter in the next basis by replacing β_r), we have

$$\mathbf{a}_k = \sum_{i=1}^{m} y_{ik}\beta_i. \tag{A.2.25}$$

As $y_{rk} > 0$, therefore expressing β_r in terms of $\mathbf{a}_k$ etc. we have

$$\begin{aligned}\beta_r &= \frac{1}{y_{rk}}\mathbf{a}_k - \left(\frac{y_{1k}}{y_{rk}}\beta_1 + \frac{y_{2k}}{y_{rk}}\beta_2 + \cdots + \frac{y_{r-1\cdot k}}{y_{rk}}\beta_{r-1} + \frac{y_{r+1\cdot k}}{y_{rk}}\beta_{r+1} + \cdots + \frac{y_{mk}}{y_{rk}}\beta_m\right)\\ &= -\sum_{\substack{i=1\\ i\neq r}}^{m} \frac{y_{ik}}{y_{rk}}\beta_i + \frac{1}{y_{rk}}\mathbf{a}_k \end{aligned}\tag{A.2.26}$$

Putting the value of β_r from (A.2.23) in $\mathbf{a}_j$ in (A.2.24) we have

$$\mathbf{a}_j = \sum_{\substack{i=1\\ i\neq r}}^{m} \left(y_{ij} - y_{rj}\frac{y_{ik}}{y_{rk}}\right)\beta_i + \frac{y_{rj}}{y_{rk}}\mathbf{a}_k = \sum_{i=1}^{m} y'_{ij}\hat{\beta}_i, \tag{A.2.27}$$

where $\hat{\beta}_i = \beta_i[i \neq r]$ and $\hat{\beta}_r = \mathbf{a}_k$.

The next basis is

$$\hat{B} = (\beta_1, \beta_2, \cdots, \beta_{r-1}, \mathbf{a}_k, \beta_{r+1}, \cdots, \beta_m).$$

Comparing (A.2.24) and (A.2.27) we get

$$\left.\begin{aligned} y'_{ij} &= y_{ij} - \frac{y_{ik}}{y_{rk}}y_{rj}, \quad i \neq r\\ \text{and} \quad y'_{rj} &= \frac{y_{rj}}{y_{rk}} \end{aligned}\right\}. \tag{A.2.28}$$

Due to the transformation the values of $z_j[j = 1, 2, \cdots, n]$ also change and the changed value

$$z'_j = z_j - \frac{y_{rj}}{y_{rk}}(z_k - c_k).$$

Proof. When $\mathbf{a}_k$ is the vector to enter in the basis by replacing β_k, we have $c'_{Bi} = c_{Bi}$ for $i \neq r$ and $c'_{Br} = c_k$.

Hence,

$$z'_j = \sum_{i=1}^{m} c_{Bi}y'_{ij} = \sum_{\substack{i=1\\i\neq r}}^{m} c_{Bi}y'_{ij} + c_k y'_{rj}$$

$$= \sum_{\substack{i=1\\i\neq r}}^{m} c_{Bi}\left(y_{ij} - \frac{y_{ik}}{y_{rk}}y_{rj}\right) + c_k\frac{y_{rj}}{y_{rk}} \quad \text{[from (A.2.27)]}$$

$$= \sum_{i=1}^{m} c_{Bi}\left(y_{ij} - \frac{y_{ik}}{y_{rk}}y_{rj}\right) + c_k\frac{y_{rj}}{y_{rk}} \left[\because\ y_{rj} - \frac{y_{rk}}{y_{rk}}y_{rj} = 0\right]$$

$$= \sum_{i=1}^{m} c_{Bi}y_{ij} - \frac{y_{rj}}{y_{rk}}\left(\sum_{i=1}^{m} c_{Bi}y_{ik} - c_k\right) = z_j - \frac{y_{rj}}{y_{rk}}(z_k - c_k). \quad \text{(A.2.29)}$$

Hence,

$$z'_j - c_j = z_j - c_j - \frac{y_{rj}}{y_{rk}}(z_k - c_k) \quad \text{(A.2.30)}$$

and again we have from (A.2.8)

$$z' = z_B - \frac{x_{Br}}{y_{rk}}(z_k - c_k) \quad \text{[putting } j = k\text{]} \quad \text{(A.2.31)}$$

Now a very interesting observation can be made. Suppose that we write

$$x_{Bi} = y_{i0}(i = 1, \cdot, m), \quad z_B = y_{m+1,0}, z_j - c_j = y_{m+1,j}$$

then all the formulae's of (A.2.31), (A.2.22), (A.2.30), (A.2.28) can be written in the compact form

$$\left.\begin{aligned} y'_{ij} &= y_{ij} - y_{ik}\frac{y_{rj}}{y_{rk}} \quad j = 0, 1, \cdots, n,\ i = 1, \cdots, m+1, i \neq r \\ y'_{rj} &= \frac{y_{rj}}{y_{rk}} \end{aligned}\right\} \quad \text{(A.2.32)}$$

[A very important property to remember.]

Now if the initial basis matrix be a unit matrix $B = (\mathbf{e}_1, \mathbf{e}_2, \cdots, \mathbf{e}_m)$ where $\mathbf{e}_i$ is a unit vector with ith component unity and $\mathbf{b} \geq \mathbf{0}$, then

$$\mathbf{x}_B = B^{-1}\mathbf{b} = \mathbf{b} \geq \mathbf{0}$$

$$\text{or,} \quad \mathbf{x}_B = [b_1, b_2, \cdots, b_m]$$

which is the initial B.F.S. to the problem and $z_B = \mathbf{c}_B\mathbf{b}$ and $\mathbf{y}_j = B^{-1}\mathbf{a}_j = \mathbf{a}_j$. Hence $y_{ij} = a_{ij}$ for all i and j as

$$\mathbf{a}_{j.} = [a_{ij}, a_{2j}, a_{ij}, \cdots, a_{mj}]$$

$$\text{and} \quad z_j - c_j = \mathbf{c}_B\mathbf{y}_j - c_j = \mathbf{c}_B\mathbf{a}_j - c_j.$$

With this knowledge, the initial simplex table can be constructed. The next table can also be obtained from the initial table by transforming it, using the formula (A.2.32) in such a way that the kth column vector $\mathbf{y}'_k$ of next table be a unit vector $\mathbf{e}_r$.